Lectins
Biology, Biochemistry, Clinical Biochemistry
Volume 1

Lectins

Biology, Biochemistry, Clinical Biochemistry

Volume 1

Proceedings of the
Third Lectin Meeting
Copenhagen, June 1980

Editor T. C. Bøg-Hansen

W DE G

Walter de Gruyter · Berlin · New York 1981

Editor
Thorkild Christian Bøg-Hansen, cand. scient., lic. techn.
The Protein Laboratory
University of Copenhagen
Sigurdsgade 34

DK-2200 Copenhagen N

CIP-Kurztitelaufnahme der Deutschen Bibliothek

Lectins, biology, biochemistry, clinical biochemistry:
proceedings of the Lectin Meeting. – Berlin; New York: de Gruyter.
Vol. 1. Proceedings of the third Lectin Meeting: Copenhagen,
June 1980. – 1981. ISBN 3-11-008483-X
NE: Lectin Meeting

 Printing: Karl Gerike, Berlin. – Binding: Dieter Mikolai, Berlin. – Printed in Germany.

PREFACE

This book contains the proceedings from the third lectin meeting in Copenhagen, June 8-10, 1980.

The first lectin meeting in Copenhagen was held on May 23, 1978 as a prelude to the 1979 meeting "Protides of the Biological Fluids" in Brussels. Fifteen communications were presented by scientists working in Copenhagen. The program was divided into two sessions: reactions of lectins with cells and reactions of lectins with proteins. About sixty people attended the meeting.

The second lectin meeting was held in Copenhagen on April 3, 1979, and I was very surprised to note that again sixty people turned up to listen to eleven lectures dealing mainly with the characterization and purification of glycoproteins. Guests from Poland and Canada gave the meeting an international flavour, and following an afternoon in the crowded lecture hall, I was convinced that there was a genuine interest in lectins and their reactions. This was proved by the third lectin meeting.

The third lectin meeting was arranged at the two previous meetings, - a few personal contacts, telephone calls and letters, and the program of some 45 lectures was put together. Last-minute changes had to be made, since some speakers did not obtain the necessary support. One speaker asked me to rob a few banks to provide him and others with travel grants! Needless to say, I did not take any excessive risks to provide friends with a free treat in Copenhagen. Nevertheless, I would like to thank the Danish Medical Research Council for their support in the form of eight speakers, and Medac for taking one participant to Copenhagen. The book of abstracts was printed at the last minute and we are grateful for the support received from Meda, Immuno, Nordmedac, and Pharmacia Switzerland. The City of Copenhagen invited the 100 participants to lunch on one of the two days of the meeting's duration.

The communications covered new ground and in addition to the old topics, dealt with the following five sessions: I. Lectins from plants; distribution; function and isolation. II. Lectins in other organisms; distribution and function. III. Methods based on lectin-reactions. IV. Glycoproteins studied by lectin-reactions. V. Alpha fetoprotein.

Recent publications raised a discussion on nomenclature, the definition of a lectin and the position of lectinology with regard to other related fields (see Fig. 1), - a discussion which is reflected in several of the papers in this volume. The papers were prepared after the meeting, and all the speakers handed in manuscripts. Furthermore, they contain a paper which could not be read at the meeting (G. Uhlenbruck and co-workers on the lectin in cow-pea and its apparent unimportance in the growth of the plant) and a paper from the second lectin meeting (I. Lorenc-Kubis and co-worker about in vitro activation of a plant enzyme by lectins). To make up for his inability to attend the meeting, G. Uhlenbruck sent the following reflections about his scientific love:

Lessons on Lectins, by G. Uhlenbruck

Lectins represent a fine *immunochemical* tool,
Who does not use them, seems a fool:
But please, don't touch my personal pool,
Because that would not leave me cool!

However must be considered on the other side:
An own collection is the master's pride,
Because it gives the advantage of priority
Towards other scientists in the majority!

The important question: *male or female?*
Has been solved with lectins in sweet detail:
As they have a strong sugar specifictiy,
They must be male with *sweetpart* recognizing affinity!

Male is also their flexible combining site,
Which is *true*, but sometimes does not right,
And shows ambition to *love* another carbohydrate:
That causes trouble in research, as papers do relate.

Lectins occur in bacteria, (in)vertebrates, plants and man,
A large field to investigate for a science fan,
And a never missing link in the evolution:
To *see* or *not to see*, is that the solution?

In spite of all experiments and many a discovery,
Lectins remain still to us being a mystery:
A challenge to science as in this book we can see,
Otherwise it would be the end of the story - unfortunately!

Other lectin meetings are being held all over the world, and the Copenhagen meeting is just one link in the communications-network which holds lectinologists together. However, it is my hope to be able to provide opportunities for further personal contacts and the exchange of new ideas. It is in this spirit that the fourth lectin meeting is being planned in Copenhagen, - June 8-12, 1981 - with two main themes: the biological functions of lectins and the clinical use of methods based on the reaction of lectins. The Proceedings arising from this meeting will be published as Volume 2 in this series of Lectins - Biology, Biochemistry, Clinical Biochemistry.

Copenhagen, April 1981

T. C. Bøg-Hansen

The relationship between immunology and lectinology some years ago: Though seated at the same table, the well-established rabbit-eating immunologist and the bean-and-snail-eating lectinologist appear differently nourished (by courtesy of H. Franz and D. Hasler, Berlin)

Contents

Part I. Lectins: Distribution, Isolation, Characterization, and Function

Part II. Methods Based on Reactions of Lectins

Part I

Lectins: Distribution, Isolation, Characterization, and Function

LECTINS - AN INTRODUCTION

H. Rüdiger

Institut für Pharmazie und Lebensmittelchemie der Universität Würzburg, 8700 Würzburg, W. Germany

Lectins are abundant in the plant and animal kingdoms. They usually are defined by their most obvious property, i. e. the ability to agglutinate red blood cells. Though the first lectins were named after the source they had been found in (as e. g. ricin from Ricinus, concanavalin from Canavalia), later on the term agglutinin or phythemagglutinin became popular. The latter name, abbreviated as PHA, is now in use for the mitogenic lectins from kidney beans (Phaseolus vulgaris). The term "lectin", taken from the Latin word for "to choose, to select", is now most firmly established. It has been introduced by Boyd and Shapleigh (1) in 1954. They supposed that many of these proteins would be able to select human red blood cells according to their blood group. Though this hope did not fulfil since only a very limited number of lectins proved to be blood group specific the term "lectin" has persisted.

Mostly lectins are traced and followed during purification by agglutination of red cells, either native ones or cells which had been enzymatically modified. Agglutination, however, is not restricted to red cells; also other cells as e. g. lymphocytes or microorganisms respond in a similar manner. Special attention has been paid to the fact that certain lectins preferentially agglutinate malignant cells. One of the most striking properties of many lectins is their ability to induce mitosis in lymphocytes, i. e. to simulate the action of antigens.

Lectins by definition interact with oligosaccharides of cell or glycoprotein surfaces. Consequently, free oligo- or monosaccharides of the appropriate specificity are able to inhibit or even to revert this interaction. It must be kept in

Lectins - Biology, Biochemistry, Clinical Biochemistry, Vol. I

mind, however, that lectins, though primarily acting by their carbohydrate binding sites, may reinforce their interaction with cells or proteins by hydrophobic effects (2). Because of their binding specificity, lectins may be grouped together with enzymes and antibodies to form the "affinitins" as it has been proposed by Franz and Ziska (3). The current interest in lectins manifests itself by several reviews during the last few years (4 - 8).

Lectins are used throughout the world in many laboratories because of their interesting properties. None of them, however, throws light on the problem which bothers all people interested in biological phenomena: What is the physiological role of lectins? Of course, teleological questions of this kind never can be answered definitely but one can try to accumulate experimental evidence in favor of one or the other hypothesis. Several proposals concerning the biological role of lectins have been made:

Janzen et al. (9) found that a beetle species cannot tolerate certain Leguminosae seeds because of their lectin content. They therefore proposed that lectins protect seeds from being eaten by insects, similar to the role ascribed to the alcaloids.

As Mirelman et al. (10) showed, wheat germ lectin which binds specifically to N-acetylglucosamine oligomers inhibits growth of molds since it blocks chitin biosynthesis. This experiment suggests the idea that lectinsprotect the young plant from infection by fungi.

Concerning the lectins of the Leguminosae, several authors (12 - 14) discuss the idea that lectins mediate the interaction between host plant and nodulating bacteria of the genus Rhizobium. In fact, specific interactions between the lipopolysaccharides of different Rhizobia cell coats and the respective lectins have been observed. Recent experiments (14 , 15) demand new discussion about this hypothesis.

Similar to the mitogenic effect which lectins exert on animal cells, they may also accelerate cell division of plants (16), a finding which has been contradicted by others (17).

Kauss et al. (18, 19) put forward the idea that lectins may help to glue cells together reversibly in such a manner that they can be shifted during extension growth. This hypothesis on the one hand is based on the observation that pH changes occur in plants during extension growth, on the other hand on the well known fact that many lectins agglutinate red cells in a highly pH-dependent manner. In this context recent observations of Etzler et al. are quite important. They had found a protein in the Leguminosa Dolichos biflorus which crossreacts with antibodies against the lectin of this plant though this cross reactive material itself is not a lectin (20). Presumably, it differs from the lectin by its extended C-terminal part. Apparently this protein is localized in the cell wall as opposed to the lectins which mostly are found in the protein bodies or in the cytoplasm. it can assume lectin-like properties under certain circumstances by a process which is unknown, and it is formed in abundant amounts if the plant is hurt (21). All these results indicate a role of the cross reactive material in cell growth.

Another very exciting proposal has been made by Hankins and Shannon (22). They suggest that lectins might act as enzymes in their natural environment. In fact, they found that the mung bean lectin copurifies with an α-galactosidase activity. As it will be outlined during this Meeting (23). also other lectin containing plants display glycosidase avtivities.

Very early, the apparent similarity of lectins with antibodies in their action led to the idea that lectins represent a chemical defense system of plants. Though this idea could not be substantiated in its original form, lectin or lectin-like proteins seem to be responsible for the incompatibility of pollen and stigma in flowering plants thus preventing self-fertilization (24).

Single cell slime molds aggregate to multicellular organisms before fruit bodies are formed. This process which represents a simple kind of differentiation is governed by lectins which are found on the cells during aggregation (25). Lectins also may take part in sexual recognition phenomena (26). Recent years brought the discovery that also animals contain lectins, e. g. in the membranes of their liver cells (27). These lectins bind to galactose residues which often constitute the subterminal sugar of serum glycoproteins. Aged glycoproteins, i. e. those which have lost their terminal sialic acid residues, are believed to be recognized by the lectins and thus removed from circulation. Lectin activity may also be released from connective tissue by proteolysis (28).

Apparently lectins from different sources may have quite different functions. Lectins are grouped together because they have in common that they agglutinate red blood cells. This of course is a totally unphysiological reaction, so the group of lectins may be regarded as artificial. Consequently a common biological role cannot be expected for them; this question has to be considered for each lectin separately.

If one inspects only the lectins of the Papilionaceae, one of the Leguminosae families (Fig. 1), evidently it is difficult to find a connection between structure or properties and taxonomy:

Only very few lectins are specific with respect to human blood groups (Vicia cracca, Phaseolus lunatus, Dolichos biflorus: blood group A; Ulex europaeus, Lotus tetragonolobus: blood group O (H); Bandeiraea simplicifolia: blood group B .

Three plants contain lectins of different specificities, one of them being blood group specific (Vicia cracca, Ulex europaeus, Bandeiraea simplicifolia).

Most lectins contain subunits which are identical or nearly identical in M_r. It has been proposed to call them "one chain lectins" (30). Most of them are tetramers, typically with a M_r of about 120 000 (4 x 30 000). There are some notable exceptions from this architecture, all of them in the Vicieae tribe. These lectins belong to the "two chain type" and are composed of two subunits of different size, forming tetramers of the formula $\alpha_2\beta_2$. A further lectin of this type from Lathyrus odoratus will be presented in this Meeting (31).

Nearly all lectins can be inhibited by simple sugars or glycosides, the blood group specific ones by the terminal sugar of the respective blood group substance. For two lectins of Fig. 1 (Phaseolus vulgaris, Vicia graminea), no simple inhibiting sugars are known. The Vicia graminea lectin will be described in detail during this Meeting (32).

At present it is difficult to assign a function to the lectins. As pointed out by P. Rougé & C. Chatelain (33), up to now not even the localization within the plant of the most common Leguminosae lectins is clear. Some authors assume lectins to be associated to the protein bodies, others find them to be evenly distributed in the cytoplasm. It is very important to clarify this question unambiguously. A second point which is difficult to reconcile with a role of the lectins is made by G. Uhlenbruck et al. (34). They found that only a few seed samples of a single plant species exhibit a mitogenic effect on lymphocytes suggesting that the plants differ in their lectin equipment.

Nevertheless, lectins must be expected to be important in plant life.

The first argument in favor of this expectation is quite indirect but nevertheless valid: Leguminosae seeds contain considerable amounts of lectins. In most cases as e. g. in pea, broad bean, lentil, about 0.1% of the seed weight is lectin, in Jack beans this amount is exceeded by a factor of 20 - 30, i. e.these seeds contain as much as up to 3% lectin. One cannot imagine that Nature accumulates one particular protein in such big amounts without necessity.

The second argument is that there exist lectin binding proteins in plants as we recently found (35, 36). In any species tested, there is one protein which is specifically recognized by the lectin among the many proteins of the plant. Lectin binding proteins will be one of the subjects of this Meeting (37).

The third argument is concerned with the lectin amino acid sequences. Lectins of different tribes of the Leguminosae are quite homologous (30, 38). A compilation of some of the Vicieae lectin sequences will be presented during this Meeting (39).

Much sequence information on proteins has accumulated during the last years. It has been found out that proteins with crucial biological functions, especially if they have to fit to a certain environment, evolve very slowly and are extremely conservative in their sequences. Therefore the inverse conclusion may also be allowed: The high degree of homology in lectin sequences reflect an important, though unknown biological function.

References

1. Boyd, W. C., Shapleigh, E.: Science 119, 419 (1954).

2. Ochoa, J.-L., Sierre, A., Córdoba, F.: These Proceedings

3. Franz, H., Ziska, P.: These Proceedings

4. Lis, H., Sharon, N.: Annual Rev. Biochemistry 42, 541 - 574 (1973).

5. Liener, I. E.: Annual Rev. Plant Physiol. 27, 291 - 319 (1976).

6. Rüdiger, H.: Naturwissenschaften 65, 239 - 244 (1978).

7. Hayes, C. E., Goldstein, I. J.: Advances in Carbohydrate Chemistry and Biochemistry 35, 127 - 340 (1978).

8. Kauss, H.: Plant Carbohydrates, Vol. 2, Springer, Berlin-Heidelberg-New York 1980, in press.

9. Janzen, D. H., Juster, H. B., Liener, I. E.: Science 192, 795 - 796 (1976).

10. Mirelman, D., Galun, E., Sharon, N., Lotan, R.: Nature 256, 414 - 416 (1975).

11. Bohlool, B. B., Schmidt, E. L.: Science 185, 269 - 271 (1974).

12. Wolpert, J.S., Albersheim, P.: Biochem. Biophys. Res. Commun. 70, 729 - 737 (1976).

13. Kato, G., Maruyama, Y., Nahamura, M.: Agric. Biol. Chem. 43, 1085 - 1092 (1979)

14. Pull, S. P., Pueppke, S. G., Hymowitz, T., Orf, J. H.: Science 200, 1277 - 1279 (1978).

15. Dazzo, F. B., Urbano, M. R., Brill, W. J.: Current Microbiol. 2, 15 - 20 (1979).

16. Howard, J., Shannon, L., Oki, L., Murashiga, T.: Exp. Cell. Res. 107, 448 - 450 (1977).

17. Vasil, I. K., Hubbell, D. H.: Z. Pflanzenphysiol. 84, 349 - 353 (1977).

18. Kauss, H., Glaser, C.: FEBS Letters 45, 304 - 307 (1974).

19. Kauss, H., Bowles, D. J.: Planta 130, 169 - 174 (1978).

20. Talbot, C. F., Etzler, M. E.: Biochemistry 17, 1474 - 1479 (1978).

21. Etzler, M.: Federation Proceedings 39, Abstr. 2698 (1980).

22. Hankins, C. N., Shannon, L. M.: J. Biol. Chem. 253, 7791 - 7797 (1978).

23. Shannon, L. M., Hankins, Ch. N., Strosberg, A. D.:

These Proceedings

24. Golynskaya, E. L., Bashkirova, N. V., Tomchuk, N. N.: Fisiol. Rast. 23, 88 - 97 (1976), cited after Chem. Abstr. 84, 161759D (1976).

25. Rosen, S. D., Kaur, J., Clark, D. L., Pardos, B. T., Frazier, W. A.: J. Biol. Chem. 254, 9408 - 9415 (1979).

26. Homan, W. L.: These Proceedings

27. Sarkar, M., Liao, J., Kabat, E. A., Tanabe, T., Ashwell, G.:J. Biol. Chem. 254, 3170-3174 (1979).

28. Chang-Fournier, F., Gerfaux. J.: These Proceedings

29. Heywood, V. H., in: Chemotaxonomy of the Leguminosae (Harborne, J. B., Boulter, D., Turner, B. L., eds.), Academic Press London - New York 1971.

30. Foriers, A., De Neve, R., Strosberg, A. D.: Physiol. Vég. 17, 597 - 606 (1979).

31. Kolberg, J., Michaelsen, T. E., Sletten, K.: These Proceedings

32. Prigent, M. J., Bourillon, R.: These Proceedings

33. Rougé P., Chatelain, C.: These Proceedings

34. Uhlenbruck, G., Karduck. D., Van Mil, A.: These Proceedings

35. Gansera, R., Schurz, H., Rüdiger, H.: Hoppe-Seyler's Z. Physiol. Chem. 360, 1579 - 1585 (1979).

36. Gebauer, G., Schiltz, E., Schimpl, A., Rüdiger, H.: Hoppe Seyler's Z. Physiol. Chem. 360, 1727 - 1735 (1979).

37. Rüdiger, H., Gansera, R., Gebauer, G., Schurz, H.: These Proceedings

38. Baumann, C., Rüdiger, H., Strosberg, A. D.: FEBS Letters 102, 216 - 218 (1979).

39. Baumann, C. M., Rüdiger, H., Strosberg, A. D.: These Proceedings

LECTINS : SUGAR-SPECIFIC OR RECEPTOR-SPECIFIC PROTEINS ?

P. Balding
Oncology Research Unit. Dept. Radiotherapy, University of Bristol, U.K.

At the turn of the century, the study of haemagglutination by plant extracts involved only a few species: Ricinus communis, Abrus precatorius, Croton tiglium and Robinia pseudoacacia (1,2,3,4). Names for these substances were derived from the plant names, i.e. ricin, abrin, croton and robin respectively. Due to their toxicity they were collectively referred to as phytotoxins. Later, non-toxic, haemagglutinating plant extracts were discovered, initially in species of the genus Phaseolus: they were therefore called phasins (5). Expressions like glabrification" and "conglutination" were used to describe their activity until Elfstrand (3) introduced the expressions "haemagglutinin" and "haemagglutination". This change in terminology was not universally accepted (6). Assman (7) quoting Eisenberg (8) attempted to solve the problem and wrote "exactly speaking agglutination is only a function of a substance and if we call the hypothetical material substrate of this function an "agglutinin" then this means only that we use a symbol in order to make logical operations with this concept easier". The wisdom of this approach was recognised and the terms have been generally accepted. Following the discovery of many more haemagglutinating plant extracts, the expression "phytohaemagglutinin" became common (9), adopting the prefix "phyto" as already used for the term "phytotoxins".

After the discovery of blood group specific seed extracts (10,11) Boyd (12,13) introduced a new name "lectin", from the latin to pick out, or to choose, with which to refer to such reagents. For some time "lectin" continued to be used only for the blood group specific seed haemagglutinins although in the etymological sense all agglutinating reagents whether of plant or animal origin could be included.

Lectins - Biology, Biochemistry, Clinical Biochemistry, Vol. I

However, at the time Boyd (13) referred to the natural heterohaemagglutinins in the haemolymph of invertebrates, which were discovered soon after the plant haemagglutinins (14), as antibody-like substances. Then, in 1964, Johnson (15) described the first blood group specific non-immune haemagglutinin not of plant origin; an anti-A_1 in the haemolymph of the clam Saxidomus giganteus. This agglutinin was similar to the anti-A specific plant lectins in that haemagglutination was inhibited by N-acetyl-galactosamine. One year later Boyd and Brown (16) and Prokop et al (17) simultaneously reported the presence of blood group A specific haemagglutinins in the albumin glands of certain snails (Otala lactea and Helix hortensis respectively). Prokop et al (18,19) used the descriptive term antibody-like substance introduced by Boyd when referring to this anti-A. They also introduced a further name "protectin" for the same group of agglutinins, derived from an assumed role in an immunological protective mechanism (19). This term has not been adopted, its use being restricted mainly to Prokop and co-workers.

Gold and Phelps (20) discussed the use of the term "antibody-like substance" concluding that the similarity of lectins to antibodies was rather superficial. Instead they introduced the purely descriptive term "Receptor-specific proteins" to cover the wide range of reagents found in viruses, bacteria, fungi, plants, sponges, snails, fishes and other lower vertebrates. The broader adjective "receptor-specific" was chosen rather than the limited "sugar-specific" used by Sharon and Lis (21) since at that time a small number of these reagents were already known which were not inhibited in haemagglutination tests by simple sugars or oligosaccharides, e.g. Agaricus campestris (22) Clostridium botulinum types C and D (23) and Mannose-resistant E.coli (24). A provisional classification of Receptor-specific proteins was proposed based on four patterns of haemagglutination-inhibition known about at that time. This classification is no longer valid since (i) the specificity of the sponge Axinella sp, reported to be of an unusual nature (25) has been investigated further, resulting in the establishment of terminal β-D-galactoside binding by the lectin (26); (ii) inhibition of the lectins of Cl.botulinum types C and D by gangliosides has been demonstrated (27), (iii) one mannose-resistant E.coli agglutinin has been shown to be specific for the amino-acid L-histidine (24,28). A

Table 1 Revised provisional classification of Receptor-specific proteins: plant and animal lectins (adapted from ref. 20).

Lectin	Inhibiting substances			
	mono -	oligo	amino acid	biological substances*
	saccharides			
Most lectins	+	+	-	+
A few lectins	(-	+	-	+
	(-	-	-	+
M.R. E.coli D520**	-	-	L-histidine	+
M.R.E.coli 9044/45**	-	-	-	-

* e.g. cyst fluids, secretions, glycoproteins, glycolipids (incl. gangliosides)
** see ref. 24, 28.
M.R. - Mannose-resistant strains
\+ - Inhibition, - no inhibition

revised version of this provisional classification is shown in Table 1.

Sharon and Lis (21) also included the snail agglutinins together with the plant agglutinins, regardless of any blood-group specificity, under the term lectin. Although using the expression " sugar-specific proteins", they pointed out that for some lectins no saccharide had been found which inhibited agglutination.

The use of the name "lectin" as an umbrella term for all cell-agglutinating or glycoconjugate-precipitating proteins or glycoproteins of non-immune origin was more fully developed by Gold and Balding (29) who urged that this usage should be universally adopted. The field of study could then be termed "lectinology" an expression introduced by Potapov (30). The receptor-specific protein nature of lectins was emphasised. However Sharon (31) and Goldstein (32) have persevered with the limited use of lectin for sugar- specific reagents only. Sharon (31) may have been unaware of the reports of other than simple saccharide specificity when he stated that "no lectin has been found that is specific for compounds other than carbohydrates". Recently (33) the expansion of the term lectin was again endorsed although as already described above, the term has for some time been used for other than plant seed carbohydrate binding proteins. Goldstein et al (33) persist in restricting the lectins to reagents with only monosaccharide or simple oligosaccharide specificity. This restriction not only excludes the _E.coli_ agglutinating proteins, but also PHA, _Agaricus campestris_, _Cl.botulinum_ types C and D and other lectins for which a simple sugar or oligosaccharide inhibitor has not yet been described. (The structure of the receptor for PHA has been reported (34,35) but it is not a simple oligosaccharide).

Attention has been drawn to the involvement of hydrophobic interactions between lectins and their receptors (36). A sugar-specific lectin has even been described which can also be inhibited by albumin (37) although this inhibition is most probably due to hydrophobic interaction, considering the concentration of albumin necessary for inhibition, rather than to specificity. Thus it is difficult at present to explicitly define a lectin in unambiguous terms as regards specificity, unless as sugar-binding proteins they relegated to a, perhaps major, section of a much larger group of receptor-specific proteins, which in turn could form part of the class of proteins referred to as affinitins (38). However, it would seem premature to exclude, for example, the L-histidine specific lectin isolated from _E.coli_ (28) which fulfills all the other criteria necessary to achieve the status of "lectin" listed by Goldstein _et al_ (33), solely on the basis of specificity; it is quite possible that many more lectins exhibiting other than simple sugar or oligosaccharide specificity will be discovered, particularly in the bacteria.

During the 22 years since their discovery, a host of different names have been proposed by various authors. But apart from the little used term "protectin" none of these names has been based on a functional role _in vivo_, which remains speculative. It is time to recognise Eisenberg's wisdom and common sense. The term lectin is only a symbol: we can use it to make logical operations with the concept easier.

1. Stillmark, H. : Über Ricin, ein giftiges Ferment aus den Samen von _Ricinus communis_ L. und einigen anderen Euphorbiaceen. Thesis, Dorpat (1888).

2. Hellin, H. : Der giftige Eiweisskörper Abrin und dessen wirkung das Blut. Dissertation, Dorpat (1891).

3. Elfstrand, M. : Über blutkörperchenagglutinierende Eiweisse. Görbersdorfer Veröffentlichungen. Editor : R.Kobert, F.Enke, Stuttgart (1898)

4. Lau, K. : Über vegetabilisch Blutagglutinine, Dissertation, Rostock (1901)

5. Wienhaus, O. : Biochem. Z., 18, 228 (1909)

6. Miessner, H. and Rewald. : Z. Immun - Forsch. 2, 323, (1907)

7. Assman, F. : Arch. ges. Physiol., 137, 489 (1911)

8. Eisenberg, P. : Zbl. Bakt., I. Abt. Orig. 41, 96 (1906)

9. Landsteiner, K. and Raubitschek,H. : Zbl. Bakt. I. Abt. Origin. 45 660 (1907)

10. Renkonen, H.L. : Ann. Med. Exp. Fenn. 26, 66 (1948)

11. Boyd, W. C. and Reguera, R.M. : J. Immunol. 62, 333 (1949)

12. Boyd, W.C. : The proteins of immune reactions. In: The Proteins, Vol. 2. p.788. Ed. Neurath. H and Bailey, K. Academic Press, New York, and London (1954)

13. Boyd, W.C. and Shapleigh, E. : Science, 119, 419 (1954)

14. Noguchi, H: Univ. Penn. Med. Bull. 15: 295 (1902)

15. Johnson, H.M.: Science, 146, 548 (1964)

16. Boyd, W.C. and Brown, R. : Nature, 208, 593 (1965)

17. Prokop, O., Rackwitz, A. and Schlesinger, D:J. Forens. Med. 12.108(1965)

18. Projop O. and Rackwitz, A.: Acta. Biol. med. germ. 15, 191 (1965)

19. Prokop, O., Uhlenbruck, G. and Kohler, W. : Dtsch. Gesundh - Wes., 23 318 (1968)

20. Gold, E..R. and Phelps, C.F. : Z. Immun. - Forsch. 143, 430 (1972)

21. Sharon, N. and Lis. H. : Science, 177, 949 (1972)

22. Sage, H.J. and Connett, S.L. : J.Biol.Chem. 244, 4713 (1969)

23. Balding, P., Gold, E..R. Boroff, D.A. and Roberts, T.A. : Immunology 25, 773 (1973)

24. Balding, P., Gold, E.R.:Z.Immun-Forsch. 146, 36 (1973)

25. Gold, E.R. Phelps, C. F., Khalap, S and Balding, P.: Ann. N.Y. Acad. Sci. 234, 122 (1974)

26. Bretting, H. In: Protides of the Biological Fluids Vol. 27 p. 409 Ed. H. Peeters, Pergamon Press (1979)

27. Balding P., Smart J.L., Roberts, T.A. Gold, E.R. : The lectins of _Cl.Botulinum_ and their relationship to toxin specificity.

In: Lectins: biology and biochemistry. Proc. 3rd. Lectin Conference, Copenhagen. Ed. T.C. Bøg-Hansen, Walter de Gruyter, Berlin, New York, 1980.

28. Balding P. In: Protides of the Biological Fluids, vol. 27 p.405, Ed. H. Peeters. Pergamon Press (1979)

29. Gold, E.R. and Balding, P. : Receptor-specific proteins: Plant and Animal lectins. Excerpta Medica, Amsterdam (1975)

30. Potapov, I.M. : Izv. Akad. Nank Kazah S.S.R. Ser. Fiziol.Med. Soviet Union, Nr. 5 712 (1968)

31. Sharon, N.: Sci. Amer. 236 (6), 108 (1977)

32. Goldstein, I.J. : Adv. Carb. Chem. 35, 127 (1978)

33. Goldstein, I.J., Hughes, R.C. Monsigny, M. Osawa, T., Sharon. N. : Nature, 285, 66 (1980)

34. Kornfeld, S. and Kornfeld R. : Proc. nat. Acad. Sci. 63, 1439 (1969)

35. Kornfeld, S. and Kornfeld R. : The structure of phytohaemagglutinin receptor sites. In Glycoproteins of Blood cells and Plasma, p.50 Ed. G. A. Jamieson and T. J. Greenwalt. Lippincott. U.S. (1971)

36. Sparrman, M. Ochoa, J. -L and Kristiansen, T. In : Protides of the Biological fluids, vol. 27, p. 431. Ed. H. Peeters. Pergamon Press (1979)

37. Ochoa, J-L, Kristiansen, T., Romero, M.L. and Rodriguez, H. In: Protides of the Biological Fluids vol. 27 p. 381 Ed. H. Peeters Pergamon Press (1979)

38. Franz, H. and Ziska, P. : Affinitins: combining sites containing proteins. In: Lectins: biology and biochemistry. Proc. 3rd. Lectin Conference, Copenhagen. Ed. T.C. Bøg-Hansen. Walter de Gruyter, Berlin New York, 1980.

AFFINITINS: COMBINING SITES CONTAINING PROTEINS

H. Franz and P. Ziska
Staatliches Institut für Immunpräparate und Nährmedien,
DDR-1120 Berlin, Klement-Gottwald-Allee 317/321

The term "lectin", introduced in 1954 by Boyd (1) is now used for the designation of a "sugar binding protein of non-immune origin which agglutinates cells and/or precipitates glycoconjugates" (2). It is the advantage of this term, that it does not anticipate a distinct provenance (in contrast to phythemagglutinin, phytagglutinin, phasin) or a certain function in vivo (like plant antibody, antibody-like substance, protectin). The last point is of particular importance because at present time we do not know enough about the function of lectins in vivo. For the understanding of the characteristics of lectins the ability to agglutinate cells of other species seems to be of secondary importance. Legume seeds-lectins e. g. cannot get into contact with animal cells under biological conditions. More essential seems to be the existence of combining sites with high affinity to certain sugars. However, combining sites with sugar specificity are also present in carbohydrate specific enzymes and antibodies. Lectins can be considered to be glycoproteins or proteins with one or more sugar-specific combining sites as far as they are not enzymes or antibodies (or perhaps some other groups of proteins). On the other hand the existence of combining sites with high affinity to distinct chemical structures implies a close relation between lectins, enzymes and antibodies. In order to underline this aspect we have proposed the term "affinitin" as collective name (3, 4).
The reaction of the thermo-labile affinitins with their

Lectins - Biology, Biochemistry, Clinical Biochemistry, Vol. I

partners in vivo is of biological relevance. It cannot be excluded that there exists an evolutionary connection between the groups of affinitins like assumed for enzymes and antibodies by Roitt (5).
On principle the classification of a certain affinitin as lectin, enzyme or antibody is possible without difficulties. Nevertheless there are interesting borderline cases.

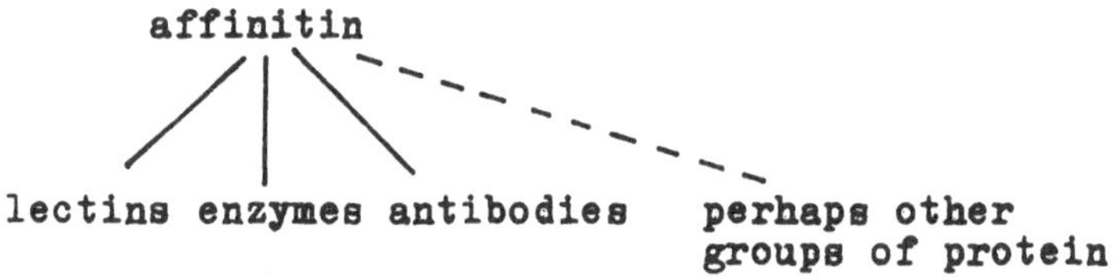

From our point of view the "affinitin", isolated by Hankins and Shannon (6) is not a lectin but a (temporary hemagglutinating) enzyme. In the same way a carbohydrate reacting antibody is an antibody and not a lectin. Foregoing considerations should be the subject of a "comparative affinitinology".

The interactions of affinitins belonging to different groups are of interest, too. Generally only one reaction partner acts as affinitin. Thus, an antibody reacts as affinitin with a lectin if it is directed against antigenic determinants of the lectin (Fig. 1a). On the other side a lectin reacts as affinitin if it combines with the carbohydrate moiety of an immunoglobulin molecule (Fig. 1b). Both reactions can take place simultaneously, too (Fig. 1c).

The term "affinitin" is strictly to distinguish from "receptor specific protein" of Gold and Balding (7) which is ultimately an other expression for lectins only. Likewise Burke a. o. (8) emphasize the necessity of the introduction of a common terminology for systems consisting of a thermo-labile "active" and a "passive" component. The

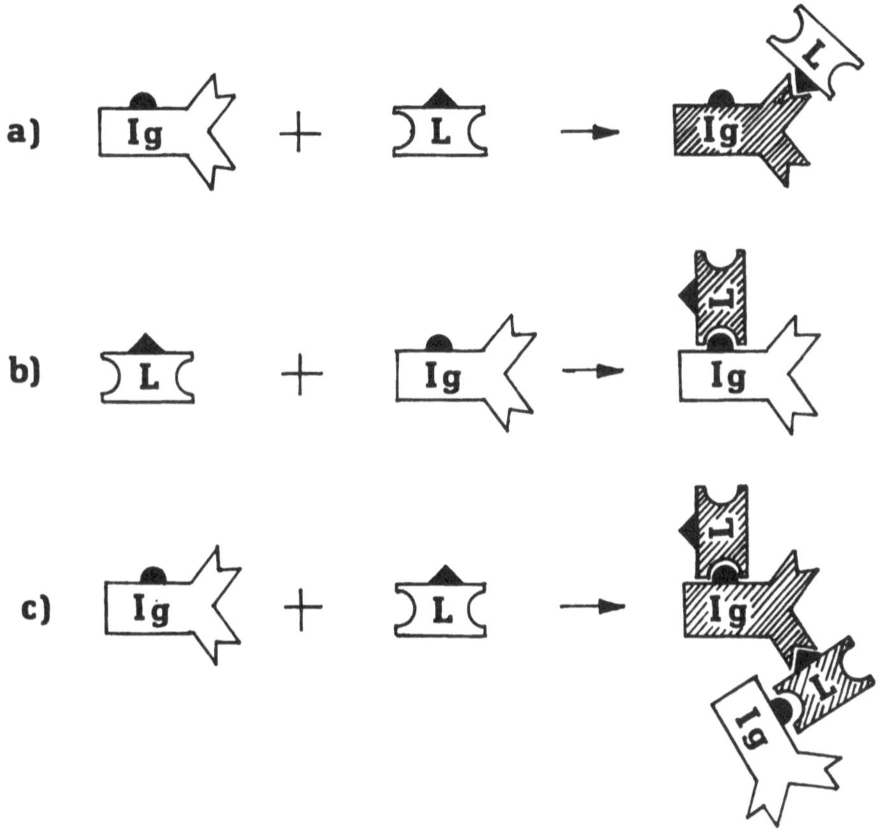

Fig. 1: Interactions between immunoglobulins and lectins

a) as an affinitin (hatched) reacts immunoglobulin
b) as an affinitin (hatched) reacts lectin
c) as affinitins (hatched) reacts immunoglobulin and lectin

= Immunoglobulin with combining site and carbohydrate moiety

= Lectin with combining site and antigenic determinant

active partner (cognor) recognizes the passive one (cognon). In contrast to the cognor-cognon system the term affinitin is not only a describing one because it is based on the community of chemical structures (combining sites). In the first line the reason to propose "affinitin" is to give the base for a systematization following Aristotle: "Ought we, for instance, to begin by discussing each separate species - man, lion, ox, and the like - taking each kind in hand independently of the rest, or ought we rather to deal first with the attributes which they have in common in virtue of some common element of their nature, and proceed from this as a basis for the consideration of them separately?" (cited after (9)). This can be illustrated by an example. All animals with a vertebral column are systematized as vertebrates. Commonly the main groups of vertebrates (fishes, amphibias, reptiles, birds and mammals) are easily to distinguish. Obviously there is an evolution from fishes to mammals. Concerning affinitins in our example combining sites correspond to vertebral column. Provided that there exists also an evolution of affinitins lectins are at the beginning and antibodies are at the end.

The classification of lectins as one of the groups of affinitins is in concordance with the establishment of an independent lectinology. Lectinology cannot be pseudo- or paraimmunology at all. The relation between lectinology and immunology consists in the fact that they both are branches of "affinitology".

References

1. Boyd, W.C., Slapeigh, E.: Science 119, 419 (1954)

2. Goldstein, I.J., Colin Hughes, R., Monsigny, M., Osawa, T., Sharon, N.: Nature 285, 66 (1980)

3. Franz, H., Bergmann, P., Ziska, P.: Histochemistry 59, 335-342 (1979)

4. Franz, H.: In "Affinity chromatography and Molecular Interactions" Colloque Strasbourg 1979, Editor J.-E. Egly Editions INSERM, Paris 1979

5. Roitt, J.: Leitfaden der Immunologie", Dr. Dietrich Steinkopf, Darmstadt 1977

6. Hankins, N.H., Shannon, L.M.: Biol. Chem. 253, 7791-7797 (1978)

7. Gold, E.R., Balding, P.: Receptor-specific proteins. Plant and animals lectins Excerpta Media, Amsterdam and American Elsevier Publishing Company Inc., New York 1975

8. Burke, D., Mendonca-Previato, L., Ballow, C.E.: Proc. Natl. Acad. Sci. USA 77, 318-322 (1980)

9. Morowitz, H.J.: Energy Flow in Biology Biological Organization as a Problem in Thermal Physics Academic Press, New York and London 1968

ISOLATION OF LECTIN-LIKE SUBSTANCES FROM CONNECTIVE TISSUE AND INVOLVEMENT IN CELL-CELL RECOGNITION AND CELL-MATRIX INTERACTIONS

F. Chany-Fournier and J. Gerfaux
Institut National de la Santé et de la Recherche Médicale, U. 43, Hôpital Saint-Vincent-de-Paul, 74 avenue Denfert-Rochereau, 75014 Paris, France

Introduction

The interactions between the cell surface and the extracellular matrix and their consequences on cell functions were explored in two subsequent studies. Firstly, water soluble proteins were isolated from loose connective tissue. They share some properties of lectins and induce rapidly the degradation of the antiviral state established in cells by interferon. Secondly, by enzymatic digestion, subunits of insoluble structures, deriving mainly from the basement membrane, were isolated. The role of these substances in cell recognition and cell adhesion was studied.

Previous observations concerning the human amniotic membrane led us to study the regulatory function of the extracellular matrix on the adjacent cells. We showed that amniotic cells embedded in the basal lamina were resistant to viral infection and unable to multiply (1). When these cells were separated by trypsinization from their natural environment, they started to multiply and became in parallel virus sensitive. We concluded from these observations that substances located in the loose connective tissue or in the basement membranes could affect the metabolism of the adjacent cells.

Lectins - Biology, Biochemistry, Clinical Biochemistry, Vol. I

A. Isolation and Biological Action of Water Soluble Proteins

We thus attempted to isolate, from different connective tissues, human fetal membranes (2, 3) and murine thoracic cartilage (4, 5), substances which could affect cell receptors and consequently modify cell functions.

As a model of defined cellular function we studied the cellular expression of interferon (Fig. 1). Interferon, a cell-coded glycoprotein synthesized after viral infection, acts as a messenger protein on target cells and induces the antiviral resistance, at least in homologous cells. The antiviral state is transitory and decreases with time. Its decay is governed

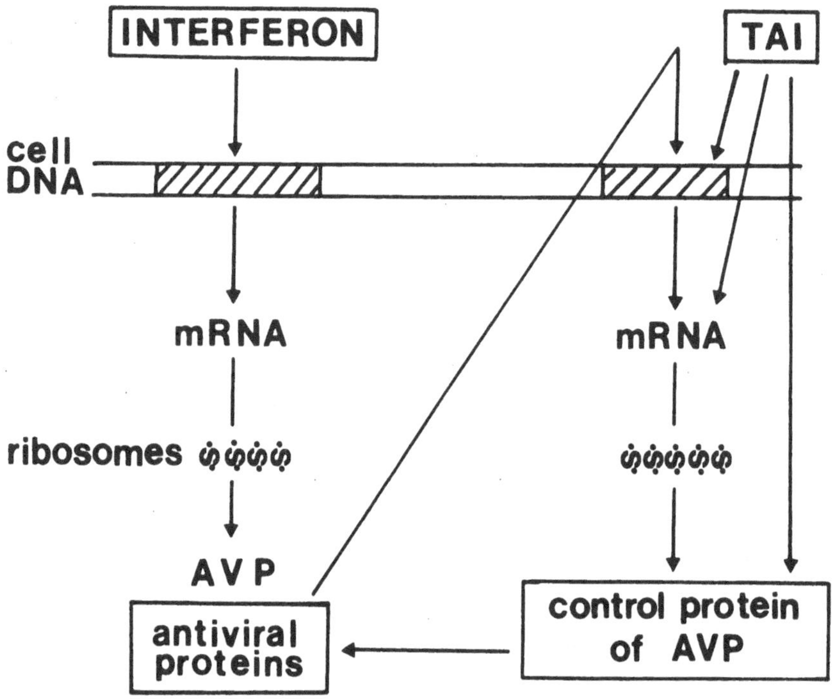

Fig. 1. Regulatory mechanism of the antiviral state induced in the cell by interferon.

by regulatory proteins which might be triggered when the concentration of antiviral factors reaches a critical level (6). Consequently, the cell recovers its primary sensitivity to viruses. The decay of the antiviral state can be increased by extracellular matrix substances acting from the outside of the cell. Because of their antagonistic effect on the interferon-induced cell, these substances were termed "TAI" (tissue antagonist of interferon) (2). We have demonstrated that the antagonistic action of TAI is indirect and could act precisely through the regulatory protein either as a co-inducer of its synthesis or as a catalyser of its action.

The biological effect of tissue extracts (TAI) on the decay of the antiviral state induced in the cells by interferon is shown in Fig. 2. The dose-response relationship between the concentration of tissue extracts and a single dose of interferon is studied. The activities of both interferon and TAI are estimated by their respective effect on the replication of vesicular stomatitis virus (VSV), used as a challenge virus, in murine L cells. In the absence of interferon treatment, tissue extracts do not affect virus multiplication which reaches 10^6 PFU/ml. In the presence of interferon, a significant inhibition (5 logs) of viral replication can be seen. In the presence of both interferon and TAI (added 5 hours later - the time necessary for the induction of the antiviral state), we observe a linear relationship between the concentration of extracts and the degradation of the antiviral state. At the maximal concentration of 1g%, TAI restores in the cell the virus sensitivity of about 4 logs.

The same chronology and experimental conditions were used to show that plant lectin, such as phytohemagglutinin (PHA) can mimic a comparable antagonistic effect on the antiviral state. This effect is similarly concentration dependent (7).

These two experimental models suggest that extracellular stimuli are in both cases responsible for the degradation of this induced cell function. The striking resemblance between the

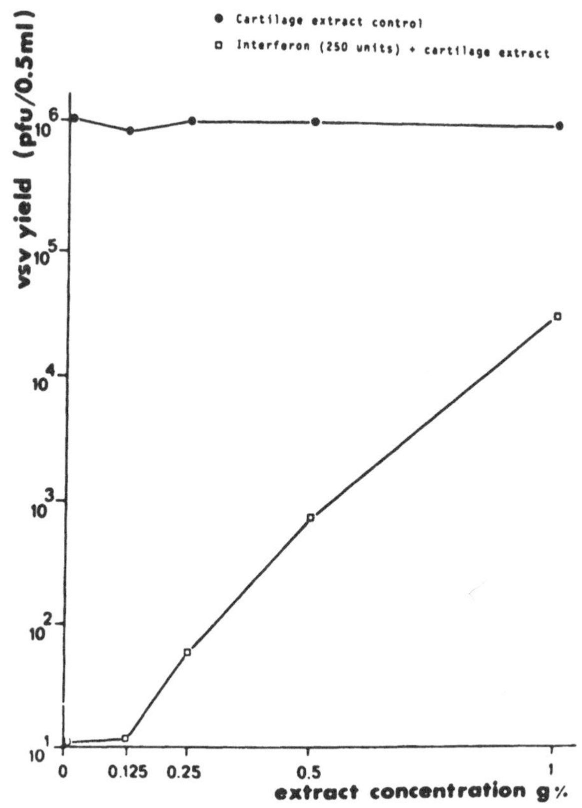

Fig. 2. Dose-response effect of increased concentrations of crude cartilage extract on the interferon-induced antiviral state.

anti-interferon effect of PHA and TAI raises the question whether TAI could share or not some properties generally attributed to lectins.

Indeed, crude tissue extracts prepared from murine thoracic cartilage agglutinate mouse embryonic fibroblasts (MEF) and murine transformed L cells (929 strain). The agglutinating titer of TAI (2g% dry weight) is usually about 100 units per ml for transformed cells and 10 times higher for fibroblasts. In contrast, purified PHA (1.7 mg/ml), used as reference, agglutinates more intensively transformed cells (the titer is about 20,000 units per ml), whereas in normal cells the activity is practically nil. In these two types of cells, the cytoagglutination due to TAI is inhibited by the following sugars: 2-deoxy D-glucose, D-glucosamine, 2-deoxy D-galactose, D-galactosamine, and also D-galacturonic acid to a lesser extent.

Because of these two properties, cell agglutination and affinity for specific saccharides, we suggest that substances present in the tissue preparations could bind, like lectins, to polysaccharidic residues, such as gangliosides or glycoproteins located in the cell membrane (5).

In Fig. 3, we summarize the involvement of extracellular matrix lectins in the modulation of the cell membrane receptors. Firstly, the interaction of interferon with membrane glycoconjugates results in a modulation of membrane receptors compatible with the establishment of the antiviral state. It has been suggested that the interferon receptors are composed of two sites; one consisting in gangliosides and the other in glycoproteins (8). For both the establishment and maintenance of the antiviral state, the integrity of this interferon receptor system is needed.

Secondly, the addition of different lectins from plants (PHA) or extracellular matrices (TAI), which present affinity for polysaccharidic structures of the cell membrane, induces a new distribution of its receptors. This modification imposed from the outside of the cell leads to the decay of the antiviral state.

Thirdly, this re-distribution of membrane receptors is reversible, since the lectins can be removed or displaced by competitive saccharides. The recovery of the initial state of the cell is checked by the rescue of the interferon receptors, as shown by the re-induction of the antiviral state.

Figure 3

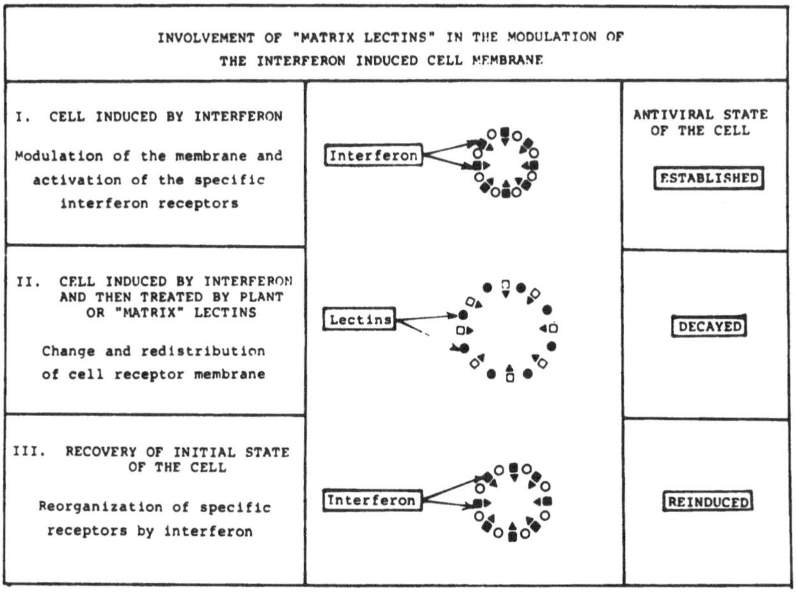

B. Study on the Insoluble Extracellular Matrix

Basement membranes are generally defined as a specialized connective tissue which separates cells from different embryological origin. In the case of the human glomerule the basement membrane separates the urinary epithelium from the vascular endothelium. This structure can be obtained free of cellular material by physical methods (such as ultrasonic treatment). The basement membrane is a heterogenous complex composed of collagen type IV associated to structural glycoproteins. The type IV collagen, genetically defined as the collagen of basement membranes, consists of three identical α (IV) chains, forming a triple helical domain. These chains are similar to the α chains of other types of collagen, but differ by the presence of globular telopeptide situated at each extremity and characterized by heteropolysaccharidic residues and disulfide bridges (9). The N-terminal telopeptide is apparently not removed before its integration into the basement membrane. For this reason, type IV collagen can be assimilated to a procollagen-like molecule.

Since these basement membrane structures are highly insoluble, their study requires enzymatic proteolysis followed by fractionation and purification. The different subunits are then studied and characterized and their functional role related to their structure (10, 11).

The cytoagglutinating activity of human glomerular basement membrane (HGBM) fractions issued from pepsin or collagenase digest is tested in the presence of human normal or transformed cells. As can be seen in Fig. 4, a limited pepsin proteolysis releases collagen moiety, consisting mainly in a triple helical material associated to the globular heteropolysaccharidic telopeptide. In this case, all the fractions are active. The agglutinating activity is obtained at concentrations ranging from 0.1 μg to 20 μg for $2x10^5$ cells per 0.5 ml. After collagenase digest, a material containing either collagen or heteropolysaccharidic components develops no activity. Adequate controls of purified collagens (types I, II, III) from different origins are similarly ineffective.

A study in SDS-polyacrylamide gel electrophoresis, under reducing conditions, indicates that the active material (from pepsin digest) contains mainly three bands of 150 K, 100 K and 70-80 K, which are characteristic of the type IV procollagen.

We then tested some sugars for their capacity to compete with the HGBM agglutinating factors for the binding sites of the cell membrane. The inhibitory sugars tested in the presence of Kb cells are the N-acetyl-osamines and the N-acetyl-neuraminic acid. The basement membrane fractions involved in cell agglutination at 4 hours induce cell attachment and spreading after an incubation period of 20 hours at 37°C. This spreading activity does not correspond to a mitogenic effect, since the

CELL AGGLUTINATING ACTIVITY OF HGBM FRACTIONS

	HGBM FRACTIONS (40 µg/ml)		HUMAN CELLS ($2x10^5$/0.5 ml) F7000	HUMAN CELLS ($2x10^5$/0.5 ml) Kb
	COLLAGENASE DIGEST			
	S.Col	soluble collagenase material	64	32
Sephadex G200	F_I	heterogenous collagen components	64	32
Sephadex G200	F_{II}	heterosaccharidic noncollagen components	<2	<2
Sephadex G200	F_{III}	collagen components	<2	<2
	F_I	(reduced alkylated)	<2	<2
Biogel P_2	Dc_{IV}	collagen peptide	<2	<2
	PEPSIN DIGEST			
NaCl Precipitation	A	soluble pepsin material	128	64
NaCl Precipitation	B ↓0.9 M	triple helical procollagen type IV	128	128
NaCl Precipitation	C ↓1.7 M	triple helical procollagen type IV	512	256
NaCl Precipitation	D ↓4.5 M	triple helical procollagen type IV	512	128
	A	(reduced alkylated)	<2	<2
	INTERSTITIAL COLLAGENS			
	Type I	bovine tendon	<2	<2
	Type II	murine cartilage	<2	<2
	Type III	calf skin	<2	<2
	Deglycosylated bovine tendon		<2	<2
	FIBRONECTIN		ND	256
	Pepsin digest fibronectin		<2	<2

Figure 4

number of cells is not modified and the incorporation of ^{3}H-thymidine is not increased.

In the case of the insoluble basement membranes, we propose in conclusion two models of cell-matrix interactions (Fig.5). Firstly, as shown by Wiestner et al.(12), the fibroblast present in loose connective tissue synthesizes a procollagen molecule. Before its integration into the loose connective tissue, the procollagen molecule loses its extension telopeptide which exerts a negative control on the biosynthesis of collagen.

Secondly, the epithelial cell is strongly attached to the basement membrane. We postulate that the conformational units of the basement membrane consist of type IV procollagen associated to heterosaccharidic components. This peculiar structure could interact directly with the cell membrane and mediate cell adhesion and cell regulation. The conformational units

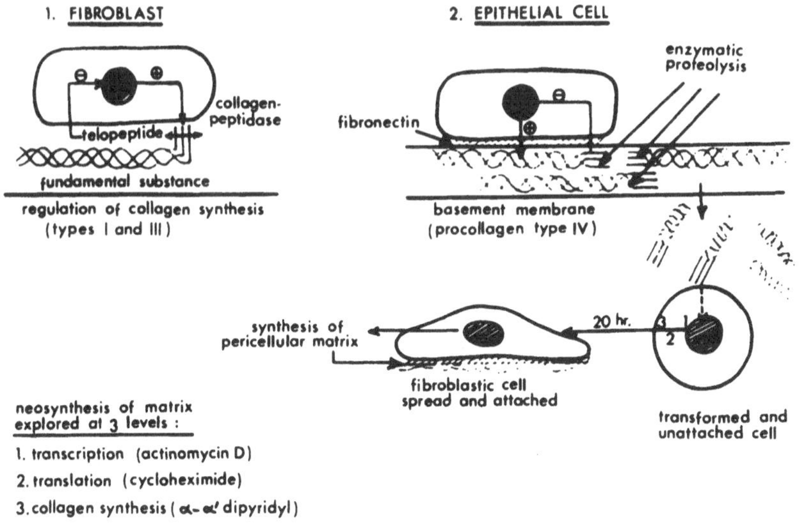

Fig. 5 Two models of cell-matrix interactions.

released by enzymatic proteolysis interact in vitro with the cell membrane in cell to cell recoqnition process which is the lectin-like mechanism (11). This transmembrane siqnal is followed by a phenotypic chanqe of the transformed cell into a fibroblastic shape, which involves intracellular events and results in attachment of the cell to the substratum followed by spreadinq. This increased adhesion of the cell could be explained by the neosynthesis of a pericellular matrix.

References

1. Chany, C., Gresser, I., Vendrely, C., Robbe-Fossat, F.: Proc. Soc. Exp. Biol. Med. 123, 960 (1966).
2. Fournier, F., Rousset, S., Chany, C.: Proc. Soc. Exp. Biol. Med. 132, 943 (1969).
3. Fournier, F., Chany, C., Sarraqne, M.: Nature 235, 47 (1972).
4. Chany-Fournier, F., Pauloin, A., Chany, C.: C. R. Acad. Sci. (Paris) 285, 941 (1977).
5. Chany-Fournier, F., Pauloin, A., Chany, C.: Proc. Nat. Acad. Sci. (USA) 75, 2333 (1978).
6. Chany, C., Fournier, F., Rousset, S.: Nature 12, 113(1971).
7. Pauloin, A., Chany-Fournier, F., Fpstein, L., Chany, C.: C. R. Acad. Sci. (Paris) 284, 1119 (1977).
8. Chany, C.: Biomedicine 24, 148 (1976).
9. Prockop, D., Kiqirikko, K., Tuderman, L., Guzman, N.: New Enqland J. Med. 301, 13-23, 77-85 (1979).
10. Gerfaux, J., Chany-Fournier, F., Bardos, P., Muh, J.P., Chany, C.: C.R. Acad. Sci. (Paris) 288, 915 (1979).
11. Gerfaux, J., Chany-Fournier, F., Bardos, P., Muh, J.P., Chany, C.: Proc. Nat. Acad. Sci. (USA) 76, 5126 (1979).
12. Wiestner, M., Krieq, T., Horlein, D., Glanville, R.W., Tietzek, T., Muller, P.K.: J. Biochem. 254, 7016 (1979).

RAT PERITONEAL MACROPHAGES EXPRESS TWO DIFFERENT LECTIN-LIKE RECEPTORS ON THE CELL SURFACE

Y. Nagamura and H. Kolb

Diabetes Research Institute, University of Düsseldorf
D-4000 Düsseldorf, Federal Republic of Germany

Introduction

Macrophages are able to differentiate between self and altered self or nonself (1, 2). The recognition mechanism involved remains unclear although many hypotheses have been put forward (3, 4). Recently we have reported that rat hepatocytes and Kupffer cells express a receptor with specificity for terminal D-galactosyl residues which is able to recognize altered self- and non self-structures in vitro (5, 6, 7). We have found a similar receptor (gal-lectin 1) on the surface of rat peritoneal macrophages (8, 9) and on the same cells a second D-galactose specific receptor (gal-lectin 2) which is expressed after induction of phagocytosis or pinocytosis. In this paper we discuss the characteristics of the two lectin-like receptors on rat peritoneal macrophages.

Materials and Methods

Animals. Osborn-Mendel rats weighing 200-250 g were used.

Substances. Orosomucoid and asialo-orosomucoid were a gift from Dr. G. Ashwell (National Institute of Health, Bethesda, USA). D-Glucosyl-bovine serum albumin (D-glucosyl-BSA) and D-galactosyl-bovine serum albumin (D-galactosyl-BSA) were a gift from Dr. Y.L. Lee (John Hopkins University, Baltimore,

Lectins - Biology, Biochemistry, Clinical Biochemistry, Vol. I

USA). Neuraminidase (Behringwerke, Marburg, Fed. Rep. Germany) was from Vibrio cholerae. Colchicine and cytochalasin B were obtained from Serva, Heidelberg, Fed. Rep. Germany) and Eagle's medium from Flow Laboratories, Bonn, Fed. Rep. Germany. Drawing ink (Gunther Wagner, Pelikan fount india) was dialysed against 0.9 % physiological saline over night and centrifuged at 200 x g for 10 minutes. The supernatant was used for injection. Polyvalent rabbit or goat anti rat IgG was purchased from Nordic, Tilburg, The Netherlands.

Peritoneal cells. Rats were injected with 0.2 ml (0.5 ml) dialysed drawing ink, 0.2 ml (0.5 ml) liquide paraffin or 0.2 ml (10 ml) thioglycollate medium intraperitoneally. After 30 minutes (2 days) peritoneal cells were harvested by rinsing the peritoneal cavity with 15 ml Eagle's medium, and washed twice.

Neuraminidase treatment of erythrocytes. Fresh rat erythrocytes were mixed with sodium citrate and washed three times with Eagle's medium. Erythrocytes were suspended in Eagle's medium at a concentration of 10^9 cells/ml. Incubation with neuraminidase (0.02 units/ml) was done at 37^oC for 30 minutes. After incubation, erythrocytes were washed three times.

Cell adhesion assay. 50 µl of peritoneal cell suspension (5×10^6 cells/ml) were mixed with an equal volume of erythrocytes (10^8 cells/ml) in plastic test tubes. When the adhesion test was performed with normal erythrocytes tubes were centrifuged at 100 x g for 15 seconds and incubated for 1 hr. at 4^oC. When the test was performed with asialo-erythrocytes centrifugation was omitted. Cells were resuspended by gentle shaking of tubes, and fixed with 0.5 % glutaraldehyde (in 0.2 % pH 7.2 phosphate buffer) for 3 minutes. Before adding Nadi esterase staining solution cells were incubated with bovine serum albumin (final concentration 2 %) for 10 min. This staining procedure could be omitted when macrophages which had phagocytosed carbon particles were used. Esterase staining and phagocytosis of carbon particles described the same cell population. Cells were counted in hemocytometer. Binding was regarded as positive when three or more erythrocytes adhered to one macrophage.

Inhibition studies. Peritoneal cell suspensions were preincubated with various saccharides, glycoproteins and with membrane fragments of erythrocytes for 5 min at 0°C. Preincubation with sodium azide, colchicine, cytochalasin B and anti-immunoglobulin serum was done for 30 min at 37°C.

Results

Table 1 Recognition of desialylated erythrocytes by peritoneal macrophages (gal-lectin 1)

Erythrocytes	Saccharides (25 mM)	Rosette Formation (% ± SD)	Inhibition (%)
Untreated	None	8.9 ± 3.6	
Desialylated	none	61.2 ± 13.0	
	D-Glucose	60.0 ± 11.3	1.6
	N-Acetyl-D-glucosamine	60.0 ± 10.0	2.0
	D-Mannose	58.4 ± 11.6	4.6
	D-Galactose	17.2 ± 3.7	71.8
	D-Galactosamine	58.0 ± 11.6	5.2
	N-Acetyl-D-galactosamine	14.8 ± 3.2	75.8
	D-Fructose	58.0 ± 10.8	5.2
	L-Fucose	22.0 ± 5.2	64.1
	D-Fucose	19.1 ± 4.5	68.8
	Lactose	21.9 ± 4.9	64.2
	Lactulose	17.4 ± 4.3	71.6
	Melibiose	30.9 ± 6.3	49.5

The spontaneous adhesion of neuraminidase-treated erythrocytes to peritoneal macrophages is shown in Tab. 1. Rosette formation between untreated or desialylated erythrocytes and macrophages is 8.9 % and 61.2 % respectively. The binding of desialylated erythrocytes can be inhibited by the addition of sugars which are related in structure to D-galactose (Tab.1)

Table 2 Glycoproteins inhibit contacts between peritoneal macrophages and asialo-erythrocytes (gal-lectin 1)

Glycoproteins		Rosette Formation (% ± SD)	Inhibition (%)
None		66.7 ± 10.2	
Orosomucoid	50µM	63.3 ± 10.7	5.1
Asialo-orosomucoid	50µM	17.7 ± 3.2	73.5
D-Galactosyl-BSA	1µM	15.0 ± 3.1	77.5
D-Glucosyl-BSA	10µM	16.1 ± 4.0	75.9
Bovine serum albumin	10µM	61.6 ± 7.7	7.6

Glycoproteins with terminal D-galactosyl residues are potent inhibitors of lectin-dependent cell contacts (Tab. 2). The lectin also recognizes D-glucosyl-BSA but with lower avidity than D-galactosyl-BSA. This crossreaction is not apparent at the level of monosaccharides (Tab. 1).

Table 3 Characteristics of peritoneal macrophages receptor for asialo-erythrocytes (gal-lectin 1)

Inhibitor		Rosette Formation (% ± SD)	Inhibition (%)
None		63.5 ± 8.3	
EDTA	2mM	7.4 ± 1.4	88.3
EGTA	2mM	6.1 ± 1.4	90.4
NaN_3	10mM	20.9 ± 5.5	67.1
Cytochalasin B	40µM	63.2 ± 8.2	0.5
Cochicine	250µM	59.9 ± 8.0	5.7
Anti-Rat IgG (Rabbit)	20µl	60.4 ± 7.3	4.8
Anti-Rat IgG (Goat)	20µl	59.1 ± 7.7	6.9

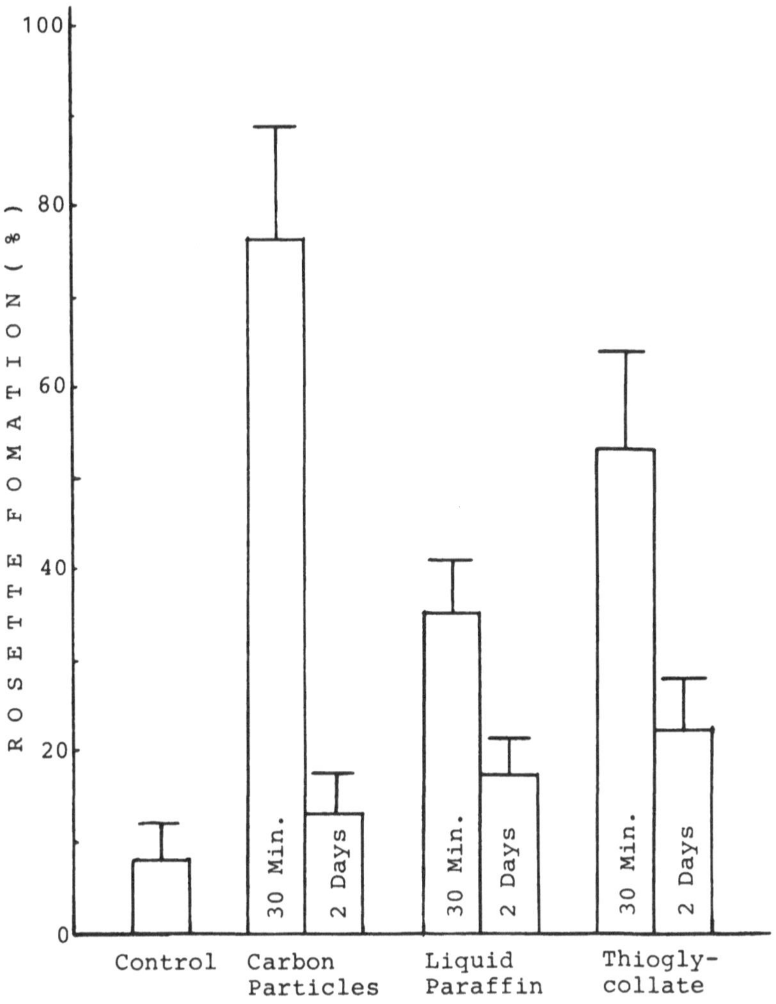

Fig. 1 Recognition of autologous erythrocytes by activated peritoneal macrophages (gal-lectin 2)
Peritoneal cells were harvested either 30 minutes or two days after i.p. injection of carbon particles, liquid paraffin or thioglycollate.

Lectin function appears to depend on the presence of calcium ions since chelating agents as EDTA and EGTA blocked rosette formation. Pretreatment of peritoneal cells with sodium azide also was inhibitory whereas the addition of cytochalasin B, colchicine or anti-rat IgG serum was of no effect. Thus microfilaments, microtubuli or antibodies are probably not involved in cell contact formation.

Fig. 1 shows that after the induction of endocytosis the percentage of autorosette forming macrophages increases strongly. The best activation is found with carbon particles, leading to spontaneous autorosettes in about 80 % of phagocytes. When macrophages are tested 2 days later the percentage of autorosette is near normal again.

Table 4 Specificity of macrophage-erythrocyte cell contacts: Inhibition by saccharides (gal lectin 2)

Saccharides (25 mM)	Rosette Formation (% ± SD)	Inhibition (%)
None	77.3 ± 11.7	
D-Glucose	77.6 ± 13.6	0.0
D-Mannose	76.8 ± 10.4	0.6
D-Fucose	70.1 ± 9.0	9.3
L-Fucose	74.7 ± 8.2	3.3
D-Galactose	74.5 ± 10.8	3.6
D-Galactosamine	72.5 ± 9.7	6.2
N-Acetyl-D-galactosamine	11.0 ± 4.8	85.8
Lactose	65.0 ± 5.9	15.9
Lactulose	71.2 ± 4.8	7.9
Melibiose	75.7 ± 7.4	2.1

Rat peritoneal macrophages were activated by phagocytosis of carbon particles as described in Materials and Methods.

Table 5 Specificity of macrophage - erythrocyte cell contacts Inhibition by glycoproteins (gal lectins 2)

Proteins	Rosette Formation (% ± SD)	Inhibition (%)
None	77.4 ± 12.6	
Orosomucoid	75.5 ± 12.5	2.5
Asialo-orosomucoid	32.1 ± 4.2	58.5
Bovine serum albumin	75.6 ± 11.8	2.8
D-Galactosyl-BSA	21.2 ± 3.4	72.6
D-Glucosyl-BSA	72.2 ± 11.2	6.7
Solubilised membrane of 10^8 rat erythrocytes	25.8 ± 5.2	66.7
Solubilised membrane of 10^8 rat asialo-erythrocytes	23.5 ± 4.4	69.4

The specificity of cell contacts was analysed by competitive inhibition experiments with monosaccharides, oligosaccharides glycoproteins and erythrocytes membrane vesicles. Of various mono- and oligosaccharides tested in 25mM concentration only N-acetyl-D-galactosamine was able to impair rosette formation (Tab. 4). Similarly, glycoproteins with terminal D-galactosyl residues blocked cell contacts, but not glycoproteins with terminal sialic acid (orosomucoid) or D-glucosyl residues (Tab. 5). For D-galactosyl BSA the 50 % inhibition concentration was less than 10^{-6}M. D-galactosyl BSA therefore appears to be a highly specific inhibitor of the peritoneal macrophage receptor. We have also tested whether membrane fragments of erythrocytes are able to interfere with autorosette formation. As shown in Tab. 5 inhibition was found with membrane fragments from normal as well as from desialylated rat erythrocytes.

Table 6 Further characteristics of cell contacts between rat peritoneal macrophages and rat erythrocytes (gal lectin 2)

Addition		Rosette Formation (% ± SD)	Inhibition (%)
None		76.1 ± 10.3	
NaN_3	10mM	72.2 ± 11.2	5.2
Cytochalasin B	0.04mM	75.0 ± 11.3	1.4
Colchicine	0.25mM	75.9 ± 8.0	0.2
EDTA	2mM	70.9 ± 12.2	6.8
EGTA	2mM	69.0 ± 10.2	9.3
Anti Rat IgG (Rabbit)	20µl	74.1 ± 7.5	2.8
Anti Rat IgG (Goat)	20µl	73.2 ± 8.4	3.8

Autorosette formation occurred independent from the metabolic state of macrophages and under condition of blocked microfilament and microtubulus function. I.e., preincubation of macrophages with NaN_3, cytochalasin B and colchicine had no effect (Tab. 6). Also, receptor function was not dependent on the presence of divalent cations. The receptor did not react with anti-immunoglobulin sera (Tab. 6).

Discussion

We have described the spontaneous adhesion of normal and of desialylated autologous erythrocytes to peritoneal macrophages. Since rosette formation is observed at 4°C enzyme activities therefore are unlikely to be involved. Specific inhibition of cell contact formation is achieved by the addition of N-acetyl-D-galactosamine, asialo-orosomucoid, D-galactosyl BSA, but not by saccharides unrelated in structure to D-galactose. Antisera against rat immunoglobulin do not affect cell adhesion. The receptors for normal and for desialylated autologous erythrocytes therefore behave functionally as lectins. The receptor mediating adhesion of desialylated erythrocytes (gal-lectin 1) is shown here to be different from the receptor mediating cell contacts with normal erythrocytes (gal-lectin 2). Gal-lectin 1 is distinguishable from gal-lectin 2 for the following reasons: (i) Activity of gal-lectin 1 is present on unstimulated macrophages as well as after induction of endocytosis. (ii) Activity of gal-lectin 2 appears only after the induction of endocytosis. (iii) Gal-lectin 1 does not recognize normal but only desialylated erythrocytes. (iv) Gal-lectin 1 reacts with D-glucosyl residues, gal-lectin 2 does not. (v) Activity of gal-lectin 1 is dependent on the presence of calcium ions. Thus the two lectin-like receptors on peritoneal macrophages appear to be different with respect to specificity and cation requirements.

Although the receptor for autologous erythrocytes could not be demonstrated on untreated macrophages a high activity of gal-lectin 2 is seen within 30 minutes after inducing phagocytosis or pinocytosis. This short time period suggests the expression of previously hidden or inactive receptor molecules. Alternatively the new receptor may be formed from elements of previously existing lectins. Another possibility would be the absorption of receptor molecules from the peritoneal fluid. Such a mechanism appears less probable since the induction of receptor expression could also be achieved in serum free tissue culture (Nagamura and Kolb, to be published).

The function of the galactose-specific receptors on macrophages is not known. Lectins may be involved in the process of self-non self discrimination or they may serve nonimmune functions as has been postulated by Ashwell (10).

Acknowledgements. We are grateful to Prof. Gries for stimulating discussions. This work was supported by the Minister für Wissenschaft und Forschung des Landes Nordrhein-Westfalen.

References

1. Metchnikoff. E.: Lectures on the comparative pathology of inflamation, Dover, New York 1968

2. Wilkinson, P.E.: Clin. exp. Immunol. 25, 355-366 (1976)

3. Aminoff, D., vor der Bruegge, W.F., Bell, W.C., Sarpolis, K. and Williams, R.: Proc. Natl. Acad. Sci 74, 1521-1524 (1977)

4. Kolb, H.: Develop. Compar. Immunol. 1, 193-206 (1977)

5. Kolb, H. and Kolb-Bachofen, V.: Biochem. Biophys. Res. Commun 85, 678-683 (1978)

6. Kolb, H., Kriese, A., Kolb-Bachofen, V. and Kolb, H.A.: Cell Immunol. 40,457-467 (1978)

7. Kolb, H., Schudt, C., Kolb-Bachofen, V. and Kolb, H.A.:Exp. Cell Res. 113, 319-325

8. Nagamura, Y. and Kolb, H.: FEBS Lett. 115, 59-62 (1980)

9. Nagamura, Y. and Kolb, H.: Hoppe Seyler's Z. Physiol. Chem. 361, 306-307 (1980)

10. Ashwell, G. and Morell, A.G.: Trends Biochem. Sci. 2, 76-78 (1977)

HOST-MICROORGANISM INTERACTION: DEVELOPMENTALLY REGULATED LECTIN-MEDIATED CAPTURE OF NEMATODES BY NEMATODE-TRAPPING FUNGI.

B. Nordbring-Hertz, E. Friman
Department of Microbial Ecology, Ecology Building
University of Lund, S-223 62 Lund, Sweden

P.A. Johansson and B. Mattiasson
Department of Pure and Applied Biochemistry, Chemical Center
University of Lund, S-220 07 Lund, Sweden

Introduction

Nematodes are common soil inhabitants. Many of them are bacterial or fungal feeders, whereas others are plant-parasitic, causing damage to agricultural crops. Nematophagous fungi are natural enemies of nematodes. These soil fungi comprise a taxonomically heterogeneous group, but they are biologically similar in that they can utilize nematodes as nutrients. Some capture their prey in sophisticated nematode-trapping devices such as adhesive networks (Fig.1), adhesive branches, adhesive knobs or constricting rings, whereas others use spores as infecting agents (for review see ref.1).

A common feature of these fungi is that the ability to attack nematodes is mostly confined to a certain stage of the fungal development: the trap or the spore. Further, in many cases the process of capture is connected with an adhesive developmental phase of the fungus.

Specificity in the choice of prey is little understood in the nematode-nematophagous fungus system, as are the recognition mechanisms behind a possible specificity. Furthermore, although different aspects of taxonomy, morphology and

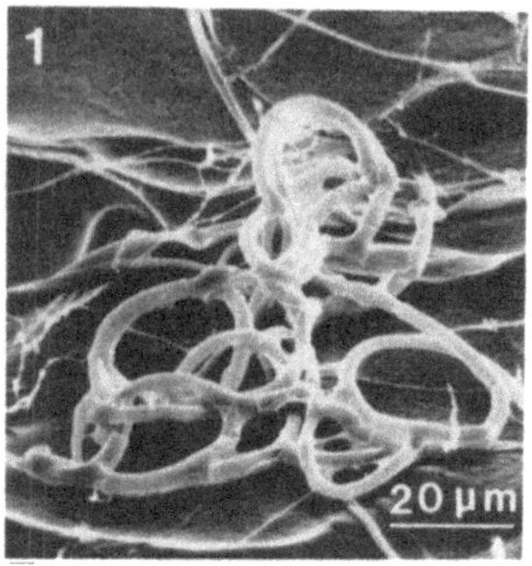

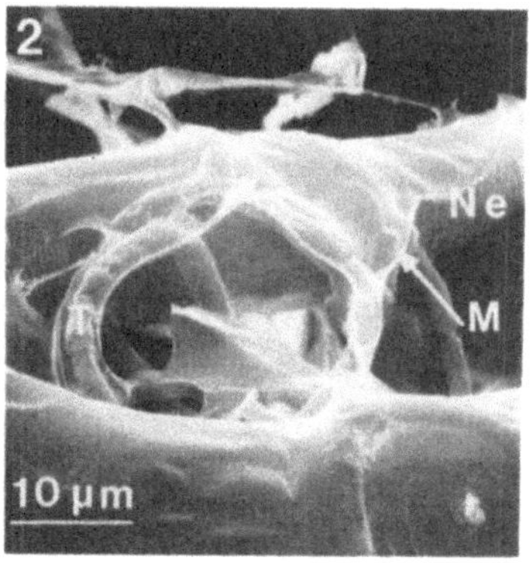

Fig.1. Adhesive network of Arthrobotrys oligospora. (From Nordbring-Hertz, B. 1972. Ref.2)

Fig.2. Nematode (Ne) captured in adhesive network (T) of A. oligospora. (From Nordbring-Hertz, B. and Stålhammar--Carlemalm, M. 1978. Ref.4)

physiology have been studied for a long time, the molecular mechanism behind the capture of nematodes has been totally obscure. Recently, however, evidence was presented for the presence of a lectin on the traps responsible for binding to surface structures of the nematode (3).

In this article we review our present knowledge of this host-microorganism relationship. The significance of our results in terms of specificity of fungus-animal interaction and the recognition mechanism behind the specificity is discussed.

Results

Arthrobotrys oligospora Fres. (ATCC 24927) is a nematophagous fungus common in all types of soil. It captures nematodes with the aid of three-dimensional structures (Fig.1) equipped with an adhesive material. Although the traps are hyphal structures, they are different from hyphae in many respects (Table 1). Thus no adhesive is present on hyphae, and the traps differ ultrastructurally from hyphae in their thicker cell wall and their organelle contents (4). When nematodes are added to a fungal culture containing traps, they are captured within a few minutes. Scanning electron microscopy (SEM) revealed increased secretion of a mucilaginous substance in the presence of prey (Fig.2). Transmission electron microscopy (TEM) showed an electron dense layer between the nematode surface and the fungal trap (Fig.3) and that penetration of the nematode cuticle (Fig.4) occurred within 60 min after capture. As the nematode was invaded the electron dense organelles in the trap gradually disappeared (4).

Thus, although many structural and developmental features were known, the mechanism of capture was obscure. How do the nematodes stick? Are lectins involved?

If lectins were involved in the attachment of nematodes to the traps, then capture should be prevented by treatment of the traps with carbohydrates. To investigate the nematode trapping ability after different treatments, A.oligospora was grown on dialysis membranes on top of agar substrates, and trap formation was induced as described previously (3). The membrane cultures were flooded with solutions of carbohydrates (20 mM dissolved in 0.01 M phosphate buffered saline) and the ability to capture the bacterial feeding nematode Panagrellus redivivus was tested after different periods of time. Capture was specifically inhibited by N-acetyl-D-galactosamine (GalNAc) at a concentration of 20 mM, whereas other galactose derivatives could prevent or at least delay capture (3) only at increased concentrations (200 mM).

This result suggested that a lectin present on the traps was involved in or responsible for the interaction. To demonstrate the presence of the lectin on the traps we used human red blood cells (RBC) as a model prey. RBC accumulated almost exclusively around the traps. There was no marked difference between the blood groups in these in vivo tests, although blood group A (GalNAc terminal carbohydrate residue) showed a tendency to attach more easily. In addition, GalNAc residues were indicated on the nematode surface with the aid of a peroxidase-lectin conjugate prepared from Helix pomatia lectin. Binding was inhibited by GalNAc (3).

There was a mutual influence of the RBC and the traps: The trapped RBC lysed within 1 hour. Immediately after RBC lysis, the traps lost their three-dimensional appearance and collapsed. Since no significant lysis occurred spontaneously this effect was attributed to the trap. Further, glutaraldehyde fixed RBC were trapped, but did not lyse.

These results indicated that a lectin is present on the traps of A.oligospora and binds to carbohydrates on the nematode surface. This binding initiates an enzymatic alteration of

Table 1

Characteristics of traps and hyphae in Arthrobotrys oligospora

Traps	Hyphae	Ref.
Adhesion of nematodes	No adhesion	1
Mucilaginous coat (SEM)	No mucilaginous coat	2,4
Electron dense organelles, thick cell wall	No organelles thin cell wall	4,v.Hofsten unpubl.
Electron dense layer between trap and nematode	No electron dense layer	4,v.Hofsten unpubl.
Penetration of nematode cuticle by traps	No cuticle penetration by normal hyphae	4
High K^+ (SEM+X-ray microanalysis)	Low K^+	5
High metabolic activity (fluorescein diacetate)	Lower metabolic activity	6
Attraction of nematodes strong	Attraction weaker	Jansson unpubl.
Presence of lectin(s)	No lectin(s) detected	3,7
Toxin?	No toxin?	

both nematode and erythrocyte surfaces leading to penetration of the nematode cuticle and lysis of the RBC.

To study further the binding properties of the supposed lectin, a ^{125}I-labelled homogenate was prepared. Fungal material containing traps, grown on dialysis membranes, was collected and washed in a buffer containing metal ions (1 mM each of Ca^{2+}, Mg^{2+}, Mn^{2+}, Zn^{2+}, K^{+} and Na^{+}) and was then

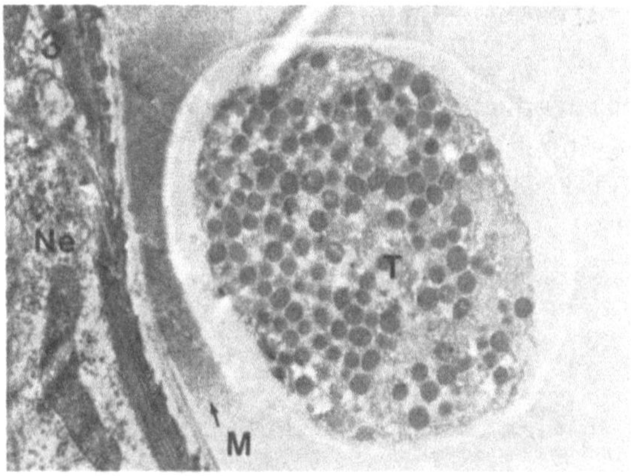

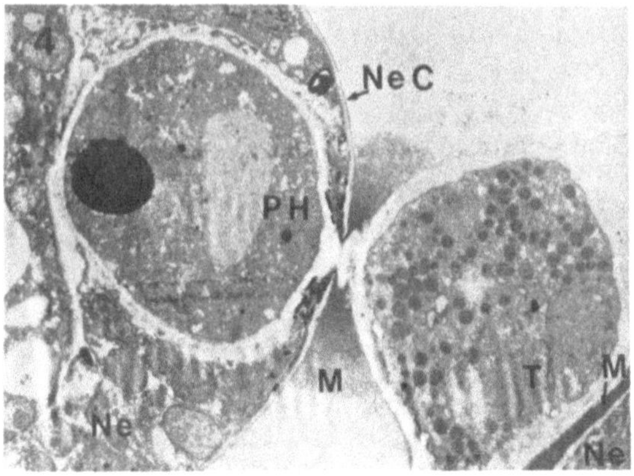

Fig.3. Section of fungal trap (T) 15 min after capture of nematode (Ne). Note contents of organelles in trap and electron dense layer (M) between trap and nematode. (From Nordbring-Hertz, B. and Stålhammar-Carlemalm, M. 1978. Ref. 4)

Fig.4. Penetration of nematode cuticle (NeC) by a penetration hypha (PH) originating from the trap (T). (From Nordbring-Hertz, B. and Stålhammar-Carlemalm, M. 1978. Ref. 4)

treated with 4-sulfo-2(^{125}iodo)-benzendiazoniumchloride (200 µCi) in the presence of 100 mM GalNAc in the same buffer. After homogenization and dialysis the homogenate was tested for binding to 1) GalNAc substituted Sepharose and 2) nematodes (7).

Binding to GalNAc-Sepharose (100 %) was reduced to 30 % when free GalNAc was present. Homogenate, pretreated with trypsin, showed 15 % binding to GalNAc-Sepharose, indicating the protein nature of the binding structure on the trap. In a blank experiment with hydrolyzed epoxi-Sepharose binding occurred to 24 %, which was reduced to 15 % in the presence of GalNAc. Thus there was a certain unspecific binding to Sepharose (Table 2A).

Table 2

^{125}I-labelled fungal homogenate bound to (A) N-acetyl-D--galactosamine-Sepharose and (B) washed nematodes (mean of 2-5 experiments)

A.	
<u>N-acetyl-D-galactosamine-Sepharose</u>	
Homogenate	100 %
Homogenate + GalNAc (0.1 M)	30-32 %
Trypsin-treated homogenate	15 %
<u>Hydrolyzed epoxi-Sepharose</u>	
Homogenate	24 %
Homogenate + GalNAc (0.1 M)	15 %
B.	
<u>Nematodes</u>	
Homogenate	100 %
Homogenate + GalNAc (0.1 M)	20-50 %
Homogenate + trypsin-treated nematodes	67 %

Binding of the homogenate to nematodes was also studied. In this case, too, free GalNAc could partially inhibit binding. Trypsin treatment of the nematode surface prior to binding studies resulted in a reduction of the binding capacity, indicating that glycoproteins are probably present on the nematode surface (Table 2B).

Discussion

To summarize our results hitherto, we have shown that the traps of A.oligospora bear a lectin binding to carbohydrate residues on the nematode surface. The lectin is GalNAc specific and metal ion dependent, and it seems to bind to a glycoprotein on the nematode cuticle. The trap lectin must be developmentally regulated, since in normal hyphae there is no capture ability, nor is lectin activity detected. The results add one more feature to the list of differing characteristics of traps and hyphae, respectively, constituting the unique nature of the trap (Table 1).

The trap lectin is another example of lectins from microorganisms. The involvement in the capture process adds one more facet to the picture of different functions of lectins in nature. The existence of developmentally regulated lectins in nematode-trapping fungi has many features in common with other microorganisms where lectin generation is physiologically regulated, e.g. the slime mold lectins (8,9) and the lectins from Myxococcus xanthus recently detected (10). In these cases the lectins appear when cells are aggregating, and there is a close connection between lectin production and cohesiveness of the cells.

The trap lectin binding to GalNAc on the nematode surface initiates a series of events which are probably responsible for the firm attachment of the nematode. This attachment leads to penetration of the nematode cuticle by a penetration hypha originating from the trap (Table 1) and digestion of the nematode. Our understanding of these further events is still poor. Recent ultrastructural studies (A.von Hofsten, unpublished) confirm the differences between traps and hyphae noted previously (4;Table 1) and suggest a possible connection between trap organelles and the adhesion to the nematode surface and digestion of the nematode.

Concerning specificity in the choice of prey, A.oligospora seems to capture all types of nematodes (but no other animals). No difference in the adhesion of nematodes was noticed between bacterial and fungal feeding or plant-parasitic nematodes (11). Parallel studies of different fungi in their natural environment and in laboratory experiments are now in progress.

This investigation was supported by a grant from the Swedish Natural Science Research Council.

1. Barron, G.L.: The Nematode-Destroying Fungi. Topics in Mycobiology No 1. Canadian Biological Publications Ltd., Guelph, Ontario, Canada 1977

2. Nordbring-Hertz, B.: Physiol. Plant. 26, 279-284 (1972).

3. Nordbring-Hertz, B., Mattiasson, B.: Nature 281, 477-479 (1979).

4. Nordbring-Hertz, B., Stålhammar-Carlemalm, M.: Can. J. Bot. 56, 1297-1307 (1978).

5. Nordbring-Hertz, B.: Trans. Br. mycol. Soc. 68, 53-57 (1977).

6. Nordbring-Hertz, B.: Nematologica 23, 443-451 (1977).

7. Mattiasson, B., Johansson, P.A., Nordbring-Hertz, B.: Acta Chem. Scand. Ser. B. In press.

8. Rosen, S.D., Kafka, J.A., Simpson, D.L., Barondes, S.H.: Proc. Natl. Acad. Sci. USA 70, 2554-2557 (1973).

9. Rosen, S.D., Simpson, D.L., Rose, J.E., Barondes, S.H.: Nature 252, 149-151 (1974).

10. Cumsky, M., Zusman, D.: Proc. Natl. Acad. Sci. USA 76, 5505-5509 (1979).

11. Jansson, H.B., Nordbring-Hertz, B.: Nematologica. In press.

SEXUAL CELL RECOGNITION IN *CHLAMYDOMONAS EUGAMETOS*

W.L. Homan
Plant Physiology Department, University of Amsterdam
1098 SM Amsterdam, The Netherlands

Introduction

Sexual reproduction in Chlamydomonas provides a relatively simple system for studying the mechanism of cell-cell recognition (1,2). For example, cells are easily cultured and gametes readily induced, and the flagella, which are involved in gamete contact, are easily isolated.

The first step in the copulation process is the recognition between gametes of opposite sexes. This recognition is highly selective, only gametes of the same species and different sex reacting. The elucidation of the molecular basis of this recognition process is the aim of the research described in this paper.

Whether cell recognition between *C. eugametos* gametes is mediated by a lectin-receptor interaction or whether other interactions are involved is as yet unknown. Some of the possibilities have been reviewed by Wiese and Wiese (3). The fact that both protein and carbohydrate moieties seem to be involved (see below) favors a lectin-receptor interaction. Many observations, however, contradict the idea of a lectin-like interaction: erythrocytes are not agglutinated by the isolated agglutination component, and, what is more important, adhesion is not inhibited by monosaccharides in concentrations, which do not impair the mobility of the gametes (1). More must be known about the molecular basis of Chlamydomonas gamete recog-

Lectins - Biology, Biochemistry, Clinical Biochemistry, Vol. I

nition before something meaningful can be said about the kind of ligands involved in this process.

Sexual reproduction in C.eugametos

The unicellar biflagellate green alga C.eugametos occurs as two sexually compatible mating types: mating type + (mt^+) and mating type - (mt^-), which are morphologically indistinguishable. A detailed discription of sexual interaction in C.eugametos has been given by Mesland (4). When gametes of the two mating types are mixed, they all instantaneously aggregate into many large clumps. The cells agglutinate together by a sex-specific adhesiveness of their flagella. Just after mixing the contacts seem to be of a random nature, but gradually the flagella reassociate so that they adhere over their whole length. After the gametes are lined up in this way, cell fusion takes place by way of newly-formed fertilization tubes that project out from between the flagellar bases. Directly after fusion, the flagella lose adhesiveness, so that the resulting vis-à-vis pairs can escape from the clumps. Eventually the cells fuse completely to form a zygote.

Flagellar agglutination

This paper describes some research aimed at understanding the mechanism of flagellar adhesion. Sexual adhesion is easily distinguished from other types of cell aggregation, such as antibody or lectin crosslinking, through a highly characteristic violent twitching of the flagella, which gives the clumps a vibrating appearance. A stable noncovalent binding of receptor to ligand seems a too simple assumption to explain flagellar adhesion. The most important observations which lead to this conclusion are the dynamic character of the agglutination process, which allows continuous reassociation, the reversibility of agglutination seen after cell fusion and the fact that at least one of the gamete types must be alive for agglutination to occur (5). For example, neither killed gametes nor isolated flagella can agglutinate killed gametes of the other mating type. Our present working hypothesis is that a weak, temporary, receptor to ligand binding is involved, and that other factors such as the distribution and turnover of the adhesion sites strongly affect the degree of binding.

Isoagglutinins

One approach to the problem was to study vesicles derived from the flagellar membrane. These vesicles are called isoagglutinins and have a less complicated structure than flagella, while the adhesive properties are the same (1,6,7,8). Mixing vesicles with gametes of the opposite mating type results in isoagglutination of these gametes. Isoagglutinins are easily isolated and purified from conditioned, cell-free medium of

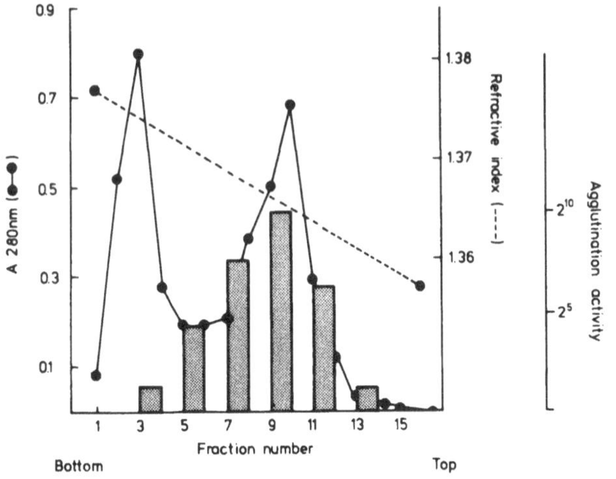

Fig.1. Purification of mt⁻ isoagglutinin on a CsCl density gradient. The sample was centrifuged for 24 h at 156,000 g in 2.8 M CsCl (8). Two bands of particles were produced. Agglutination activity (histogram) was restricted to the band with a refractive index of 1.3655, while the denser, inactive band, whose refractive index was 1.3745 consisted of cell wall material that contaminated the isoagglutinin preparation.

gametes by differential centrifugation, followed by CsCl density centrifugation. Figure 1 shows the distribution of isoagglutinin activity in a CsCl gradient. The biologically active band has a density of 1.32 gram per ml. The other band is inactive and consists of contaminating cell wall material (8).

Composition of isoagglutinins

Agglutination clearly involves components exposed at the outside of the flagellar or isoagglutinin membranes, so it is reasonable to assume that either lipids or membrane proteins are involved. Wiese (1) found that $\underline{mt}^{-}$ agglutination was sensitive to proteolytic enzymes, while $\underline{mt}^{+}$ gametes were resistant. $\underline{Mt}^{+}$ gametes, on the other hand, were sensitive to Con A treatment (9,10). That carbohydrates are important in $\underline{mt}^{+}$ agglutination was confirmed by the fact that α-mannosidase treatment rendered them inactive (1).

Since the proteins in isoagglutinin vesicles are all glycosylated (8),our attention has been mainly directed to the glycoproteins. Figure 2a shows periodic acid Schiff (PAS) stained 5 % polyacrylamide SDS gels of purified isoagglutinins. In this gel system (11) only one high molecular weight protein, on the top of the gel, is stained by Coomassie Brilliant Blue (not shown). This is a glycoprotein, as can be seen in the PAS-stained gels, in which other PAS stainable bands are also present. These bands are numbered PAS 1 to PAS 9, as illustrated. When isoagglutinins are electrophoresed in a 4 % SDS polyacrylamide system (12), PAS 1 is resolved in two bands numbered PAS 1.2 and PAS 1.4 (Figure 2b).

A comparison of the gel pattern of $\underline{mt}^{+}$ and $\underline{mt}^{-}$ isoagglutinins shows that there are no detectable differences (8). All glycoproteins present in $\underline{mt}^{+}$ flagellar membranes are also present in those of $\underline{mt}^{-}$. The amino acid composition of $\underline{mt}^{+}$ and $\underline{mt}^{-}$

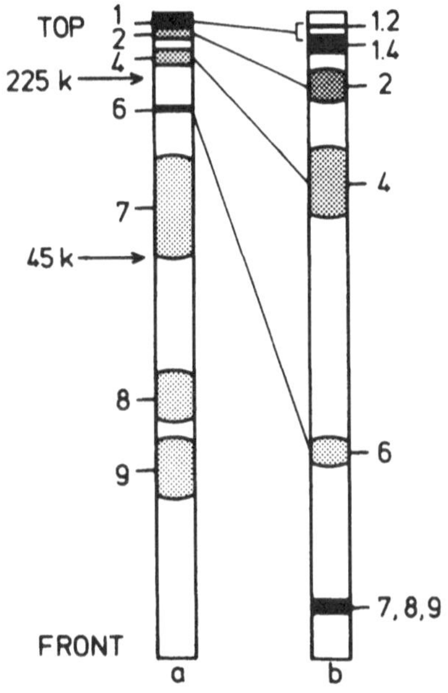

Fig.2. PAS stained sodium dodecyl sulfate polyacrylamide gels of mt⁻ isoagglutinin of C.eugametos in: a) a 5 % acrylamide gel (11) and b) a 4 % acrylamide gel (12).

isoagglutinins only differs in hydroxyproline content. The carbohydrate composition, however, differs remarkably. Both mating types appear to contain arabinose, rhamnose, xylose, mannose, galactose, glucose and N-acetyl-glucosamine, and in addition unidentified sugars unique to either mt^+ or mt^- (unpublished). These sex specific differences are also found on analyzing separate fractions from polyacrylamide gels. But whether these differences have any bearing on the specificity of the recognition process is not known.

The mt^- agglutination component

We suppose that flagella or isoagglutinin vesicles are capable of cross-linking flagellar surfaces of the opposite mating type because they are multivalent. So it should be possible to isolate these components as monovalent binding sites. When mt$^-$ isoagglutinin was extracted with a number of chaotropic reagents or treated with ultrasound, isoagglutinin activity was reduced. In the lipid-free supernatant another sex-specific activity appeared. The appearence of this activity was correlated with a diminished activity in the isoagglutinin pellet (13). This characteristic activity can best be described as a violent twitching of the flagella, reminiscent of that observed during sexual agglutination, but now the cells remained separate and tended to accumulate at the surface of the fluid. On gelelectrophoresis of the solubilized material of mt$^-$ isoagglutinins the only components present were the PAS 1 to PAS 6 bands (13). Since this fraction only caused twitching and not isoagglutination it was considered probable that it contained monovalent factors, adsorbing to the flagellar membrane of the opposite mating type. Isoagglutinin vesicles could thus be considered as particles deriving their multivalency from a multiplicity of such factors in or on the lipid bilayer of each vesicle. This is supported by the fact that vesicles rendered non-agglutinative by extensive sonication could be reactivated by incubation with a twitch-effect containing solution, probably by rebinding of these monovalent agglutination components to the vesicle (unpublished).

In order to further purify the agglutination component, a

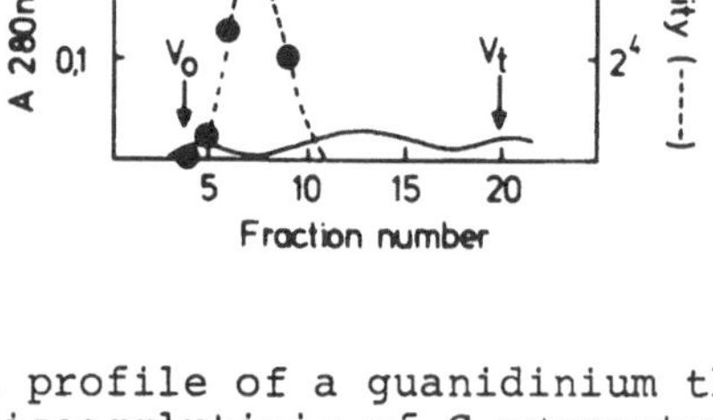

Fig. 3. Elution profile of a guanidinium thiocyanate (GTC) extract of mt$^-$ isoagglutinin of *C.eugametos* from a Sepharose 2 B Cl column eluted with 2 M GTC.

mt⁻ isoagglutinin extract was fractionated in a Sepharose 2 B column equilibrated in 2 M guanidinium thiocyanate (Figure 3). Activity was only found in the fractions just after the void volume of the column. On electrophoresis, only PAS 1.2 was found to be present in these fractions.

So it seems that at least for mt^- isoagglutinins, the active component in agglutination is the high-molecular weight PAS-1.2 glycoconjugate. It should be kept in mind that these results are only valid for mt^- material. When mt^+ isoagglutinin was extracted, no activity could be detected.

References

1. Wiese, L.: Annals N.Y. Acad. Sci. 234, 383-395 (1974).
2. van den Ende, H.: In "Sexual Interactions in Eukaryotic Microbes" (O'Day, D. Horgen, P.A. eds.) Acad. Press, Inc., New York (1980), in press.
3. Wiese, L. and Wiese, W.: In "Cell Interactions", Symp. Soc. Exp. Biol. (A.S.G. Curtis, ed.) 32, 83-104. Cambridge University Press, Cambridge, London, New York, Melbourne (1978).
4. Mesland, D.A.M.: Arch. Microbiol. 109, 31-35 (1976).
5. Goodenough, U.W.: In "Receptors and Recognition" (J.L. Reissig, ed), Series B, 3, 323-351. Chapman and Hall, London (1977).
6. Bergman, K., Goodenough, U.W., Goodenough, D.A., Jawitz, J. and Martin, H.: J. Cell Biol. 67, 606-622 (1975).
7. Snell, W.J.: J. Cell Biol. 68, 48-69 (1976).
8. Musgrave, A., Homan, W.L., van den Briel, M.L., Lelie, N., Schol, D., Ero, L. and van den Ende, H.: Planta 145, 417-425 (1979).
9. Musgrave, A., van der Steuyt, P. and Ero, L.: Planta 147, 51-56 (1979).
10. Wiese, L. and Wiese, W.: Dev. Biol. 43, 264-276 (1975).
11. Weber, K., Pringle, J.R. and Osborn, M.: In "Methods of enzymology" (Hirs, C.H.W., Timasheff, S.N. eds.) 26, 3-27 Acad. Press, New York (1972).
12. Laemmli, U.K.: Nature (London) 227, 680-682 (1970).

13. Homan, W.L., Musgrave, A., Molenaar, E. and van den Ende, H.: Arch. Microbiol. (1980), in press.

THE LECTINS OF CLOSTRIDIUM BOTULINUM AND THEIR RELATIONSHIP TO TOXIN SPECIFICITY.

P. Balding

Oncology Research Unit, Department of Radiotherapy, University of Bristol U.K.

J. L. Smart and T. A. Roberts

Meat Research Institute, Langford, Bristol U.K.

E. R. Gold

Department of Child Health, University of Bristol, U.K.

Introduction

The anaerobic, gram-positive bacterium Clostridium botulinum produces an intracellular toxin/haemagglutinin complex whose characteristics and properties, with particular reference to toxin, have been reviewed by Lamanna (1), and Schantz and Sugiyama (2). The clinical aspects of botulism, the condition resulting from ingestion of the toxin have been reviewed by Koenig (3).

Seven types of Cl.botulinum designated A, B, C, D, E, F and G are universally recognised, each producing an antigenically distinct neurotoxin. Types A to G also produce haemagglutinins. (1, 4, 5, 6). Types A, B, E and F cause poisoning in man. Although cases of human botulism caused by types C and D have occasionally been reported (7, 8, 9), confirmation was generally inaccurate and some doubt remains whether man is sensitive to toxins of types C and D (10, 11). Type C commonly causes botulism in wild water fowl which feed on dead invertebrates and type D most often affects cattle. Type E is probably of marine origin (12).

There is evidence for the involvement of bacteriophage in the toxigenicity of types C and D (13, 14, 15, 16).

Lectins - Biology, Biochemistry, Clinical Biochemistry, Vol. I

In 1948, Lamanna (4) first reported that addition of crystalline or amorphous preparations of type A toxin to a suspension of animal red cells (chicken, guinea pig, rabbit, sheep and man) resulted in agglutination. Alteration of the erythrocyte surface by enzyme activity of the toxin was ruled out from the results of absorption and elution experiments. Later a haemagglutinating molecule, which could be separated from the toxin, was discovered (17). Lowenthal and Lamanna (18, 19) investigated the factors affecting haemagglutination and toxicity of type A toxin preparations and the relationship between the two properties, and established the separate identities of toxin and haemagglutinin. The effects of temperature, salt and pH on both haemagglutination and absorption of haemagglutinin to erythrocytes were investigated. They suggest that at acid pH values a toxin-haemagglutinin complex formed, which resulted in simultaneous absorption of both haemagglutinin and neurotoxin from acid solution by chicken erythrocytes. In alkaline solution the complex dissociated permitting absorption of haemagglutinin without appreciable absorption of neurotoxin.

Results and Discussion

Figure 1 shows a comparison of the effect of pH on absorption of type A haemagglutinin to red cells (from 19) and the effect of pH on agglutination. There is close agreement between the results: maximum agglutination and absorption occurred between pH 6.5 and pH 8.0. The haemagglutinating activities of types A, C and D were affected similarly by pH and temperature (Figs. 2, 3 and 4).

Haemagglutination by types B and E was reported by Lamanna and Lowenthal (17) and by type C by Lamanna and Jensen (20). Stern (21) confirmed the activity of type C and compared agglutination by types C and D. No further work on haemagglutination by Cl.botulinum was reported until our study (22) in which haemagglutination tests using normal and enzyme treated cells and inhibition tests with a panel of sugars, amino acids, sera and other biological substances confirmed agglutination by types A, C and D but not by type E. Allowing for the different agglutinin concentrations, the haemagglutinating activities of types A and B were later shown to be

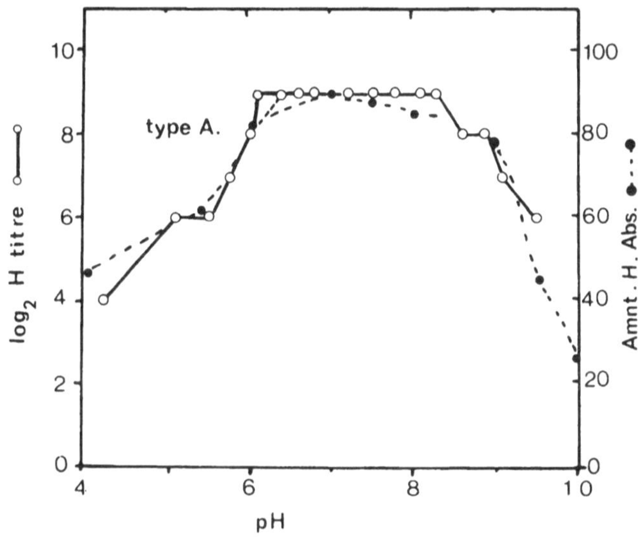

Figure 1 The effect of pH on haemagglutination (H) titre and absorption to red cells (from ref. 19) of Cl.botulinum type A

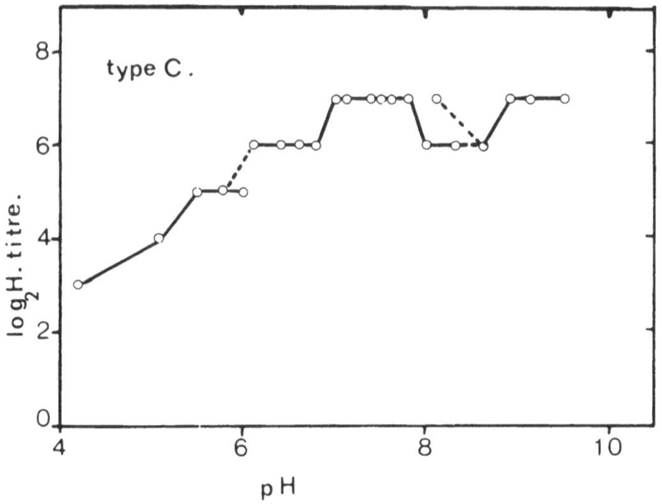

Figure 2 The effect of pH on H titre of Cl. botulinum type C

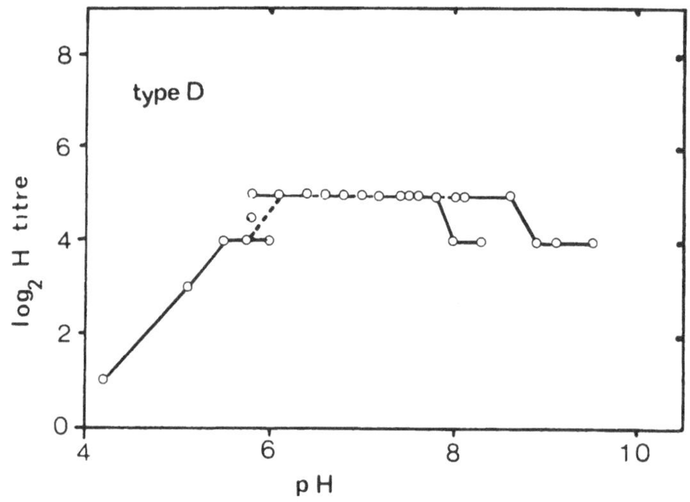

Figure 3 The effect of pH on H titre of *Cl. botulinum* type D

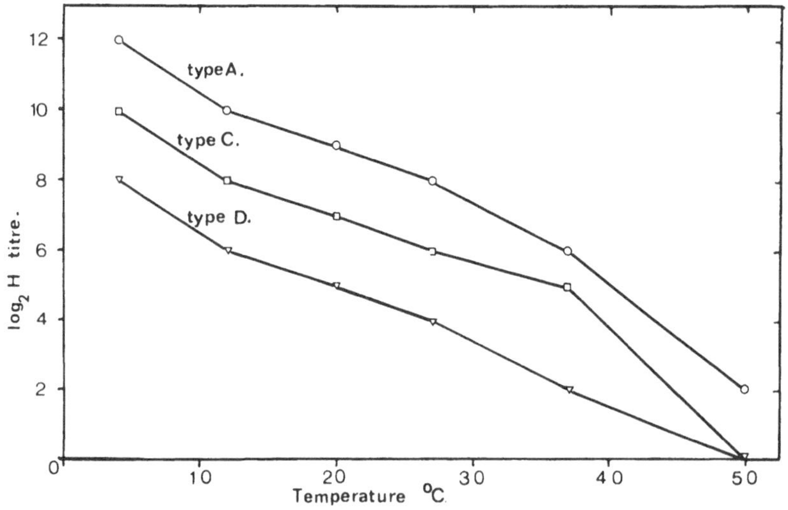

Figure 4 The effect of temperature on H titre of *Cl. botulinum* types A, C and D.

Table 1. A comparison of the haemagglutination titres of Cl.botulinum types A and B with native and enzyme treated human and animal red cells.

	Native cells		Papain-treated cells		N-ase treated cells	
Cl.botulinum	Type A	Type B	Type A	Type B	Type A	Type B
Red cell species						
Human	512	64	2048	64	5000	512
Rabbit	512	16	256	16	1024	256
Guinea Pig	128	4	512	8	2048	256
Pig	256	64	64	64	512	256
Chicken	128	2	128	4	256	16

Table 2 Examples of the major differences in specificity of the lectins of Cl. botulinum types A, C and D.

Test System	Reactivity		
	A	C	D
1. H. with Nase treated human O cells *	5000	0	0
2. ** Inhibition of sheep cell agglutination by sheep serum	8	16	n.i.
3. ** Inhibition by human milk	8	n.i.	16
4. Inhibition by β-Methyl-D-Gal	2.5 μM	n.i.	n.i.
5. Inhibition by Pn XIV	0.75 μg	n.i.	n.i.

* a similar difference in haemagglutination (H) was shown by erythrocytes from mink, guinea pig, horse, sheep, pig, ox, chicken, trout.

** Inhibition - number indicates greatest dilution of inhibitor to give complete inhibition of a dilution of lectin with a titre of 1 in 4 with a 3% erythrocyte suspension.

n.i. no inhibition

similar (23), but not identical (Table 1). Cl.botulinum can be grouped according to biochemical properties (10) and antigenic composition (24) into (i), types A, B and F, (ii) types C and D, (iii) type E. In the absence of data on type F, haemagglutination tests appear to group Cl. botulinum in the same way. Some major differences between types A, C and D (Table 2) suggest that typing by haemagglutination may be possible although type B has not yet been investigated in sufficient detail to distinguish it from type A.

Table 3 Results of inhibition studies using Cl.botulinum lectins

Inhibitor (uM)	Quantity of inhibitor to completely inhibit agglutination[1] by Cl.botulinum type			
	A	B	C	D
α-Methyl-D-Gal	5.0	10.0	n i	n.i
β-Methyl D-Gal	2.5	2.5	n.i	n.i
Lactose	2.5	2.5	n.i	n i
Pn XIV (µg)	0.75	40.0	n.i	n.i
Ganglioside [2]				
G.Gn.S.L.C.	5.6	-	0.7	1.4
S.G.Gn.S.L.C.	3.0	-	3.0	3.0
G.Gn.S.S.L.C.	2.5	-	3.1	6.2
S.G.Gn.S.S.L.C	3.0	-	3.0	1.5
G.Gn.L.C.	3.0	-	250.0	250.0
Sialoglycoproteins				
1% sialic acid (µg)	0.4	-	100.0	100.0
15% sialic acid (µg)	25.0	-	50.0	50.0
MN active (µg)	50.0	-	6.0	12.0

1 lectin titre 1 in 4 with a 3% erythrocyte suspension
2 alternative nomenclature in Figure 5
n.i no inhibition

Haemagglutination-inhibition studies are illustrated in Table 3. Whereas type C and D were not inhibited by any single sugar or oligosaccharide, types A and B were inhibited by α and β - D - galactoside and by Pneumococcus type XIV polysaccharide (Pn XIV). The β - D - galactoside

Figure 5 Gangliosides most commonly found in mammalian cell membranes.

1. Basic Structure

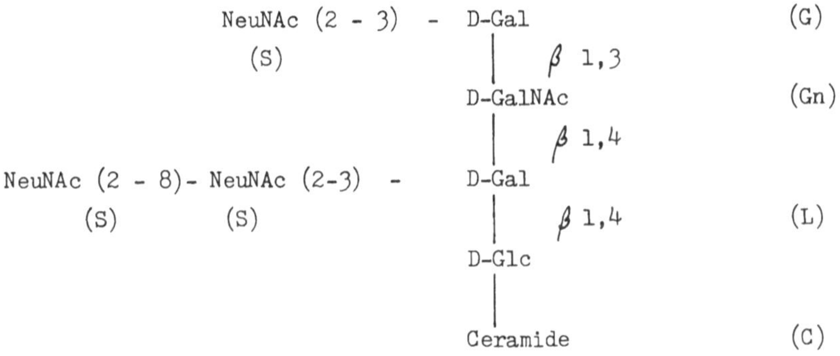

2. Nomenclature according to :

Kuhn and Wiegandt (41)	Svennerholm (42)	McCluer (27)
-	G_{M3}	SLC
GO	G_{M2}	GnSLC
GI	G_{M1}	GGnSLC
GII	G_{D1a}	SGGnSLC
GIII	G_{D1b}	GGnSSLC
GIV	G_{T1}	SGGnSSLC
Tay Sachs	G_{M2}	GnSLC

specificity of types A and B has been confirmed (25). Additional inhibition studies showed that the specificity of types C and D was directed towards an oligosaccharide unit containing N-acetyl-neuraminic acid (26). Gangliosides were a good source of such sugar chains and there were differences in the reactivities of the different gangliosides tested. The most strongly inhibiting ganglioside was GI, a monosialoganglioside with a sialic acid molecule on the internal galactose moeity. Using McCluer's nomenclature (27) this would be G.Gn.S.L.C. (See Figure 5).

Asialoganglioside did not inhibit haemagglutination. Blood group specific glycoproteins from secretions containing increasing concentrat-

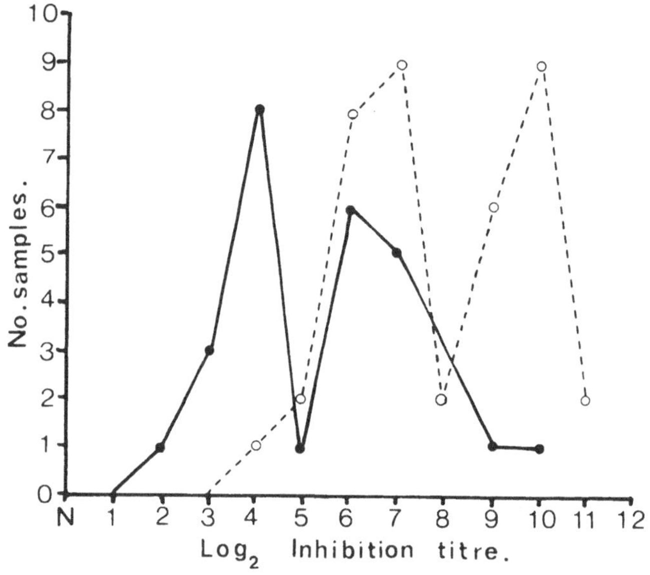

Figure 6 Inhibition titres with Cl. botulinum type C and unrelated saliva samples. (No inhibition by three Caucasian samples or one Bantu sample).
(o - - - o = Caucasian ; •———• = Bantu)

ions of sialic acid inhibited in proportion to their sialic acid content, but were more than 100 times less effective inhibitors than gangliosides. An MN active glyco protein isolated from erythrocytes was a much stronger inhibitor and inhibition of type A was related to the presence of terminal β - D - galactoside in these glycoproteins and glycolopids.

During the course of the inhibition studies using types C and D lectins, we found that most, but not all, samples of human saliva inhibited agglutination. Inhibition was unrelated to ABH, Lewis or Sid secretor status, sex or ethnic origin (26). In Figure 6 the number of saliva samples giving a particular inhibition titre with type C has been plotted against that inhibition titre. There was no inhibition by three Caucasian or one Bantu sample. The distribution of titres was bimodal, perhaps corresponding to genetic differences: approximately 54% were homozygous positive, 39% were heterozygous and 7% were homozygous negative individuals.

Table 4 Intraspecies differences of native and papain-treated rabbit red cells revealed by Clostridium botulinum lectins type C and D.

	haemagglutinin titre			
	Type C		Type D	
Rabbit red cell sample	Native	Papainised	Native	Papainised
1	256*	64	2	16
2	128	16	8	1
3	128	4	0	0
4	512	8	0	0
5	256*	8	0	0
6	0	0	0	4
7	128*	8	32	16
8	0	1	32	1
9	nt	nt	4	16

*Marked "prozone" effect

The distribution of inhibiting salivas in family studies strongly suggests that the inhibiting substances is a dominant inherited character (see ref. 26).

All human red cell samples tested were agglutinated by type C and type D, but samples from a variety of animals revealed intraspecies differences e.g. rabbit red cells (Table 4) unrelated to sex or strain of rabbit. Titres ranged from 1 in 512 to complete lack of agglutination. Papain treatment of erythrocytes from different rabbits altered agglutination titres to different extents (see Table 4). In collaboration with Professor C. Cohen, Illinois University, association with rabbit blood groups was investigated but although significant differences in agglutination were revealed there was no apparent resemblance of the specificity of types C and D to any known rabbit blood group antibodies, therefore suggesting that a previously unrecognised "antigenic" system in the rabbit has been recognised. The inhibition results using human saliva samples suggest that the system could be inherited (26), and the ganglioside inhibition results indicate that the structural basis of the system could be ganglio-

Table 5 Agglutination of Rat Cerebral Cortex Fractions.

Fraction	Agglutination titre Type C	Type D
Synaptic Membranes	16	8
" Vesicles	2	2
Myelin	4	2
Microsomes	0	0
Mitochondria	0	0
Control (RBC)	32	16

Table 6 Neutralization of Toxicity (10 MLD Toxin, 20 hr. 30° C) *

GANGLIOSIDE (100 nM)	Type A	Type C	Type D
G.Gn.S.L.C. (GI)	-	P	-
S.G.Gn.S.L.C. (GII)	SS	P	-
S.G.Gn.SS.L.C. (GIV)	-	P	-
G.Gn.SS.L.C. (GIII)	SS	SS	-
Pn XIV (40 µg)	-	P	-
H_2O (Control)	SS	P	-

- = survived
P = pinched
SS = severe symptoms
* = no inhibition of toxicity when 10 MLD toxin incubated for 2 hr. at 30°C, or 300 MLD toxin incubated for 30 min. at 40°C.

side or ganglioside-like oligosaccharide chains. The difference in erythrocyte surface structure may involve a change in ganglioside accessibility or the deletion of one ganglioside type which is compensated by increased production of another. Involvement of a sialoglycolipid, rather than sialoglycoprotein, in the erythrocyte receptor is supported by a strong agglutination of En (a-) erythrocytes (23) which lack the major sialoglycoprotein molecule (28).

The lectin in the toxin complex may be responsible for achieving close association between the active site of the toxin and its receptor, i.e. it may function as a binding agent (29). A similar mechanism has been described for some viruses (30), for the toxic lectins abrin and ricin (31) and tetanus and cholera toxins (32, 33). The receptor for the lectin haptomer of abrin or ricin is β - D - Gal, the same receptor as for the lectins of Cl.botulinum types A and B, and the choleragenoid and tetanus receptor is ganglioside with which the lectins of types C and D react. The analogy between myxoviruses and types C and D lectins appears to be significant since treatment of cells with neuraminidase abolishes haemagglutination. It is generally accepted that the site of action of the toxin is the synaptic region of myoneural junctions but it may act presymaptically (33, 34). Acetylocholine is stored within synaptic vesicles and its release is prevented by the action of the toxin which alters transport of Ca^{++} and perhaps other inorganic ions, across the synaptic nerve cell membrane (35). Gangliosides may facilitate the transport of ions such as calcium which is released into synaptic vesicles by means of a sequence of reversible and cyclic reactions with receptor consequent upon the action of serotonin whose receptor may also be ganglioside (S.S.L.C.) (32). These associated events provide evidence that interference with normal nervous function, such as occurs in botulinal poisoning, may be brought about by an interaction between the toxin and a ganglioside-related receptor in nervous membranes of the synaptic cleft.

In an attempt to investigate the relationship of the lectins to toxicity, their interaction with isolated fractions of nervous cell membranes was studied using synaptosomes i.e. pinched off nerve endings which retain their in situ characteristics. Synaptosomes, synaptic membranes and myelin, absorb and are agglutinated by type C and type D (Table 5), supporting the demonstration of gangliosides in these three fractions (36). Binding of type A toxin to synaptosomes was demonstrated by Habermann (37) who used ^{125}I labelled toxin. Binding could be prevented or reversed by neuraminidase treatment of synaptosomes confirming that sialic acid residues are involved in the interaction. Later work has confirmed these observations (33, 35).

Although agglutination by both types C and D is inhibited by gangliosides, no inhibition of the toxicity of type D could be demonstrated (Table 6). The effects of type C toxin were not reduced by the gangliosides which inhibited agglutination; the ganglioside with terminal galactose and two sialic acid residues (GIII) slightly enhanced the effect of the toxin. However, Pn XIV polysaccharide and gangliosides GI and GIV at relatively high concentration (100 nM) and after 20 hours incubation prevented the severe symptoms of botulism caused by type A toxin in control mice. Simpson and Rapport (38) reported rapid inactivation of type A toxin by ganglioside GIV, over a wide temperature range. We, and other groups (34, 39) have not been able to confirm this work which assessed toxicity by intravenous assay rather than by the intraperitoneal assay we used.

The result of haemagglutination - inhibition studies reported here support the view expressed by Simpson and Rapport (38) that inactivation of toxin is a consequence of what they term "non specific binding" but which could be binding of lectin to receptor, since they found that inactivation is reversible, whereas binding of toxin to its receptor in vivo is not reversible. Also, Simpson and Rapport found that galactocerebroside antagonised the ganglioside effect; this would be expected if the interaction between ganglioside and lectin, rather than ganglioside and neurotoxin were being studied.

On the other hand, Kitamura et al (40) have confirmed the interaction between type A neurotoxin and ganglioside S.G.Gn.S.S.L.C. Although toxicity was reduced to 0.03% of control values in the presence of ganglioside, binding of neurotoxin to synaptosomes was only reduced to 70% of control levels compared to a complete reversal of neurotoxin binding after neuraminidase treatment of synaptosomes (36). Thus the site of action of the toxin may not be ganglioside, but be in close proximity to ganglioside molecules in the synaptic region. The lectin part of the haemagglutinin - toxin complex may facilitate its binding to a specific receptor.

References

1. Lamanna, C.: Science, 130, 763-772 (1959)
2. Schantz, E. J. and Sugiyama, H. : Essays on Toxicology, Ed. W. J. Hayes. 5, 99 (1973)

3. Koenig, M.G. : Neuropoisons, Ed. L. L. Simpson, 1, 283 (1971)

4. Lamanna, C. : Proc. Soc. exptl. Biol. Med., 69, 332-336 (1948)

5. Ciccarelli, A.S. and Gimenez, D.F. : Infect. Immun. 5, 985-986 (1972)

6. Sakaguchi G., Ohishi, I. Kozaki, S., Sakaguchi, S. and Kitamura, M. : Jap. J. med. Sci. Biol. 27, 95-99 (1974)

7. Meyer, K.F., Eddie, B., York, G.K., Collier, C.P. and Townsend, C.T.: Proc. 6th. Int. Congr. Microbiol., Rome, 4, 123-124 (1953)

8. Prévot, A.R., Terasse, J., Daumail, J. Cavaroc, M., Riol, J. and Sillioc, R. : Bull. nat. Acad. Med. 139, 355-358 (1955)

9. Demarchi, J., Mourques, C., Orio, J. and Prévot, A.R. : Bull. nat. Acad. Med. (Paris) 142, 580 (1958)

10. Holdeman, L.V. : J. Wildlife Diseases, 6, 205-210 (1970)

11. Roberts T. A. and Gibson, A. M. : J. Fd. Technol., 14, 211-216 (1979)

12. Dolman, C.E. : Jap. J. med. Sci. Biol. 10, 383 (1957)

13. Inoue, K. and Iida, H. : Jap J. med. Sci. Biol. 24, 53-56 (1971)

14. Eklund, M.W., Poysky, F.T. and Reed, S.M: Nature New Biol. 235, 16 (1972)

15. Eklund, M.W. and Poysky, F.T.: Appl, Microbiol., 27 251-258 (1974)

16. Oguma, K., Iida, H. and Shiosake, M. : Infect. Immun. 14, 597-602 (1976)

17. Lamanna, C. and Lowenthal, J.P. : J. Bact. 61, 751-752 (1951)

18. Lowenthal, J. P. and Lamanna, C. : Amer. J. Hyg., 54, 342-353 (1951)

19. Lowenthal, J. P. and Lamanna, C. : Amer. J. Hyg. 57, 46-59 (1958)

20. Lamanna, C. and Jensen, W. I. : Bact. Proc. M71, 106 (1952)

21. Stern, D. M. : Science, 119, 440 (1954)

22. Balding, P., Gold, E.R., Boroff, D.A. and Roberts, T.A. Immunology 25 773-782 (1973)

23. Balding, P. Haemagglutinins of Bacteria. Ph.D Thesis. University of Bristol (1977)

24. Walker, P. D. and Batty, I.J. appl. Bact. 27, 140 (1964)

25. DasGupta, B.R. and Sugiyama, H. : Can. J. Microbiol. 23, 1257-1260 (1977)

26. Balding, P. and Gold, E.R: J. med. Genetics, 10, 323-327 (1973)

27. McCluer, R. H: Chem. Phys., Lipids, 5, 220 (1970)

28. Tanner, M.J.A. and Anstee, D. J. : Biochem. J. 153, 271 (1976)

29. Boroff, D.A., Nyberg, S. and Hoglund, S. : Infect. Immun. 6, 1003-1007 (1972)

30. Andrews, P: Biochem. J. 91, 222 (1964)

31. Olsnes, S., Refnes, K and Pihl, A. : Nature, 249, 627 (1974)

32. Van Heyningen, W.E. : Nature 249, 415 (1974)

33. Haberman, E. : Arch. Pharmacol., 293, 1-9 (1976)

34. Mellanby, J., Thompson, P.A. and Hampden N. : Arch. Pharmacol. 276, 303 (1973)

35. Wonnacott S., Marchbanks, R. M. and Fiol. C.J: Neurochem. 30, 1127-1134 (1978)

36. Hamberger, A. and Svennerholm, L.J: Neurochem. 18, 1821 (1971)

37. Habermann, E. : Arch. Pharmacol. 281, 47 (1974)

38. Simpson, L. L. and Rapport, M.M. : J. Neurochem. 18, 1751 (1971)

39. Van Heyningen, W.E. and Mellanby, J. : Arch. Pharmacol. 276, 297 (1973)

40. Kitamura, M., Iwamori, M., Nagai, Y. : Biochem. Biophys. Acta 628, 328-335 (1980)

41. Kuhn, R. and Wiegandt, H. : Chem. Ber. 96, 866 (1963)

42. Svennerholm, L. : J. Neurochem. 10, 613 (1963)

ON THE SPECIFICITY AND HYDROPHOBICITY OF LECTINS

Ochoa, J.-L., Sierra, A., and Córdoba, F.

CIB-Biological Research Center. Box 128,
La Paz, B.C.S. Mexico

For almost nearly a century plants have been known to contain substances that are capable of inducing the agglutination of suspended animal cells (1). The best term for these substances is probably plant agglutinins, or "phytagglutinins" (2), although some of them may be classified as lectins (3). In other words, lectins are just a small group of protein agglutinins with the remarkable characteristic of being carbohydrate-inhibitable in a similar fashion to glycosidases and other sugar-binding proteins.

It has been possible to show that most lectins possess no common feature that could be regarded as the key of their binding characteristics (4). Lectins are neither chemically, nor structurally alike. Moreover, even those belonging to the same family, or genus, differ in their physico-chemical and biological properties (5-8). A question one may ask is whether the binding characteristics of these particular group of agglutinins is a mere accident, or a result of an unknown function. An answer based on the studies performed with only a few lectins could hardly be found satisfactory (5-6). As a matter of fact, lectin's specificity is a relative term. Most research workers in the field when refering to a lectin

Lectins - Biology, Biochemistry, Clinical Biochemistry, Vol. I

inhibitor speak often about the "best" one, thus meaning that there may be other inhibitors with quite a different structure, although less efficient. This problem can be of the utmost importance. For example, Phaseolus vulgaris agglutinin is known to be inhibited by N-Acetyl-D-Galactosamine (9). Curiously, the erythrocyte receptor for this lectin lacks such carbohydrate in its structure (10), and furthermore a core of three mannose residues was found to be relevant for binding. These observations have been confirmed by other authors (11-12). Therefore, one may also ask whether the N-Acetyl-Galactosamine residues of glycoproteins are involved in the Phaseolus vulgaris agglutinin-glycoprotein recognition process at all. Apparently, and according to (10),

THE CARBOHYDRATES STRUCTURE OF PHASEOLUS VULGARIS INHIBITORS

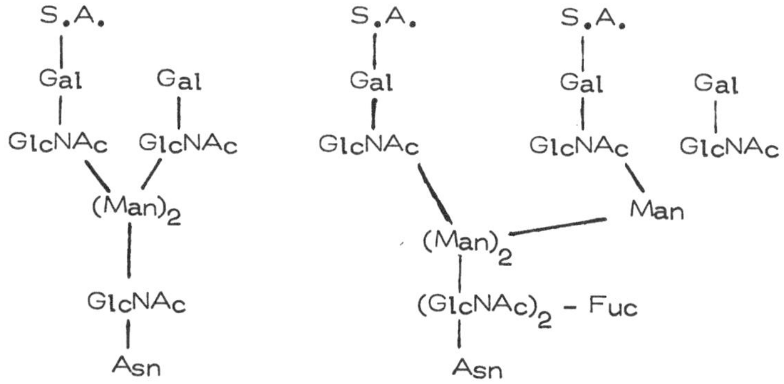

Erythrocyte (Ref 10)

Porcine thyroglobulin (Ref 12)

the O-linked N-Acetyl-Galactosamine residues, in the case of the erythrocyte membrane glycoprotein, can be removed without significant variation in its inhibitory capacity towards this lectin. In another case, (13) however, a similar treatment carried out with a glycoprotein from saliva, or from the ovarian cyst fluid, causes the complete loss of the inhibiting activity against Phaseolus vulgaris agglutinin.

It has also been pointed out that other interactions than carbohydrate-mediated ones may participate in the recognition of glycoproteins by lectins (10,14). In recent papers, the contribution of hydrophobic bonding on such phenomena has become increasingly evident (15-20). For example, interferon adsorption to Concanavalin A-agarose columns is attributed to a carbohydrate-mediated interaction that is reinforced by a hydrophobic bonding occurring between both entities (16). This assumption is supported by the fact that the displacement of interferon from Concanavalin A-agarose columns requires the use of both, the corresponding competitive sugar and an organic solvent (16). Such observations are the same in the case of Influenza virus adsorption to Vicia ervilia-agarose columns (20). In this latter instance, however, the use of a mixture of carbohydrate and detergent was found most appropriated for elution (Table 1). In fact, detergents solutions have been used before to diminish the undesirable non-specific interactions between lectins and glycoproteins that are often a problem in affinity chromatography systems involving immobilized lectins (18,24).

TABLE I

EFFECT OF THE COMPETITIVE INHIBITOR, A REDUCING POLARITY AGENT, AND A DETERGENT IN THE ADSORPTION OF INFLUENZA VIRUS TO *Vicia ervilia* AGGLUTININ - SEPHAROSE*

Adsorption Conditions	Percent of Virus Adsorbed
PBS	75
PBS, IM Glucose	18
PBS, 50% Ethylene Glycol	68
PBS, 1% Triton X-100	60
PBS, 1% Triton X-100, IM Glucose	4

*0.5 ml (98 hemagglutinating units) of influenza virus A England 42/72 purified by gradient centrifugation in Sucrose, followed by gel filtration on Sepharose 2B, was applied to (0.5 × 5 cm) *V. ervilia*-Sepharose 2B columns, (29), equilibrated with the corresponding buffer at 20° C and a flow rate of 40ml/h. After passing 2 bed volumes, the hemagglutinating activity was calculated from the eluant.

It has been suggested that the oligosaccharide portion of glycoproteins may play a role in the exposure of its hydrophobic residues (21-22). If interferon glycosylation is specifically inhibited during biosynthesis, with tunicamycin for example (25,26), not only its capacity to bind to Concanavalin A-agarose columns is lost, but also its ability to adsorb to Phenyl-agarose (16). Therefore, it appears that interferon ability to bind to immobilized lectins and hydrophobic gels depends on its content in carbohydrate.

The role of the carbohydrate moiety on the hydrophobicity of a number of glycoproteins has been estimated by studying their interaction with a number of hydrophobic fluorescent probes (27). As it turned out, the most hydrophobic glycoproteins studied were those possessing the largest portion of carbohydrate residues. For lectin-glycoprotein interactions, these observations may be considered meaningful.

The hydrophobicity of lectins, by another part, has been studied by looking at their chromatographic behaviour on hydrophobic derivatives of agarose gels (14,20). In the case of Concanavalin A, for instance, it was found that the lowest member of the series of alkyl-agarose derivatives showing an appreciable affinity for this lectin was hexyl-agarose, with a degree of substitution of 60mMoles substituent/Mol galactose. Adsorption of Concanavalin A to hexyl-agarose, and other higher members of the series, increases with temperature and salt concentration, in agreement with the theory of Hydrophobic-interaction chromatography (23).

In conclusion, it seems to us that the participation of hydrophobic bonds in lectin-glycoprotein complexes may be rather the rule than the exception. Non-specific adsorption, or low recoveries in affinity chromatography systems involving lectin-glycoprotein associations may be related to the hydrophobic character of the participating entities. It is therefore advisable to consider the use of low polarity agents, and/or mild detergents, in order to abolish such drawbacks in this type of chromatography.

TABLE II

ADSORPTION* OF CONCANAVALIN A TO HYDROPHOBIC GELS

	Temperature (C°)			Salt Concentration (Moles of NaCl added to PBS				
Type of Adsorbent	4	20	37	0	0.5	1.0	2.0	4.0
Hexyl-agarose	10	22	22	0	10	21	40	60
Octyl-agarose	20	40	40	22	50	60	65	65
Decyl-agarose	50	50	50	50	60	65	65	65

*35 μg of ^{125}I-Concanavalin A (2×10^5 cpm), (19), in 0.5 ml of PBS (Phosphate saline buffer) was applied to columns (0.5×2.5 cm) equilibrated with the same buffer (except in the case of increasing salt concentration) at a flow rate of 10 ml/h. Thereafter, the columns were washed using 10 ml of the adequate buffer and the radioactivity thus adsorbed was measured directly, pouring the gels in Searle gamma vials with a Beckman LS 100C Scintillation counter. All experiments were run by triplicate in a temperature chamber at 20° C, if nothing else is stated. The reported values, in percent of total material, are averages with a standard deviation of about 5%. The degree of substitution of the hydrophobic gels ranged between 50-60 mMol of substituent / Mol galactose (for the preparation of the gels see Ref 28).

References

1. Stillmark, H.: Uber Rizin, ein giftiges Ferment aus dem Semen Von Ricinus Communis L. und einiges anderen Euphorbiaceen" Inaug. Diss., Dorpat (1888).

2. Mäkelä, O.: Ann. Med. Exp. Biol. Fenn. 36, Suppl. 11 (1957).

3. Boyd, W.C., Sharpleigh, E.J.: Science 119, 419 (1954).

4. Ochoa, J.L.: Path. et Biol. 27, 103-113.

5. Goldstein, J.J., Hayes, C.E.: Adv. Chem. Biochem. 35, 127-340 (1978).

6. Lis, H., Sharon, N.: In: The Antigens. (Sela, M., Ed.) Vol. IV, p. 429-529, Acad. Press N. Y. (1977).

7. Hankins, C.N., Kindinger, J.J., Shanon, L.M.: Plant Physiol. 64, 104-107 (1979).

8. Foriers, A., Baumann, C., Lieber, S., R. de Neve, Strosberg, A.D.: In: 27th Coll. on Prot. of the Biol. Fluids (H. Peeters, Ed.) p. 443-446. Pergamon Press, Oxford (1979).

9. Borberg, H., Woodrof, J., Hirschhorn, R., Gesner, B., Misecher, P., Silber, R.: Science. Vol. 154, 1019-1020 (1966).

10. Kornfeld R., Kornfeld, S.: J. of Biol. Chem. 245, 2536-2545 (1970).

11. Debray, H., Decout, D., Strecker, G., Montreuil, J., Monsigny, M.: In: 27th Coll. on Prot. of Biol. Fluids (H. Peeters, Ed.) p. 451-454. Pergamon Press, Oxford (1979).

12. Toyoshima, S., Fukuda, M., Osawa, T.: Biochemistry 11, 4000-4005 (1970).

13. Strauchen, J.A., Moldow, C.F., Silber, R.: J. of Immunol. 104, 766-768 (1970).

14. Ochoa, J.L., Kristiansen, T.: FEBS Letters 90, 145-148 (1978).

15. Davey, M.W., Sulkowski, E., Carter, W.A.: Biochemistry 15, 704-713 (1976).

16. Miztahi, A., O'Malley, J.A., Carter, W.A., Takatsuki, A., Tamuri, G., Sulkowski, E.: J. Biol. Chem. 253, 7612-7615 (1978).

17. Poretz, R.D., Goldstein, I.J.: Biochem. Pharmacol. 20, 2727-2739 (1971).

18. Lotan, R., Beattie, G., Hubbell, W., Nicolson, G.L.: Biochemistry 16, 1787-1794 (1977).

19. Ochoa, J.L., Kristiansen, T., Påhlman, S.: Biochim. et Biophys. Acta 577, 102-109 (1979).

20. Sparrman, M., Ochoa, J.L., Kristiansen, T.: In: 27th Coll. on Prot. of Biol. Fluids (H. Peeters, Ed.) Pergamon Press Oxford. p. 431-434 (1979).

21. Leavitt, R., Schlesinger, S., Kornfeld, S: J. Biol. Chem. 252, 9018-9023 (1977).

22. Hickmann, S., Kulczycki, A. Jr., Lynch, R.G., and Kornfeld, S.: J. Biol. Chem. 252, 4402-4408 (1977).

23. Ochoa, J.L.: Biochimie 60, 1-15 (1978).

24. Zanetta, J.P., Gombos, G.C.: In: Con A as a tool, (Bittinger, H., Schnebli, H.P., Eds.) p. 389-398, Wiley and Sons, London (1976).

25. Takatsuki, A., Kawamura, K., Okina, M., Kodama, Y., Ito, T., Tamuya, G.: Agr. Biol. Chem. 41, 2307-2309 (1977).

26. Waechter, C.J., Lennarz, W.J.: Annu. Rev. Biochem. 25, 95-112 (1976).

27. Sachdev, G.P., Zodrow, J.M., Carubelli, R.: Biochim. et Biophys. Acta 580, 85-90 (1979).

28. Hjertén, S., Rosengren, J. and Påhlman, S., : J. of Chromatogr., 101, 281-288 (1974).

29. Kristiansen, T.: In: 23rd. Coll. on Prot. of the Biol. Fluids (H. Peeters, Ed.), p. 663-665 (1975).

ENZYMIC PHYTOHEMAGGLUTININS: THEIR RELATION TO "CLASSIC" LEGUME PHYTOHEMAGGLUTININS

Leland M. Shannon, Charles N. Hankins
Biochemistry Department, University of California
Riverside, California 92521 USA

A. Donny Strosberg
Institut de Recherche en Biologie Moléculaire
CNRS, Université Paris VII, France

Introduction

Many lectins have been described from both plant and animal sources (1-3). Collectively, these proteins display tremendous variations in chemical and physical properties and thus, it is doubtful that all lectins have a common function. There is, however, a group of plant lectins that share many properties and which may have a common function(s). These lectins, which we refer to as "classic" lectins, are those found in relatively large concentrations in the seeds of many dicotyledonous plants, especially those of the legumes (1,4). Many of these lectins have been studied intensively with repect to their physical properties and effects on animal cells (2,5), but relatively little is known about their synthesis, distribution, or function(s) within plants.

While a large number of legumes possess hemagglutinins, it has been difficult to develop generalities about these

Lectins - Biology, Biochemistry, Clinical Biochemistry, Vol. I

proteins. The reason for this is because their physiological function is totally unknown. Furthermore, information required to establish the relationships between different lectins has not been available. It would be useful to know how many physiologically or evolutionarily distinct classes of proteins the lectins represent, and what, if any, relationship exists between each class.

Amino acid sequence analysis is required for unambiguous evidence of structural relationships among proteins. When proteins are found to be structurally similar, there is a strong probability of both an evolutionary and functional relationship. Foriers and Strosberg (6-7) have reported extensive homologies among the segments of the first 25 residues of amino terminal sequences of a large number of legume lectins, Figure 1.

Since the relationships among hemagglutinins from different legume species has only recently come under careful scrutiny, it is not possible at this time to state how many different functional classes these proteins comprise. The recent discovery (8-9) that certain legume enzymes possess hemagglutinin properties has led to an exciting area of research that provides new avenues of approach to the general study of plant hemagglutinins. In this report we will attempt to summarize what is now known about enzymic phytohemagglutinins and how they are related to the "classic" legume phytohemagglutinins.

Results and Discussion

Working with seed extracts of mung bean, a very potent agglutinin of trypsinized rabbit erythrocytes was observed. The mung bean extract displayed two unique characteristics.

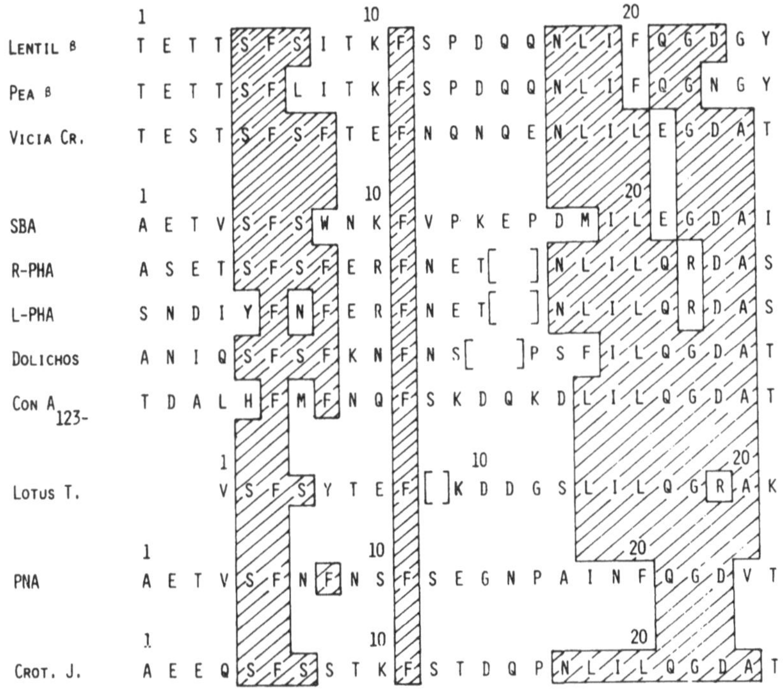

Figure 1. A comparison of the amino terminal sequences of 11 different lectin chains prepared from the seeds of leguminous plants. Shadded areas cover residues identical in six or more proteins.

Firstly, the hemagglutinin activity was inhibited by xylose and inositol, in addition to galactose, and thus possessed a specificity distinct from previously described hemagglutinins. Secondly, the extracts displayed an interesting property referred to as "clot dissolving activity". Apparently, under the appropriate conditions, this protein is capable of enzymically altering those erythrocyte receptors with which it interacts, resulting in a dissolution of cell aggregates. The disaggregated erythrocytes are permanently altered and are no longer agglutinable by the mung bean extract (although they remain agglutinable by many other legume lectins). Since these observations could be explained by an enzymatic alteration of receptors on the erythrocytes, we examined the mung bean extract for various hydrolytic activities and found a very strong α-galactosidase. During purification we followed both hemagglutin and α-galactosidase activities.

Figure 2 depicts the gel filtration elution profile from a Sephacryl S-200 column, the final purification step. Note that both hemagglutinin and α-galactosidase activities elute at coincident positions. <u>Insert A</u> shows the appearance

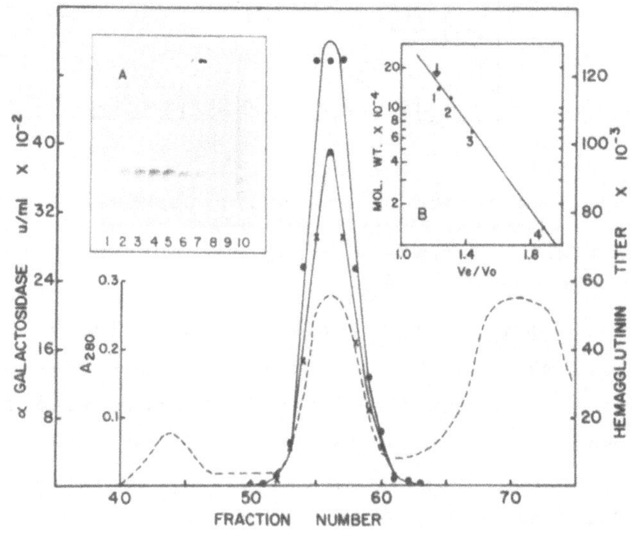

Figure 2. Sephacryl S-200 elution profile. Hemagglutinin (●—●); α-galactosidase (X—X); A_{280} (----). <u>Inset A</u> shows the appearance of SDS gels of 100 μl portions of fractions across the peak. <u>Inset B</u> shows the calibration curve of column.

of SDS gels across the peak. A single band is noted, the intensity of which coincides with A_{280} and activity peaks. Insert B shows calibration curve of the column. The extrapolated native molecular weight for α -galactosidase-hemagglutin was 160,000. The molecule appeared to be a tetramer composed of a single type of subunit (45,000 MW). All attempts to separate the α -galactosidase from heamgglutinin activity were unsuccessful. What was observed instead, was that these two activities exactly co-purified by every separation method tested. Attempts to selectively inactivate one of the activities by the use of heat or PCMB were also unsuccessful. Both activities were equally sensitive to these treatments and were equally protected from inactivation by the hapten sugars galactose, xylose and inositol (8).

A number of carbohydrates were examined for their ability to inhibit the agglutination and α -galactosidase reactions. These results are summarized in Table 1. The only compounds showing significant inhibitory properties were pNPG(α), galactose, xylose, and inositol. All sugars which inhibited agglutination either inhibited the α-galactosidase or

Table 1.

Sugar inhibition of mung bean α -galactosidase-hemagglutinin

SUGAR	HEMAGGLUTININ* (50% INHIB)	GALACTOSIDASE** Ki
	mM	mM
pNPG(α)	0.3	--
GALACTOSE	1.0	0.7
XYLOSE	4.0	5
INOSITOL	20	20
$GalNH_2$	>50	>50
GalNAc	>50	>50
RAFFINOSE	>100	>100
STACHYOSE	>100	>100
pNPG(B)	>100	>100

* Minimum concentration causing 50% inhibition of 4 hemagglutinin units.

** Ki determinations were performed using pNPG (α) as the substrate.

functioned as substrate. In every case, the concentration of sugar causing 50% inhibition of enzyme activity was very similar to that causing 50% inhibition of hemagglutination. Data presented thus far indicate that hemagglutinin and α-galactosidase activities co-purify to yield a single size class of polypeptide. The fact that both activities are simultaneously bound to erythrocytes and are simultaneously released, are equally heat labile (in presence and absence of protective sugars), possess indistinguishable substrate and inhibitor specificity--These facts, we believe, provide compelling evidence that both the hemagglutinin and α-galactosidase activities reside within a single protein species.

The implications of these findings could be very important. The fact that a phytohemagglutinin may possess hydrolytic activity offers a completely new dimension to our thinking about the function of lectins. For this reason it is important to know if the mung bean protein is some freak enzyme that just happens to possess hemagglutinin activity, or if it is genuinely homologous to other legume seed lectins. In our search for additional α-galactosidase hemagglutinins we

Table 2.

Kinetic studies of α-galactosidase activity using pNPG(α) as substrate.

	Km	Ki*		
		GAL	XYL	INOS
	mM	mM	mM	mM
Soybean	0.33	1.06	3.8	12.9
Lima bean	0.38	0.33	2.0	15.0
Lupine	0.39	0.30	1.5	7.5
Thermopsis	0.20	0.40	1.6	8.8
Pueraria	0.42	0.90	2.9	8.8
Mung bean	0.29	0.75	5.2	20.0

*Carbohydrates which were noninhibitory at 100 mM included galactosamine, N-acetylgalactosamine, raffinose, stachyose, glucose, mannose and fucose.

examined seed extracts from 26 genera of legumes (10). We detected clot-dissolving-enzymatic-phytohemagglutinins in 5 additional species, Table 2. The kinetic and hemagglutinin properties are virtually identical with each other and with the mung bean protein. They possess practically identical K_m values and each extract is inhibited by the same sugars

Table 3.

Kinetic studies of hemagglutinin and non-hemagglutinin forms of α-galactosidase using pNPG (α) as substrate.

	Km	Ki*		
		GAL	XYL	INOS
	mM		mM	
Non-Hemagglutinin α-Galactosidases				
Amorpha	1.2	0.5	2.9	12.6
Bandeiraea	1.1	0.6	4.4	14.7
Bauhinia	1.0	0.5	3.1	15.8
Caragana	0.8	0.6	3.8	9.5
Cercis	1.8	0.5	3.4	9.0
Colutea	0.4	0.4	3.2	17.3
Conavalia	1.2	0.2	1.7	9.5
Cytisus	1.0	0.3	1.8	6.1
Dolichos	1.2	0.5	3.6	--
Genista	0.6	0.3	1.6	11.0
Laburnum	1.4	0.5	2.7	14.7
Lathyrus	1.3	0.6	3.5	8.7
Lens	1.2	1.2	3.3	6.9
Lespedeza	0.5	0.3	2.7	11.7
Mimosa	0.7	0.4	2.2	11.6
Phaseolus	0.7	0.5	3.1	12.8
Sophora	0.6	0.3	2.2	9.0
Spartium	1.7	0.4	3.5	11.4
Ulex	0.7	0.4	1.7	10.0
Wistaria	0.7	0.6	3.6	17.8
Hemagglutinin α-Galactosidases				
Soybean	0.2	1.1	3.8	12.9
Lima bean	0.4	0.3	2.0	15.0
Lupine	0.2	0.3	1.5	7.5
Thermopsis	0.4	0.4	1.6	8.8
Pueraria	0.4	0.9	2.9	8.8
Mung bean	0.3	0.7	5.2	20.0

* Ki determinations were performed using pNPG(α) as the substrate.

with virtually identical Ki values from plant to plant. The same three sugars inhibited hemagglutinin activity with virtually identical Ki values from plant to plant. The soybean α-galactosidase-hemagglutinin has been purified to apparent homogeniety and appears to be immunologically identical to the mung bean protein (11). The α-galactosidase-hemagglutinin has also been isolated from soybean leaves indicating that this protein is not limited to seed tissues. The α-galactosidase-hemagglutinin in soybean, and lima, are distinct proteins from the well characterized "classic" lectins present in these species. The two groups of proteins can be separated from each other by affinity chromatography or by immunoprecipitation, they have different sugar specificities, and they can exist simultaneously in the same tissue.

The above survey detected clot-dissolving α-galactosidase-hemagglutin activity in only 6 genera. α-galactosidase activity, however, was present in every legume examined (12). The kinetic properties of hemagglutinin and non-hemagglutinin forms of α-galactosidase are presentd in Table 3. The α-galactosidase activities are virtually identical. They possess the same K_m for pNPG(α) and each extract was inhibited by the same sugars with virtually identical Ki from plant to plant. Each extract displayed Michaelis-Menten substrate utilization with competitive inhibition by galactose, xylose and inositol.

We next challenged each of the above extracts with antibody raised against mung bean α-galactosidase-hemagglutinin. Each extract contained materials which cross reacted with the mung bean antibody, i.e., each extract showed CRM - even though the extract did not contain a clot dissolving α-galactosidase-hemagglutinin. This observation led us to inquire if the α-galactosidase activity present in each extract was responsible for the CRM. To answer this, IgG was purified from mung bean antibody and from normal sera. Each IgG was then coupled to Sepharose and extracts from the different legume species, passed over the columns. In each case virtually 100% of the α-galactosidase activity could be specifically adsorbed from the extract by antisera raised against the mung bean protein. No specific binding was seen for B-galactosidase, B-glucosidase, or α-mannosidase. These immunoprecipitation studies indicate that the non-hemagglutin α-galactosidases are immunologically closely related to the mung bean α-galactosidase-hemagglutinin and were at least partially responsible for the CRM. This result, coupled with the similarities in their physical and kinetic properties, suggests that both types of α-galactosidases are members of one specific functional class of protein, i.e., they are homologues.

The evidence collected to date indicates that the α-galactosidase-hemagglutinins and the "classic" legume hemagglutinins are members of distinct classes of protein: they can be separated from each other; they can exist simultaneously in the same tissue; and they have different sugar specificity. Immunological studies reveal, however, that these two classes of proteins share at least some antigenic determinants (13). Antisera raised against several "classic" hemagglutinins cross react with mung bean α-galactosidase-hemagglutinin and antisera to the mung bean protein cross react with several "classic" hemagglutins. Thus, while the α-galactosidase-hemagglutinins and the "classic" legume hemagglutinins appear to be members of distinct classes of proteins, immunochemical data suggest they may be evolutionarily related.

Summary

Evidence at this time suggests that there are at least two evolutionarily related, but distinct classes of legume proteins which possess hemagglutinin properties: these are the "classic" lectins and the α-galactosidase-hemagglutinins. All legumes appear to contain a specific α-galactosidase, but only rarely does this enzyme possess hemagglutinin activity. Hence, we suspect the hemagglutinin activity associated with the α-galactosidase-hemagglutinin molecule may not be important with respect to the general functioning of this class of protein. It is clear that a great many questions about these interesting and yet elusive proteins remain to be answered. We hope the observations reported here will stimulate additional work and help provide the foundation from which the physiological roles of lectins will ultimately emerge.

References

1. Liener, I. E.: Annu. Rev. Plant Physiol. 27, 291-319 (1976).
2. Lis, H., Sharon, N.: Annu. Rev. Biochem 42, 541-574 (1973).
3. Goldstein, I. J., Hayes, C. E.: Adv. in Carb. Chem. Biochem. 35, 127-340 (1978).
4. Toms, G. C.: Chemotaxonomy of Legumes, (Harbourne, J. B., Boulter, D., Turner, B.L. eds.) pp. 367-462 Academic Press, New York (1971).
5. Nicolson, G. L.: Int. Rev. Cytol. 39, 89-190 (1974).
6. Foriers, A., de Neve, R., Strosberg, A. D.: Physiol. Vég. 17, 597-606 (1979).
7. Foriers, A., Wuilmart, C., Sharon, N., Strosberg, A. D.: Biochem. Biophys. Res. Commun. 75, 980-986 (1977).
8. Hankins, C. N., Shannon, L. M.: J. Biol. Chem. 253, 7791-7797 (1978).
9. Paus, E., Steen, H. B.: Nature 272, 452-454 (1978).
10. Hankins, C. N., Kindinger, J. I., Shannon, L. M.: Plant Physiol. 65, 618-622 (1980).

11. del Campillo, E.: Legume hemagglutinins with α-galactosidase activity. Ph.D. Thesis. Department of Biochemistry, University of California, Riverside.
12. Hankins, C. N., Kindinger, J. I., Shannon, L. M.: Plant Physiol. in press (1980).
13. Hankins, C. N., Kindinger, J., Shannon, L. M.: Plant Physiol. 64, 104-107 (1979).

Acknowledgment

This research was supported in part by grant PCM 7901450-01 from the National Science Foundation and a Short Term Fellowship to LMS by European Molecular Biology Organization.

A COMPARISON OF THE TWO LECTINS FROM VICIA CRACCA

C.M. Baumann, H. Rüdiger
Institut für Pharmazie der Universität Würzburg
D-8700 Würzburg, Germany

A.D. Strosberg
Molecular Immunology, Institute of Molecular Biology,
University Paris VII, Place Jussieu 2, Paris, France

Lectins occur in the plant as well as in the animal kingdom. Mostly lectins of only one sugar specificity can be isolated from one species. *Vicia cracca*, the common vetch, is an exception in this respect. It contains two lectins of different specificities. We isolated and characterized both because we think that it is important to compare them with respect to their structures and interactions (1, 2).

The lectins were isolated from a crude seed meal extract with Tris buffered saline of pH 8. Purification was achieved using affinity chromatography. One of the lectins was adsorbed to a CH-Sepharose 4B column to which galactosamine was covalently linked. Elution with a pH-gradient yielded two active fractions (1). Both of them agglutinate specifically erythrocytes of human blood group A1. This interaction can be inhibited by sugars. The best inhibitor is GalNAc; half-maximal inhibition takes place at 0.35 mM (Table 1).

The other lectin is purified on Sephadex G-100. It is desorbed by glucose or by acid. This lectin is unspecific towards human erythrocytes.

The unspecific lectin does not show any pH-dependence in its agglutination titre, whereas the titre of the A-specific lec-

Lectins - Biology, Biochemistry, Clinical Biochemistry, Vol. I

tin is highly pH-dependent (1). Below pH 6 no agglutination occurs (Figure 1). Furthermore, the specific activity of the A-specific lectin is about forty times higher than that of the unspecific one.

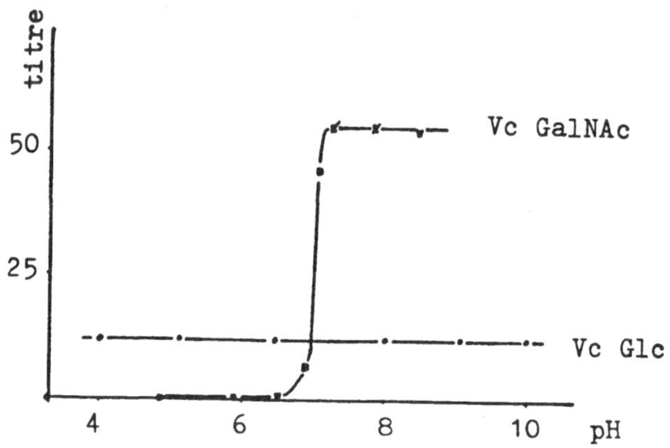

Figure 1: pH-dependence of the agglutination by the Vicia cracca A-specific (Vc GalNAc) and unspecific (Vc Glc) lectins.

Sugars	Concentration needed for 50 % inhibition of agglutination	
	Vc Glc	Vc GalNAc
D-Glucose	5 mM	› 200 mM
D-Mannose	1.6 mM	› 200 mM
D-Galactose	› 200 mM	30 mM
N-Acetyl-D-galactosamine	› 200 mM	0.35 mM
D-Maltose	4 mM	› 200 mM
D-Melibiose	› 200 mM	6 mM
D-Melizitose	1.0 mM	› 200 mM

Table 1: Agglutination inhibition of both Vicia cracca lectins by various sugars. The following sugars do not inhibit: allose, glucuronic acid, cellobiose, sorbose, ribose, xylose and sorbitol.

The agglutination of erythrocytes by the unspecific lectin can be inhibited by various sugars (Table 1). The best inhibiting monosaccharide is mannose. Among the disaccharides tested, maltose inhibits agglutination best. The best inhibitor among all sugars tested is the trisaccharide melizitose.

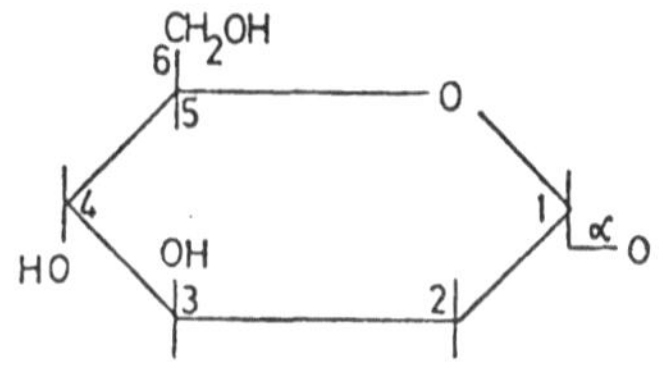

Figure 2: Sugar specificity of the Vicia cracca unspecific lectin.

A sugar inhibiting the glucose specific lectin from Vicia cracca must provide the following structural features (Fig. 2):

1) a hydroxymethyl group at C5: glucuronic acid or arabinose are not inhibitors;
2) equatorial hydroxyl groups at C3 and C4: allose and galactose which have axial hydroxyl groups at C3 or C4, respectively, do not show any inhibition;
3) the glycosidic hydroxyl group in α-configuration: cellobiose does not inhibit, whereas maltose does;
4) either configuration of the hydroxyl group at C2 is accepted: mannose and glucose inhibit nearly equally well.

Thus the glucose or mannose specific lectin from Vicia cracca closely resembles the lectins from Pisum sativum (3), Lens culinaris (4) and Canavalia ensiformis (Con A) (5, 6) with respect to its sugar binding site.

Gel filtration on Sephadex G-100 of the GalNAc specific lectins yields a molecular weight of 125 000. Polyacrylamide electrophoresis in presence of SDS shows a single band corresponding to a molecular weight of 33 000. Thus the native lectin consists of four subunits identical in molecular weights. The

lectin	molecular weight total	molecular weight subunit	isoelectric point	K_{Ass} (M^{-1})	
Vc GalNAc I & II	125 000	4 x 33 000	5.2 - 5.6	6 x 10^3	a
				25 x 10^3	b
				2.9 x 10^3	c
Vc Glc	43 000	2 x 5 800	6.1 6.7	0.2 x 10^3	a
		2 x 15 000 - 18 000	4.6 4.8	1.4 x 10^3	b
				0.2 x 10^3	c

Table 2: Properties of the two Vicia cracca lectins. Association constants were determined by: a) difference spectroscopy, b) equilibrium dialysis, c) inhibition of agglutination.

isolectric points vary between pH 5.2 and 5.6 showing the existence of isolectins (Table 2).

The glucose specific lectin is much smaller. Gel filtration on Biogel P 60 reveals a molecular weight of only 43 000. In SDS electrophoresis we found two bands corresponding to molecular weights of 5 800 and 15 - 18 000. Apparently the native lectin consists of two small subunits - called α-chains - and two large subunits - called ß-chains.

The association constants of both lectins were determined for GalNAc or Glc, respectively (Table 2). Difference spectroscopy, equilibrium dialysis or inhibition of agglutination yield values of the same order of magnitude. They are in good agreement with values obtained for other lectins (3, 6, 7, 8). The association constants of the GalNAc specific lectins are about twenty times higher than those of the glucose specific lectin.

From equilibrium dialysis we derived the number of sugar binding sites per lectin molecule. The A-specific lectin possesses four binding sites, the glucose specific lectin only two.

One of the best methods to compare different proteins is to

	1 10 20
Vc Glc	V T S Y T L S D V V P L K D V V P E W V R I G F S A T
Vs	S - - - - G - - A - - - - - - - - - - - - - - - - - - -
Vf	L - G - - - - E - - - - - - - - - - - - - - - - - - -
Ps	- - - - - - - - - - S - - - - - - - - - - - - - - - -
Lc	- - - - - - N E - - - - - - - - - - - - - - - - - - -

	30 40 50
Vc Glc	P G A E Y A A H E V L S W S F H S E L S G T S S K Q
Vs	T - D - - - - - - - - - - - - - - - - G - - - -
Vf	T - - - - - T - - - - - - T - L - - - T - P - N
Ps	T -
Lc	T - - - F - - N - - H - - - - N - Q - G H - - K S

<u>Figure 3:</u> Sequence of the α-chain from Vc Glc lectin compared with the sequences from: Vs = Vicia sativa, Vf = Vicia faba, Ps = Pisum sativum, Lc = Lens culinaris.

align their amino acid sequences:

The α-subunit of the glucose specific lectin from Vicia cracca consists of 53 amino acid residues (Figure 3). It is totally devoid of the sulfur containing amino acids methionine and cysteine. The tetrapeptide DVVP (position 8 - 11) is repeated in position 14 - 17. Similar to the Vicia cracca glucose specific lectin, the glucose binding lectins from other Vicieae, Vicia faba (8), Vicia sativa (9), Pisum sativum (10) and Lens culinaris (11) consist of two different subunits. They are also very similar to the Vicia cracca lectin in their sequences. The sequence of the glucose specific lectin from Vicia cracca differs in only two positions from that of Pisum sativum and in six from that of Vicia sativa (Figure 3).

		1									10										20					
Vc Glc		T	E	T	T	S	F	L	I	T	K	F	S	K	D	Q	Q	N	L	I	F	Q	G	D	G	Y
Vs		-	-	-	-	-	-	-	-	-	-	-	-	P	-	E	-	-	-	-	Y	-	-	-	-	-
Vf	T	D	-	I	-	-	-	S	-	P	-	-	R	P	-	-	P	-	-	-	-	-	-	G	-	-
Ps		-	-	-	-	-	-	-	-	-	-	-	-	P	-	-	-	-	-	-	-	-	-	N	-	-
Lc		-	-	-	-	-	-	S	-	-	-	-	-	P	-	-	-	-	-	-	-	-	-	-	-	-
Vc GalNAc		-	-	S	-	-	-	S	F	-	E	-	N	Q	N	-	-	-	-	-	L	-	-	-	A	T

Figure 4: N-terminal sequences of the ß-chains from two chain lectins and N-terminal sequence of the Vicia cracca GalNAc specific lectin.

N-terminal sequence studies of the ß-chains of these lectins also show striking homology (Figure 4). Within the first 25 residues we found only two deviations from Pisum sativum and Lens culinaris and three compared to Vicia sativa. We also established the N-terminal sequence of the Vicia cracca GalNAc specific lectins (2). The two isolectins are identical up to the 25th residue. There is about 60 % of homology between the Glc specific lectin and the GalNAc specific lectin from Vicia cracca. Thus, though the lectins with different specificities from the same plant (Vicia cracca) resemble each other, the similarity between lectins of the same specificity but from different plants is much higher.

These results were confirmed by immunochemical studies. We raised antibodies against different lectins in rabbits using Freund's adjuvant. We also resolved some two chain lectins in order to get antibodies against both chains separately. We succeeded in getting anti-ß-antisera, but we did not obtain antibodies against the α-subunits. Immunodiffusion experiments were performed in agar plates containing sugars (Gal, Glc, Fuc, GalNAc,GlcNAc, 10 mM each) in order to prevent the well known lectin-glycoprotein interactions (12). As expected, each lectin formed a single precipitin line with its antiserum. Some

Lectins	Antisera								
	Vc Glc	VcßGlc	VcGalNAc	Lc	ßLc	Ps	ßPs	Con A	SBA
Vicia cracca Glc	+	+	-	+	+	+	+	-	-
Vicia cracca ß Glc	+	+	-	+	+	+	+	-	-
Vicia cracca GalNAc	-	-	+	-	-	-	-	-	-
Vicia sativa	+	+	-	+	+	+	+	-	-
Vicia faba	+	+	-	+	+	+	+	-	-
Lens culinaris	+	+	-	+	+	+	+	-	-
ß Lens culinaris	+	+	-	+	+	+	+	-	-
Pisum sativum	+	+	-	+	+	+	+	-	-
ß Pisum sativum	+	+	-	+	+	+	+	-	-
Canavalia e.(Con A)	-	-	-	-	-	-	-	+	-
Glycine max (SBA)	-	-	-	-	-	-	-	-	+
Arachis hypogaea(PNA)	-	-	-	-	-	-	-	-	-
Crotalaria juncea	-	-	-	-	-	-	-	-	-
Lotus tetragonol.	-	-	-	-	-	-	-	-	-

Table 3: Immunological crossreactions of some lectins.

crossreactions, however, could be observed. In all these cases a single spur was present indicating that the crossreacting lectin was immunologically related but not identical to the homologous lectin (Table 3).

All two chain lectins (Vicia cracca Glc specific, Vicia sativa, Vicia faba, Lens culinaris and Pisum sativum) show crossreactions. The isolated ß-chains react like the intact molecules. On the other hand, the one chain lectins (Vicia cracca GalNAc, Concanavalin A, Soybean and Peanut agglutinin) crossreact neither which each other nor with the two chain lectins (13). It should be pointed out that though the glucose specific lectin from Vicia cracca is related to the two chain lectins from other plants, it does not possess a common antigenic determinant with the GalNAc binding lectin from the same plant.

Both Vicia cracca lectins interact with each other. We suspected such an interaction since the yield of the glucose binding lectin was significantly increased, if the GalNAc binding lectin had been removed by affinity chromatography. Furthermore, if one of the lectins is attached covalently to CH-Sepharose 4B it is able to retard the other lectin from a solution. After mixing both lectins the interaction can be visualized either by the spectral changes in the protein absorption range or by a turbidity which can be assayed by nephelometry.

At present we do not know, if this interaction also occurs in vivo. To answer this question it is necessary to localize both lectins in the seed and in the plant.

References

1. Rüdiger, H.: Eur. J. Biochem. 72, 317-322 (1977).
2. Baumann, C.M., Rüdiger, H., Strosberg, A.D.: FEBS Lett. 102, 216-218 (1979).
3. Trowbridge, I.: J. Biol. Chem. 249, 6004-6012 (1974).
4. Ticha, M., Entlicher, G., Kostir, J., Kocourek, J.: Biochim. Biophys. Acta 221, 282-289 (1970).
5. Goldstein, I.J., Reichert, C.M., Misaki, A., Gorin, P.A.J.: Biochim. Biophys. Acta 317, 500-504 (1973).
6. Edelman, G.M., Cunningham, B.A., Reeke, G.N., Becker, J.W., Waxdal, M.J., Wang, J.L.: Proc. Nat. Acad. Sci. US 69, 2580-2584 (1972).
7. Stein, M.D., Howard, I.K., Sage, H.J.: Arch. Biochem. Biophys. 146, 353-355 (1971).
8. Hemperly, J., Hopp, T., Becker, J.W., Cunningham, B.A.: J. Biol. Chem. 254, 6803-6810 (1979).
9. Gebauer, G., Schiltz, E., Rüdiger, H.: unpublished.
10. Richardson, C., Behnke, W.D., Freisheim, J.H., Blumenthal, K.M.: Biochim. Biophys. Acta 537, 310-319 (1978).
11. Forier, A., De Neve, R., Strosberg, A.D.: unpublished.
12. Bøg-Hansen, T.C., Takeo, K.: J. Electroph. 1, 67-71 (1980)
13. Forier, A., Baumann, C.M., Lieber, S., De Neve, R., Strosberg, A.D.: Protides of the Biological Fluids, Vol. XXVII, 439-442 (1979); (Peeters, H. ed.) Pergamon Press., Headington Hill Hall, Oxford, England.

PURIFICATION AND CHARACTERIZATION OF VICIA GRAMINEA LECTIN. INTERACTION OF LABELLED LECTIN WITH NATIVE AND ENZYME-MODIFIED HUMAN M AND N ERYTHROCYTES.

M.J. Prigent and R. Bourrillon
Laboratoire de Biochimie, INSERM U. 180, Faculté de Médecine, 45 Rue des Saints-Pères, 75006 Paris, France.

Introduction

Lectins constitute a group of cell-agglutinating proteins that are generally found in plants, particularly among the Leguminosae, in lower invertebrates and even in some higher animals (1). While these proteins differ from one another in their specificities for carbohydrate moities on the cell surface, they can be divided in two main categories. Those which agglutinate erythrocytes regardless of blood group type (2) and those whose specificity is directed toward a blood group system (3). The presence of group N-specific lectin has been reported in Vicia graminea seeds (4). It then appeared that the Vicia graminea lectin was the only lectin to bind the N blood group antigens. The present report describes the purification and characterization as well as physicochemical and biological properties of V.graminea lectin. It also seemed interesting to investigate its interaction with erythrocyte surfaces.

Material and Methods

- Purification and homogeneity of Vicia graminea lectin

The Vicia graminea seeds, cultivated in a Phytotron at Gif-sur-Yvette were ground in a with 0.02 M of phosphate buffer 0.1 M NaCl, pH 6.8 and allowed to stand overnight at 4°C with continuous stirring. To the clear supernatant fraction obtained by centrifugation was added solid ammonium sulfate to give 20 % saturation. After

Lectins - Biology, Biochemistry, Clinical Biochemistry, Vol. I

centrifugation the supernatant collected was brought to 50 % saturation. The 20 to 50 $(NH_4)_2SO_4$ precipitate was dialyzed and fractionnated by chromatography on DE_{52} Whatman cellulose. The column was eluted with increasing molarities of phosphate buffer (from 0.01 M to 0.1 M). All fractions were analyzed for absorbance at 280 nm and for agglutinating activity. After concentration, the main fractions containing hemagglutinating activity were layered on a Sephadex G-150 column. The homogeneity of the lectin was checked by disc electrophoresis on polyacrylamide gel at two different pH, by analytical ultracentrifugation and by immunoelectrophoresis against rabbit serum and Vicia graminea crude extract.

- Molecular weight

The molecular weight of the native lectin was estimated by gel filtration. The Sephadex G-150 column was calibrated according to the procedure of Fish et al. (5) using a mixture of proteins of which the molecular weights are known as markers. Blue Dextran was used to determine the void volume of the column. The molecular weight of the lectin was also measured by high speed sedimentation equilibrium in phosphate buffer, pH 7.4, at 20°C, according to the method of Lamm (6).

The molecular weight of the subunit was estimated by gel filtration on a column of Sephadex G-150 equilibrated with guanidine hydrochloride and β-mercaptoethanol. The column was calibrated with a mixture of molecular weight markers and operated at room temperature. Measurements of the sedimentation velocity were performed in phosphate buffer 7.4 at 20°C with different concentrations of lectin. The $S^{o}_{20,w}$ was obtained by extrapolation to zero concentration.

- Polyacrylamide gel electrophoresis and isoelectric focusing

Disc gel electrophoresis was carried out at pH 10.2 and pH 8.6. To determine the molecular weight of the subunit, polyacrylamide gel (10%) electrophoresis was used as described by Weber and Osborn (7). Electrophoretic mobilities were calculated after staining with Coomassie blue. Analytical isoelectric focusing was carried out as described by Doerr and Chrambach (8) in 5% polyacrylamide gels containing 2% Ampholines (pH 3 - 10) and (pH 4 - 6).

- Chemical analysis and amino-acid sequence determination

The carbohydrate composition was determined by gas-liquid chromatography. The amino-acid composition was determined with an amino-acid Analyzer Geol after 17 h hydrolysis under vacuum at 106°C in 6 M HCl. 200 nmol of Vicia graminea lectin was sequentially degraded in a Beckman sequencer by the method of Hermodson (9) with the NN'dimethyl benzylamine program. The resulting amino-acid phenylthiohydantoïn derivatives were identified by t.l.c. in chloroform/methanol 99/1 and 88/1 r (v/v) and by high pressure liquid chromatography.

- Biological properties

Hemagglutination assays were carried out in cook's plate using MM, NN human erythrocytes. The minimum hemagglutinin concentration of lectin against human erythrocytes for 30 min at room temperature was taken as one unit.

The mitogenic activity of the lectin was assayed by studying [^{3}H]-thymidine incorporation in human peripheral lymphocytes.

- Iodination and binding of Vicia graminea lectin

The lectin was iodinated with ^{125}I by the method of Hunter and Greenwood (10). Specific activities ranged from 20 000 cpm to 30 000 cpm per µg of protein. The reaction mixtures contained 0.4 ml of phosphate buffer, pH 6.8 different concentrations (from 0.02 to 70 µg) of ^{125}I V.graminea lectin and 4×10^7 MM and NN human erythrocytes .After incubation either 15 min at room temperature or 15 min at 4°C, the cells were washed and the amount of bound ^{125}I lectin was determined. Non specific binding was assayed by incubation M and N red cells first with an excess of unlabelled lectin followed by addition of ^{125}I lectin. The number of receptor sites per cell and the association constant for V.graminea lectin were calculated as described by Scatchard (11) and by computer analysis.

- Sialidase treatments of erythrocytes

To 1 ml of packed M and N human erythrocytes suspended in 0.01 M $CaCl_2$- 0.15 M NaCl ajusted to pH 7.0 with 0.1 M $NaHCO_3$, was added 0.8 unit of neuraminidase (Vibrio cholerae one unit/ml). The suspension was gently shaken at 37°C for 1 hour.

Human asialoerythrocytes in a 4% suspension were used for the binding of ^{125}I Vicia lectin. The Warren procedure (12) was used to determine the sialic content of human erythrocytes.

Results

- Purification and homogeneity of *V.graminea* lectin

The hemagglutinating activity recovered in the 20 to 50% $(NH_4)_2SO_4$ precipitate was dialyzed and applied to a DE_{52} Whatman column. Then the main fractions containing the biological activity (10% of the total proteins) were layered on a Sephadex G-150 column. After two gel filtrations, the lectin was eluted as a single symetrical peak. Homogeneity was checked by various techniques. Through the sedimentation run, lectin sedimented as a single peak. By disc electrophoresis on polyacrylamide gel, a single band was observed at two different pH. In immunoelectrophoresis, the lectin gave only one precipitation line with rabbit serum anti-*V.graminea* extract. Isoelectric focusing showed a single peak with an isoelectric point of pH 5.3.

- Physical features

A molecular weight of 120 000 has been obtained by high-speed sedimentation equilibrium whereas gel filtration on Sephadex G-150 gave a molecular weight of 105 000 thus in good agreement with the former value. The sedimentation coefficient $S^{o}_{20,w}$ calculated from sedimentation velocity data was 5.3 S $\pm$ 5%.

The subunit molecular weight determined by polyacrylamide gel electro phoresis in sodium dodecyl sulfate and by gel filtration in dissociating medium was 25 000 $\pm$ 10% (Table I).

TABLE I - Properties of *Vicia graminea* lectin

Molecular weight	
Gel filtration	120 000
Sedimentation equilibrium	105 000
Polyacrylamide gel electrophoresis (Sodium dodecyl sulfate)	25 000
$S^{o}_{20,w}$ (S)	5.3
$\overline{V}$ (calculated)	0.73
pHi	5.3
Total carbohydrate content (%)	7.3
Minimum concentration giving erythroagglutination (µg/ml)	$\leqslant$ 0.8
Minimum concentration giving $[^3H]$ thymidine incorporation (µg/ml)	$>$ 2 000

When submitted to polyacrylamide gel electrophoresis at two different pH in the absence or in the presence of S.D.S.Vicia graminea seemed homogeneous. However four major bands were seen by isoelectric focusing on a pH 4-6 Ampholine gradient (13). Under such conditions purified peanut and soybean lectins and concanavalin A exhibit similar microheterogeneity (14, 15).

- Amino-acid sequence determination

The presence of a single type of subunit as indicated by S.D.S.polyacrylamide gel electrophoresis was confirmed by the determination of the partial N-terminal sequence of the lectin. Analysis showed only isoleucine in the N-terminal position per subunit. Determination of the first nine amino-acid showed a single N-terminal sequence, no microheterogeneity being noted for these first nine residues(Table II).

TABLE II - Comparison of the N-terminal sequence (first nine residues) of the *Vicia graminea* agglutinin with those of other lectins.

Lectin	Residue N° : 1	2	3	4	5	6	7	8	9	10
Vicia graminea	Ile	Asp	Tyr	Ile	Phe	Gln	Gly	Asp	Ala	—
L-PHA(Phaseolus vulgaris lectin)	Ser	Asn	Asp	Ile	Tyr	Phe	Asn	Phe	Glu	Arg
Concanavalin A	Ala	Asp	Thr	Ile	Val	Ala	Val	Glu	Leu	Asp
Ricin B-subunit	Ala	Asp	Val	Thr	Gln	Asp	Pro	Glu	Pro	Ile
Ricin A-subunit	Ile	Phe	Pro	Lys	Met	Tyr	Pro	Ile	Ile	Ile

- Chemical composition

Its chemical composition is characterized by a high proportion of aspartic and glutamic acids : glycine, leucine, isoleucine and serine residues. A low percentage of methionine, tyrosine, cysteine and basic amino-acid residues was noted.

Being a glycoprotein, it contains 7.3% carbohydrate including mannose (5 moles per mole subunit) in the major constituent and N-acetylglucosamine (2 moles per mole subunit) the only osamine. It contains no sialic acid (Table III)

- Biological properties

Down to a concentration of 0.8 μg/ml, this lectin is a potent activator of human N-erythrocytes, but it is 50 times less active on M erythrocytes. However it has no mitogenic activity on normal human erythrocytes.

- Iodination and binding of Vicia graminea lectin

A direct method involving fixation of ^{125}I labelled lectin to N or M erythrocytes was used. A curvilinear relationship is obtained which could be analyzed by graphic procedure. The apparent affinity constants between this lectin and normal N-erythrocytes was $7.1 \times 10^7 M^{-1}$, with 8.5×10^3 receptor sites per cell at low saturation and $2 \times 10^6 M^{-1}$ with 2.1×10^5 receptor sites per cell at high saturation (Table IV). The association constants and the number of receptor sites per cell were lower with M than with N blood cells. Neuraminidase treatment of M and N cells resulted in a stronger fixation of the lectin in both cases and was accompanied by an increase of the number of receptor sites from both types of cells(Table IV).

TABLE III - Chemical composition of purified V.graminea lectin

Constituents	Mol/mol of lectin	g/100 g of glycoprotein
Mannose	20.3	3.6
Glucose	2.2	0.4
Galactose	3.0	0.5
Fucose	4.7	0.8
Glucosamine*	9.1	2.0

TABLE IV - V.graminea lectin binding parameters on human M and N normal erythrocytes obtained by Scatchard procedure and effect of neuraminidase treatments of human M and N erythrocytes on V.graminea lectin binding parameters (See "Experimental procedure").

Human erythrocytes	K Association constants($10^7 M^{-1}$)		Sites/cell x 10^5	
	Scatchard plot		Scatchard plot	
NN 15 min at 4°C	High	Low	High aff.	Low aff.
- Normal erythrocytes	7.07	0.2	0.085	2.1
- Neuraminidase treated	7.42	0.58	0.8	6.25
MM 15 min at 4°C				
- Normal erythrocytes	6.8	0.15	0.014	0.2
- Neuraminidase treated	4.6	0.14	0.51	4.8

TABLE III : * N-acetylglucosamine.

Discussion

No lectin specific for MN system has been so far isolated. Therefore the _Vicia graminea_ lectin is the first anti-N lectin to be purified and characterized. In terms of overall composition, the _V.graminea_ lectin is similar to other purified lectins. Generally, they are glycoproteins of which mannose and N-acetylglucosamine are the major constituents, as is the case for this lectin. _Vicia graminea_ , like other lectins, has a subunit structure. It is a tetramer of four similar subunits. The fact that the subunits are dissociated by S.D.S. irrespective of the presence of a reducing agent indicates that the subunits are held together by non covalent bonds. Their subunit structure may play a role in determining their biological and particularly their agglutinating properties (16). The presence of a single type of subunit was confirmed by the determination of the partial N-terminal sequence of the lectin. As it has been suggested by Foriers the genes providing the code for the various leguminous plant lectins may have evolved from a common ancestral gene (17). Therefore we compared the N-terminal sequence of the _Vicia graminea_ lectin with that of other lectins. There are thus only limited similarities between the N-terminal sequence of the different lectins (Table II). However _Vicia graminea_ showed four bands by isoelectric focusing. Different explanations can be proposed for this apparent microheterogeneity.Multiple bands could conceivably arise from heterogeneity in the oligosaccharide chain or from differences in the amide content of the asparagine or glutamine residues (18). However the most satisfactory explanation could seem to lie in the fact that the _Vicia graminea_ seeds used were not selected on a genetic basis. Possibly the variants could be the products of several nearly identical genes. If there were slight differences between individual seeds, pooling of the seeds would resulted in the pattern observed.

The existence of two types of sites suggesting the presence of a heterogeneous population of binding sites which differ in their affinity for Vicia graminea lectin or of a hemogeneous population of receptors where the binding of this lectin no affects these antigens as to change the affinity of unfilled binding sites for the lectin. But site heterogeneity cannot be distinguished from negative cooperativity by binding measurements alone. Furthermore, after removal of sialic acid from human erythrocytes there is still two types of sites It has been assumed that the affinity of the lectin for M and N antigens is dependent on the degree of sialylation since the biological activity of V.graminea lectin is better expressed in the absence of terminal sialic acid.

The saccharide specificity of many lectins is well established, but for some lectins, like V.graminea no monosaccharide is a good inhibitor. Probably these lectins require a complex structure for efficient interaction. The presence of branched structures containing gal-galNAC seems necessary for the anti-N activity of Vicia. Furthermore the affinity of the lectin for M and N antigens is dependent on the degree of sialylation. It would appear that the combining sites of Vicia lectin may be larger than a disaccharide, probably a complex oligosaccharide. Its biological activity may be determined by the sequence of the sugars as well as by the linkages between them and also the possible steric arrangement of the amino groups of the peptide chains.

References

1. Sharon, N., Lis, H. : Science 177, 949-959 (1972).
2. Miller,J.B., Hsu,R., Heinrikson,R., Yachnin,S.: Proc.Natl.Acad. Sci. U.S.A.72, 1388-1391 (1975).
3. Carter,C.G., Etzler,M.E.: Biochemistry 14, 2685-2689 (1975).
4. Prigent,M.J.,Bourrillon,R.: Biochim.Biophys.Acta 420,112-121(1976)
5. Fish,W.W., Mann,R.G., Tanford,C.: J.Biol.Chem.244,4989-4994(1969).
6. Lamm,O.: Zur Theorie und Methodik der Ultrazentrifugierung, Z. Physik. Chem. (Leipzig) A, 143-177 (1929).
7. Weber,K., Osborn,M.: J. Biol.Chem. 24, 4406-4412 (1969).
8. Doerr,P., Chrambach, A.: Anal. Biochem. 42, 96-107 (1971).
9. Hermodson,M.A., Ericsson,L.H., Titani,K., Neurath,H., Walsh,K.A.: Biochemistry 11, 4493-4502 (1972).
10. Hunter,W.W., Greenwood,F.C.; Nature, 660-672 (1962).
11. Scatchard,G.: Ann.N.Y. Acad. Sci. 51, 660-672 (1949).
12. Warren,L. : J. Biol. Chem. 234, 1971-1975 (1959).

13. Prigent,M.J., Bourrillon,R.: Biochem.J. 189, 185-188 (1980).

14. Agrawal,B.B.L., Cuatrecasas,P., Pollard,H.H.: J.Biochem.Biophys. 124, 218-229 (1971).

15. Lis,H., Friedman,C., Sharon,N., Katchalski,E.: Arch.Biochem.Biophys. 117, 301-309 (1966).

16. Funatsu,G.,Ueno,S.,Funatsu,M.: Agric.Biol.Chem.41, 171-174(1977).

17. Foriers,A., Wuilmart,C., Sharon,N., Strosberg,A.D.: Biochem. Biophys. Res. Commun. 75, 980-986 (1977).

18. Spencer,E.M., King,T.P. : J.Biol.Chem. 246, 201-208 (1971).

STUDIES OF THE LATHYRUS ODORATUS (LATH) LECTIN

J. Kolberg, T.E. Michaelsen
Department of Immunology, National Institute of Public Health, SIFF, Oslo 1, Norway

K. Sletten
Department of Biochemistry, University of Oslo, Blindern, Oslo 3, Norway

The haemagglutinating activity in the seed extracts of Lathyrus odoratus, (sweet pea), belonging to the Leguminosae family, has been known since the beginning of this century (1). We have isolated a lectin from these seeds by fractional precipitation with ammonium sulphate followed by chromatography on DEAE- and CM-Sepharose (2).The lectin agglutinated human erythrocytes of different ABO groups equally well, and was also found to be mitogenic to human lymphocytes.In terms of maximal thymidine incorporation it was as potent as PHA, but the weight concentration giving maximum effect was much higher than for PHA (2). The activation of the following human lymphocyte subpopulations has been investigated (3):

1. T-enriched lymphocytes prepared by nylon wool filtration.
2. Non-T lymphocytes obtained by depletion of sheep erythrocyte rosette forming T-cells.
3. Monocyte depleted lymphocytes.

Compared with the LATH activation of unfractionated lymphocytes, the lectin stimulation of the T-enriched and the non-T cell fraction was 14 and 12%, respectively (4). Removal of monocytes from the unfractionated lymphocytes caused only a slight reduction in the response. A full LATH response could be restored by mixing T-enriched and non-T cells. The LATH

Lectins - Biology, Biochemistry, Clinical Biochemistry, Vol. I

lectin therefore appears to require both T- and B-cells for maximum stimulation, whereas the monocytes seem to be of minor importance (3).

With regard to the ligand specificity of LATH, the mitogenic activation was strongly inhibited by α-methyl-D-mannoside, to a lesser degree by D-glucose and N-acetyl-D-glucosamine (3). These results indicate that the LATH lectin belongs to group 8 according to the classification of Kristiansen (5), together with lectins such as Con-A, pea and lentil. The LATH lectin contained 0,5% carbohydrates (2).

The LATH preparation used in our first studies was isolated by conventional chromatography (2). SDS-gel electrophoresis revealed one major band of mol. wt. 19,000 and a faint band of smaller molecular weight. Multiple peaks were obtained by isoelectric focusing on column. The mol. wt. of the untreated lectin was estimated to 37,000 by gel filtration under neutral conditions (2). It was therefore originally suggested that the lectin was composed of two subunits of 19,000 presuming that the faint band was a contaminant.

To clarify whether the minor component was a contaminant or an integral part of the lectin molecule, affinity chromatography was performed. However, the LATH preparation isolated by adsorption to Sephadex-G-100 and specific elution with 0,1M glucose also showed, two bands upon SDS-gel electrophoresis. Similar bands were obtained for the lentil lectin run in parallel. Structural studies of lectins purified from leguminous plants have revealed the existence of two groups of proteins:

1. Composed of one type of polypeptide chains.
2. Composed of two types of chains, the light α chains and the heavy β chains.

Since the lentil lectin is known to be composed of two light (α) and two heavy (β) chains (6), the SDS gel pattern might indicate that the LATH lectin has a similar structure.

To further evaluate this, the LATH lectin was dissolved in 2 M acetic acid and gel filtered through Sephadex G-100 in this solution. The lectin was resolved into one major peak (mol. wt. 19,000) and one smaller peak of low mol. wt. The materials in the individual peaks as well as mixtures of the two peaks had no haemagglutinating activity after stepwise dialysis back to neutral conditions (7). A similar loss of activity has also been reported for the lentil lectin (8).

Amino acid analysis of the whole lectin and the components obtained by gel filtration in 2 M acetic acid neither showed presence of methionine nor cysteine(7). The analysis gave no clear information whether the minor component could be a subunit, or that it was a contaminant(s) or fragment(s) of the major polypeptide chain.

N-terminal sequence analysis by automated Edman degradation of residues 1-25 of the heavy and light components clearly demonstrated that the LATH lectin belongs to the 'two chain' lectin group. The sequence of the first 25 amino acids of the LATH α chain (7) was completely identical with the corresponding chain of the lentil lectin (6), while the first 25 amino acids in the LATH β chain (7) showed only one difference compared to the lentil β chain (6): In position 7 serine is reported for lentil (6) while we found leucine for LATH. Comparison with the sequences of other glucose-mannose binding lectins reveals extensive homologies also with the pea and *Vicia cracca* lectins (7).

The lectins with α and β chains are assosiated *in vivo* to form $(\alpha\beta)_2$. By sedimentation equilibrium ultracentrifugation the mol. wt. of the whole LATH lectin was found to be 40,000 while the mol. wt. of the β and α subunits were 17,000 and 4,400, respectively (7). On the basis of these data a stochiometry of $(\alpha\beta)_2$ would give a mol. wt. of 42,800 of the whole lectin.

According to normal taxonomy Lathyrus as well as Lens, Vicia and Pisum belong to the Vicieae tribe of the Leguminoseae family. The close relationship between Lathyrus odoratus and other spesies in the Vicieae tribe has been confirmed by similarities in subunit structure and N-terminal sequences of their lectins.

Refrences

1. Gold, E.R., Balding, P.: Receptor-Specific Proteins. Plant and Animal Lectins, Excerpta Medica, Amsterdam 1975
2. Kolberg, J.: Acta path. microbiol. scand. Sect. C. 86, 99-104, (1978)
3. Kolberg, J., Michaelsen, T.E.: Acta path.microbiol. scand. Sect. C. 87, 275-279 (1979)
4. Kolberg, J., Michaelsen, T.E.: Protides biol. Fluids 27, 577-580 (1979)
5. Kristiansen, T.: Meth. Enzymol. 34, 331-341 (1974)
6. Foriers, A., De Neve, R., Kanarek, L., Strosberg, A.D.: Proc. Natl. Acad. Sci. USA 75, 1136-1139 (1978)
7. Kolberg, J., Michaelsen, T.E., Sletten, K.: FEBS Lett. (In press) (1980)
8. Fliegerová, O., Salvetová, A., Tichá, M., Kocourek, J.: Biochim. Biophys Acta 351, 416-426, (1974)

MISTLETOE LECTINS: CHEMICAL MODIFICATION AND CARBOHYDRATE INTERACTION OF THE D-GALACTOSE SPECIFIC LECTIN I

P. Ziska and H. Franz
Staatliches Institut fur Immunpraparate und Nahrmedien,
DDR-1120 Berlin, Klement-Gottwald-Allee 317/321

Introduction

Using affinity chromatography we have isolated three lectins from the mistletoe (Viscum album L.). Lectin I reacts with a galactan gel prepared by acid treatment of agarose or Sepharose (1). Neither lectin II nor lectin III react with this gel but react with the carbohydrate moieties of immunoglobulins bound to Sepharose. From there they were eluted by 0.1 M D-galactose solution (lectin II) and by glycine-HCl buffer pH 2,8 (lectin III) respectively (in press).
Lectin I is a glycoprotein with a molecular weight of 115.000 and consists of two subunits, which are not covalently bound. Each subunit consists of two chains with molecular weights of 29.000 and 34.000 bound by disulphide bridges. It was found that the reaction of mistletoe lectin I with carbohydrate-containing macromolecules was inhibited by D-galactose. It was of interest to study the lectin-carbohydrate interaction with a view to ascertain the configuration features that the lectin and the carbohydrate-containing molecules must possess in order to react together.

Chemical modification of mistletoe lectin I

Using specific chemical modification techniques the role of amino, sulphydryl, disulphide, carboxyl, phenolic, imidazole and indole groups of the mistletoe lectin I has been determined with respect to hemagglutination (2,3). Acetylation phenolic hydroxyl groups was performed with N-imidazole and ring substitution of phenolic groups with tetranitromethan. Modification of tryptophan residues was carried out using 2-hydroxy-5-nitrobenzylbromide (Koshland's reagent) and with N-bromosuccinide.

Lectins - Biology, Biochemistry, Clinical Biochemistry, Vol. I

Amino groups were modified by reaction with diethylpyrocarbonate, fluorescein-isothiocyanate and formaldehyde. Carboxyl groups were converted to amides through reaction with glycine ethylester in the presence of 1-ethyl-3-(-3-diethylaminopropyl) carbodiimide. Photo-chemical oxidation was performed in the presence of methylene blue. Substitution of sulphydryl groups was carried out with 5,5-dithio-bis (2-nitrobenzoic acid) (Ellman's reagent). The results of the experiments are summarized in Table 1.

Cleavage of the disulfide bridges with sodium borohydride resulted in reduction of the hemagglutinating ability. The reaction of the lectin with N-acetylimidazole to form O-acetyltyrosine derivatives and nitration of the lectin to form nitro-tyrosine derivates resulted also in reduction of the biological activity.

D-galactose, lactose, -methyl-D-galactoside and β-methyl-D-galactoside did not protect the lectin against modification with tetranitromethane in a final sugar concentration of 0.2 M. But the modification of the tyrosyl groups with N-acetylimidazole was protected by above sugars and also by D-glucose, D-mannose, D-mannitol and D-galactitol. Thus we assume that this is a non-specific protection of acetylation or a reaction between the N-acetylimidazole and each of the tested sugars or sugar alcohols. The data indicate that tyrosyl residues are associated with the carbohydrate binding sites of the lectin. But further studies are needed to clarify the nature of the lectin-carbohydrate interactions and to provide additional insight into the role of tyrosyl residues.

There are several publications about chemical modification of other lectins. For example the reaction of concanavalin A with tetranitromethan and N-acetylimidazole results in a decrease in complexing ability with glycogen (4). In addition, evidence is presented that the radical ions $(Br)_2$ and $(SCN)_2$ react with tyrosine and tryptophan residues in concanavalin A, which is also accompanied with loss of glycogen precipitation activity (5).

Table 1

Chemical modification of the lectin I from Viscum album

Modification by	Modified residues	Hemagglutinating activity	Titer
Photochemical oxidation	histidine	unchanged	1:256
Diethylpyrocarbonate	histidine	unchanged	1:256
Koshland's reagent	tryptophan	unchanged	1:256
N-bromosuccinimide	tryptophan	unchanged	1:256
Ellmann's reagent	-SH	unchanged	1:256
Formaldehyde	$-SH, -NH_2$	unchanged	1:256
FITC	$-NH_2$	unchanged	1:256
Carbodiimide, amine	-COOH	unchanged	1:256
Sodium borohydride	-S-S-	reduced	1:8
N-acetylimidazole	tyrosine	reduced	1:8
Tetranitromethan	tyrosine	reduced	1:8

For hemagglutination a solution of 3 mg lectin I from Viscum album per ml saline with a titer of 1:256 against a 1% suspension of human erythrocytes was used.

Modification of concanavalin A using 2-nitrophenyl-sulfenyl chloride and N-bromosuccinimide (reagents considered to be specific for tryptophan residues in proteins) is accompanied by loss of activity which parallels the degree to which tryptophan is modified (6). Acetylation of tyrosyl groups with N-acetylimidazole strongly reduced the toxicity of abrin and ricin, the toxins from Abrus precatorius and Ricinus communis, as well as their ability to induce hemagglutination. Attempts

Table 2

Inhibition power of various D-galactose derivates

Inhibiting compound	μmol carbohydrate /ml needed for complete inhibition of four hemagglutination units.	
D-Galactose	12	
Methyl-α-D-galactopyranoside	12	
Phenyl-α-D-galactopyranoside	6	
Phenyl-β-D-galactopyranoside	3	
Lactulose (β 1-4)	1,5	
Lactose (β 1-4)	3	
Melibiose (α 1-6)	3	
Raffinose (α 1-6; β 1-2)	12	
2 Deoxy-D-galactose	12	
3-O-Methyl-D-galactose	200	(0)
3,6-Anhydro-D-galactose	200	(0)
4-O-Methyl-D-galactose	200	(0)
4,6 -O- Ethylidene-D-galactose	200	(0)
4,6-O-Benzylidene-D-galactose	200	(0)
1,2; 3,4-O-Diisopropylidene-D-galactose	200	(0)
6-Deoxy-D-Galactose	12	
6-O-Methyl-D-galactose	12	
6-O-D-Benzyl-D-galactose	3	
Galactan from Lupinus albus (β 1-4)	10 mg	(0)
Galactan from Helix pomatia (β 1-3; β 1-6)	10 mg	(0)

Where numbers are given in paranthese, inhibition was not attained with 0,2 m carbohydrate solution.

to modify tryptophan groups by treatment with N-bromosuccinimide resulted also in strongly reduced toxicity and hemagglutinating activity (7). Modification of the lectin from Pisum sativum employing acetic anhydride results in acetylation of free amino groups as well as the hydroxyl groups of tyrosine, threonine, and serine accompanied with the loss of hemagglutinating activity. But the acetylated lectin binds specifically to Sephadex (8). On the other hand acetylation of the lectin from Bandeiraea simplicifolia did not show any loss in precipitation of guaram. Modification of the carboxyl groups employing glycine and a water soluble carbodiimide indicates that these groups participate in the carbohydrate binding phenomenon (9).

Chemical modification studies confirm also the involvement of carboxyl groups in the binding of carbohydrates by concanavalin A (10) but not that of amino groups (11). Depending on the reaction conditions acetylation of the tyrosine residues and amino groups of the lectin from Lens culinaris results in a modified lectin without hemagglutinating activity. However, it retains the ability to interact with polysaccharides and to bind to erythrocytes (12). Under defined mild conditions the reaction of the lectin from Pisum sativum with 2-nitrophenylsulfenyl chloride results in sulfenylation of only two of the ten tryptophan residues of the molecule with simultaneous loss of biological activity (13).

Interaction of mistletoe lectin I with carbohydrates

Using several carbohydrates, disaccharides, derivatives of D-galactose and galactans we have studied the lectin-carbohydrate interaction quantitatively by means of the hemagglutination inhibition test (1). Derivatives of D-galactose were prepared by standard methods in our laboratory (14,15). Galactans were isolated from Lupinus albus (16) and Helix pomatia (17).

Our inhibition experiments confirm the original obsevation of Pardoe made with crude extracts of mistletoe that D-galactose and lactose are good inhibitors of the lectin hemagglutination activity (18). We have extended the study to some other carbohydrates and have found that the following carbohydrates do not inhibit hemagglutination of human erythrocytes in a concentration of 0.2 M solution: D-glucose, D-glucosamine, N-acetyl-D-glucosamine, D-mannose, D-mannosamine, N-acetyl-D-mannosamine, D-galactosamine, N-acetyl-D-galactosamine, D-talose, D-tagatose, dulcit, maltose, and sucrose. Among D-galactosides differing only in the configuration about anomeric C-1, there seems to be a slight preference for the β anomer. Aromatic aglycons of D-galactosides increased the binding of the corresponding sugars by the lectin, which do not exclude the existence of a hydrophobic region adjacent to the sugar binding site (table 2). In this respect it is similar to plant lectins such as concanavalin A and soy bean agglutinin. N-acetyl-galactosamine, D-galactosamine, 2-deoxy-D-galactose, D-talose, D-tagatose were not inhibitors indicating a requirement for a free equatorial hydroxyl group at C-2. The configuration at C-3 is also important for binding since neither 3-O-methyl-D-galactose, 3,6-anhydro-D-galactose, nor galactan from the snail Helix pomatia (β 1-3, β 1-6 linkage) were inhibitors. Also an equatorial hydroxyl group at C-4 is necessary for the lectin carbohydrate interaction since neither D-glucose, 4-O-methyl-D-galactose, 4,6-ethyliden-D-galactose, 4,6-O-benzylidene-D-galactose, nor galactan from Lupinus albus (β 1-4 linkage) were inhibitors. But a C-6 hydroxyl group seems not to be essential for binding as 6-deoxy-D-galactose, 6-O-methyl-D-galactose and 6-O-benzyl-D-galactose were inhibitors.

Inhibition data demonstrate that substitution of any of the hydroxyl groups at C-2, C-3 or C-4 results in a complete loss of activity. Therefore unmodified hydroxyl groups at the C-2, C-3, and C-4 position of the D-galactopyranosyl ring system are essential for the carbohydrate-lectin interaction. Similar results were reported about the lectin from Ricinus communis, although these experiments were performed with crude lectin preparations (19). The lectin from Arachis hypogaea with

a specificity for D-galactose, too, needs an equatorial free hydroxyl or an a amino group at C-2 and unmodified hydroxyl groups at C-4 and C-6 (20). Whereas the Bauhinia purpurea lectin reacts also with a hydroxyl group at C-2, which was substituted by an acetyl group, it needs a free hydroxyl group at C-4 for lectin carbohydrate interaction (21).
A comparison of the inhibitory activity of various sugars on the lectin from Canavalia ensiformis, Lens culinaris, Pisum sativum and Vicia faba indicates that concanavalin A differs

Table 3

Essential hydroxyl and acetamido groups for lectin-carbohydrate interaction

lectin	carbohydrate specificity	configuration			
Canavalia ensiformis	Glc		C-3	C-4	C-6
Lens culinaris	Glc			C-4	C-6
Pisum sativum	Glc			C-4	C-6
Vicia faba	Glc			C-4	C-6
Viscum album	Gal	C-2	C-3	C-4	
Ricinus communis	Gal	C-2	C-3	C-4	
Arachis hypogaea	Gal	C-2		C-4	C-6
Bauhinia purpurea	Gal	C-2		C-4	
Bandeiraea simplicifolia	GlcNAc	C-2	C-3	C-4	C-6
Triticum vulgaris	GlcNAc	C-2	C-3	C-4	
Helix pomatia	GalNAc	C-2		(C-4)	
Dolichos biflorus	GalNAc	C-2		C-4	
Phaseolus lunatus	GalNAc	C-2		C-4	
Glycine max	GalNAc	C-2		C-4	C-6

from the other three D-glucose (D-mannose) specific lectins. Whereas all four lectins are inhibited by -methyl glucosides and this inhibitory activity is abolished by substitution of the hydroxyl groups at C-4 or C-6, they differ in their response to 3-O-substituted monosaccharides (22). Whereas concanavalin A is not inhibited by 3-O-substituted monosaccharides the other three lectins are at least 10 times more strongly inhibited by 3-O-substituted monosaccharides than by the parent sugars. The N-acetyl-D-glucose specific lectin from Bandeiraea simplicifolia requires unmodified hydroxyl groups at C-3, C-4 and C-6 like the lectin from Canavalia ensiformis (23). Substitution of the 3-hydroxyl group of N-acetyl-D-glucose the lectin from Triticum vulgaris completely abolishes binding capacity. Although substitution of the 4-hydroxyl group does not affect binding, the configuration of that hydroxyl group is important, as N-acetyl-D-galactosamine is a weak inhibitor (24). The combining site of the four lectins from Helix pomatia, Glycine max, Dolichos biflorus and Phaseolus lunatus corresponds to the size of a monosaccharide. The finding of the sugar inhibition studies indicates that an axially oriented hydroxyl group at the C-4 position of N-acetyl-D-galactose is absolutely essential for binding to Glycine max, Dolichos biflorus and Phaseolus lunatus lectins. For Helix pomatia lectin, axial orientation of the C-4 hydroxyl is preferred but not absolutely essential (25). Two contact groups in the N-acetyl-D-galactosamine molecule, probably involved in hydrogen-bound formation with the Helix pomatia binding site, were identified: The carbonyl oxigen on the C-2 acetamido group and the O atom on the C-4 hydroxyl group. For the Glycine max lectin three contact groups were identified: The carbonyl oxygen on the C-2 acetamido group and the H atoms on the C-4 and C-6 hydroxyl groups. For Dolichos biflorus and Phaseolus lunatus lectin only the C-2 acetamido group in N-acetyl-D-galactosamine has been identified as being in contact with the binding sites of these lectins (25). The results of some sugar-lectin inhibition studies to estimate the essential configuration of hydroxyl groups are summarized in table 3. Lectins of the same carbohydrate specificity do not need in all cases the same hydroxyl groups for carbohydrate interaction.

References

1. Ziska, P., Franz, H. and Kindt, A.: Experientia 34, 123-124 (1978).
2. Means, G.E. and Feeney, R.E.: Chemical modification of proteins Holden-Day Inc., San Francisco 1971, 214-230.
3. Ziska, P., Eifler, R. and Franz, H.: Acta biol. med. germ. 38, 1361-1363 (1978).
4. Doyle, R.I. and Roholt, O.A.: Life Sci. 7, 841-846 (1978).
5. Davies, J.V., Moore, J.S. and Mudher, S.: Int. J. Radiat. Biol. 33, 11-19 (1978).
6. Hassing, G.S. and Goldstein, I.J.: Biochim. Biophys. Acta 271, 288 (1972).
7. Sandvig, K., Olsnes, S. and Pihl, A.: Eur. J. Biochem. 84, 323-331 (1978).
8. Trowbridge, I.S.: Proc. Nat. Acad. Sci. USA 70, 3650-3654 (1973).
9. Lönngren, I. and Goldstein, I.J.: Biochim. Biophys. Acta 439, 160-166 (1976).
10. Hassing, G.S., Goldstein, I.J. and Marini, M.: Biochim. Biophys. Acta 243, 90-97 (1971).
11. Agrawal, B.B.L., Goldstein, I.J. Hassing, G.S. and So, L.L.: Biochemistry 7, 4211 (1968).
12. Vancurová, D., Ticha, M. and Kocourek, J.: Biochim. Biophys. Acta 453, 301-310 (1976).
13. Cermakova, M., Entlicher, G. and Kocourek, J.: Biochim. Biophys. Acta 420, 236-245 (1976).
14. Bell, D.J.: Adv. Carbohydr. Chem. 6, 11-25 (1951).
15. Mahler, G.G.: Adv. Carbohydr. Chem. 10, 273-282 (1955).
16. Jones, J.K.N. and Tanaka, Y.: Meth. Carbohydr. Chem. 5, 132-134 (1965).
17. Horstmann, H.J.: Biochem. Z. 340, 548-551 (1964).
18. Pardoe, G.I., Bird, G.W., Uhlenbruck, G., Sprenger, I. and Heggen, M.: Z. Immunfschg. 140, 374-394 (1970).

19. Drysdale, R.G., Herrick, P.R. and Franks, D.:
Vox sang. 15, 194-202 (1968).

20. Lotan, R., Skutelsky, E., Davon, D. and Sharon, N.:
J. biol. Chem. 250, 8518-8523 (1975).

21. Osawa, T., Irimura, T. and Kawaguchi, T.:
Methods in Enzymology 50, 367-372 (1978).

22. Allen, A.K., Desai, N.N. and Neuberger, A.:
Biochem. J. 155, 127-135 (1976).

23. Ebisu, S., Iyer, P.N.S. and Goldstein, I.J.:
Carbohydr. Res. 61, 129/138 (1978).

24. Allen, A.K., Neuberger, A. and Sharon, N.:
Biochem. J. 131, 155-162 (1973).

25. Hammerström, S., Murphy, L.A., Goldstein, I.J. and Etzler, M.E.: Biochemistry 16, 2750-2755 (1977).

MITOGENIC LECTINS OCCUR ONLY IN FEW SEED SAMPLES OF VIGNA (COW PEA)

G. UHLENBRUCK, D. KARDUCK and A. VAN MIL
Medical University Clinic Cologne, Department of Immunobiology
D-5000 Köln 41, West Germany

Our work on β-galactosido specific lectins (see Table 1) has a triple origin:

a) The isolation of a galactose-rich glycoprotein from bovine erythrocytes (1), which had many lectin receptors (2), and from which a disaccharide, β-Gal-(1-6)-Gal, could be isolated (3).
b) The investigation of subterminal carbohydrate structures in neuraminidase-treated erythrocyte glycoproteins, which led to detection of the specificity of the peanut lectin, when we tried to define the so-called Thomsen-Friedenreich receptor (4).
c) Studies on galactans containing the above-mentioned disaccharide and on antigalactans from invertebrates: Lectin molecules from different Tridacnid clams, from Cerianthus membranaceus, from Axinella and Geodia sponges (5,6).

The aim of this paper here is not only to describe another β-galactosido specific plant lectin, namely the lectin from Vigna unguiculata, but to show methods for purification and testing of these lectins, and also to demonstrate that from the same species, only a few samples had a mitogenic lectin activity, whereas for other species, such a lectin does not seem to be essential for growth and development.

Results and discussion

Most of the anti-β-galactosyl lectins and antibodies (Table 1) can be easily purified using 'acid activated' Sepharose (7).

Lectins - Biology, Biochemistry, Clinical Biochemistry, Vol. I

Table 1. Lectins with anti-D-galactose specificity (mainly β-linked, anti-galactans and also some antibodies), which can be purified in an one-step procedure, when using acid-treated Sepharose (7 and unpublished)

Origin and Source	Lectin (or antibody)
Bacteria	Pseudomonas aeruginosa lectin
Sponges	Axinella polypoides Geodia cydonium
Invertebrates	Tridacna maxima, squamosa and other Tridacnin lectins Cerianthus membranaceus
Plants	Crotolaria juncea Ricinus communis Viscum Album Abrus precatorius Arachis hypogaea Agaricus bisporus Erythrina and Vigna Lathyrus clymenum Maclura aurantiaca Phaseolus vulgaris Sophora japonica
Vertebrate Sera	Anti-Thomsen-Friedenreich antibodies (anti-TF) Galactan-specific myeloma proteins
Vertebrate membrane-integrated lectins	Lectins with anti-galactosyl specificity
Toxins	Tetanus toxin

We attempted to classify the β-galactosido specific lectins on basis of their specificity, see Table 2.

According to haemagglutination-inhibition and to precipitin reactions in agar with galactans and asialoserum glycoproteins (see Fig. 1 and Table 3) our Vigna lectin belongs to groups II and III.

Table 2. Classification of β-D-galactosidic specific lectins and antibodies.

I.	Galactose Group	Ricinus Communis (different sorts)	β-Gal
II.	Galactan Group	Tridacnins, Axinella, Cerianthus spec., Geodia, (Myeloma proteins) Arachis hypogaea,	β-Gal-(1-6)-Gal β-Gal-(1-3)-Gal
III.	Lactosamine Group	Abrus precatorius, Tridacnins, Phaseolus, Viscum album (Pneumococcus Type XIV antiserum)	β-Gal-(1-3)-GluNAc β-Gal-(1-4)-GluNAc
IV.	Peanut Receptor Group	Arachis hypogaea Agaricus bisporus (Thomsen-Friedenreich antibody)	β-Gal-(1-3)-GalNAc
V.	Lactose Group	Some Tridacnins, Cerianthus, (antibodies to artificial lactosyl-antigens)	β-Gal-(1-4)-Glu

A similar overlapping specificity is also known for the peanut lectin from Arachis hypogaea (IV and III) cf. (8) and for the Tridacnins from different Tridacnid clams. Except for group I, all lectins in Table 2 are mitogenic, especially to human or animal T lymphocytes. This is true for the Vigna lectin too, as can be seen from Fig. 2.

Some critical remarks on the specificity of the peanut lectin (PNA) must be made in this context because most of the galactose-specific lectins react with a galactose-plus-x sturcture, where x is the next subterminal sugar.

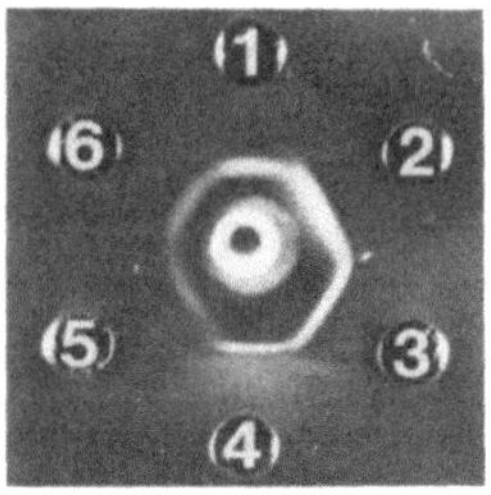

Fig. 1: Precipitation experiment

The numbers give the position in the gel (1 = 12 o'clock). Double diffusion in agarose-gel. The Vigna lectin is placed in the middle (5% conc).
The numbers refer to numbers in Table 3.

Table 3. A list of asialoglycoproteins and other clycoconjugates which precipitate in agar doubble diffusion or inhibit haemagglutination with Vigna unguiculata lectin.

Asialo-serumglycoproteins	Other glycoconjugates
IgA, IgG, IgM	Helix lucorum snail
IgD, IgE,	galactan
alpha-2-Hs-glycoprotein	edible bird nest
3.8-S-alpha-2-glycoprotein	glycoprotein
beta-2-glycoprotein	Littorina littorea
alpha-1-beta-glycoprotein	glycoprotein
transferrin	bovine pneumogalactan
thyroxin-binding globulin (5)	gum arabic galactan
alpha-2-macroglobulin	rabbit erythrocyte
prothrombin	glycoprotein
histidin-rich glycoprotein (1)	
alpha-1-antitrypsin (6)	
coeruloplasmin	
8-S-alpha-3-glycoprotein (3)	
9.5-S-alpha-1-glycoprotein (4)	
alpha-1-antichymotrypsin	
haemopexin	
leucin-rich 3.1-S-alpha-2-glycoprotein	
secretory component	
haptoglobin	

Serumglycoprotein samples were kindly and generously supplied by Behring-Werke, Marburg.

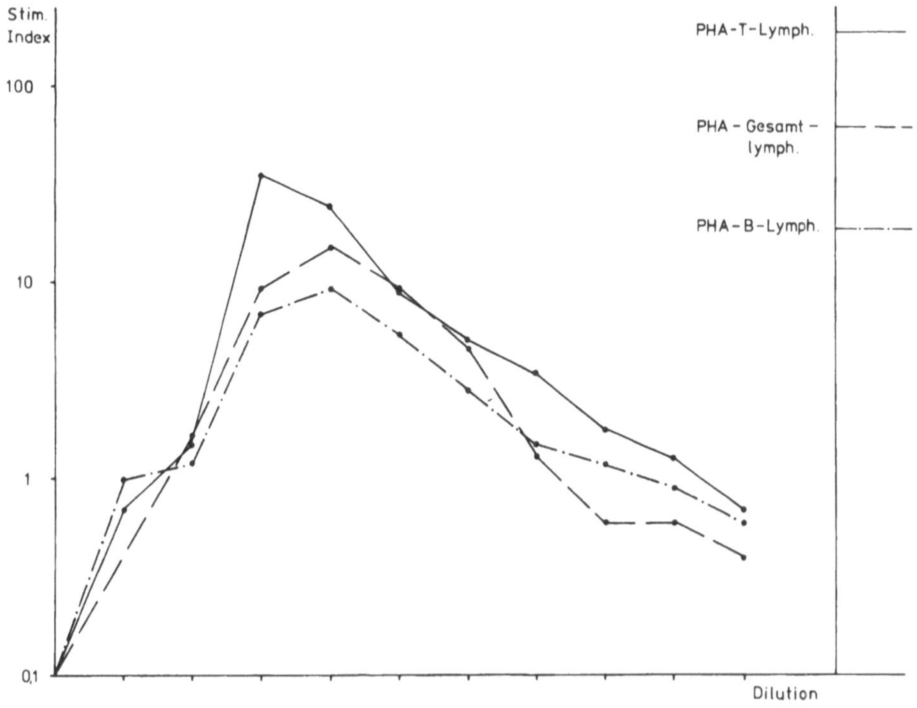

Fig. 2. Dose-dependant stimulation of human lymphocytes by a lectin from Vigna unguiculata (unpurified extract). Stimulation of total lymphocytes (---), T lymphocytes (92%, ——) and B lymphocytes (67%, -·-·-). Stimulation was measured in micro-cultures (0,2 ml with 2 x 10^5 lymphocytes), incubated for 72 hrs, including ^{3}H Thymidine incorporation for 3 hrs (spec. act. 2 Ci/mml; 0,5 uCi/culture).

Control in medium TC 199 with 15% human AB serum: Total lymphocytes 143 cpm, (T = 73 cpm, 8 = 84 cpm). PHA (Phaseolus) control 1 ug/0,1 ml culture: Total lymphocytes = 9.019 cpm, T = 14.285 cpm, 8 = 1.634 cpm).

The cpm values are expressed as stimulation index, which results from the quotient of the active sample and the AB medium control. This stimulation index is shown in the above Figure and compared with that of PHA (right side of the picture). Neuraminidase-treatment did not enhance the response.

(i) There was a slight, but definite inhibition of PNA by Helix pomatia galactan and by Pneumococcus Type XIV polysaccharide, substances lacking the PNA receptor disaccharide shown in group IV (4). The same holds for acid-hydrolyzed Sepharose, which binds specifically to the PNA.

(ii) Not only alpha-1-acid glycoprotein (orosomucoid) from serum, but also a mucus glycoprotein from edible swallows' nest and a glycoconjugate from the albumin gland of the snail Achatina fulica strongly precipitated the peanut lectin, although they do not have the peanut receptor disaccharide.

(iii) Finally, haemagglutination tests with different red cells from various sources of animals showed that some of them were clumped by PNA without neuraminidase-treatment, a fact, which was not in full agreement with the definition of the PNA receptor as being the immunodominant disaccharide of the so-called Thomsen-Friedenreich receptor which is linked alkali-labile to the glycophorin type of red cell glycoproteins. Cells that were agglutinated without neuraminidase-treatment were from guinea pig, rabbit and rat (pronase-treated).

In order to find out, whether on such erythrocytes the free PNA receptor disaccharide was responsible for this unexpected reaction, or whether the 'N-acetyl-lactosamine' type (group III) disaccharides were the cause for the 'odd' haemagglutination by PNA, we chose rabbit erythrocytes and isolated the glycoprotein fraction by phenol-saline extraction. To our surprise, 'free' PNA receptor could not be found serologically or by GLC, but only after treatment with neuraminidase. This fact favors the 'lactosamine' cross-reactivity in those cases and is supported by the reaction of group III specific lectins in this example, because these cells do react with 'N-acetyl-lactosamine' specific lectins and anti Pneumococcus type XIV antisera (unpublished results).

Various seeds of the Vigna genera grown at different locations were tested for mitogenic and haemagglutinating lectins. We found that mitogenic and haemagglutinating activity occurred simultaneously. Seed samples from Vigna capensis (L.) Walp. 1040, Vigna ssp. cylindrica (L.) van Eseltine ex Verdc. 1042, and Vigna unguiculata (L.) Walp. 1041, were positive.

The following seed samples of Vigna unguiculata were negative:

(L.) Walp. cv. Argobba Giant WP 2349, WP 3187 G, WP 3188 D, WP 3188 F, WP 4039 F, WP 7949 and WP 7964;
(L.) Walp. cv. Babile Mocaic WP 2551 D, WP 2556 C2, WP 3186 C and WP 3186;
(L.) Walp. cv. Black Lady WP 2551 H, WP 2555 G, WP 3231 A, WP 3381 and WP 7970;
(L.) Walp. cv. Brown Eye WP and SL 1516;
(L.) Walp. cv. Empress Menen WP 92 B, WP 2374, WP 2551 I, WP 2556 D, WP 3186 B, WP 3186 J, WP 3187 B2, WP 4039 E and WP 8592;
(L.) Walp. cv. Gidole WP 92 A, WP 2551 E, WP 2556 A, WP 2997 A, WP 2554 B and WP 2555 E;
(L.) Walp. cv. Gidole WP 3186 E, WP 3186 F, WP 3221 A, WP 3231 B, WP 3231 D, WP 3492 I, WP 4039 B, WP 7967, WP 8637 and WP 8674;
(L.) Walp. cv. Red Harrar WP 2555 F, WP 3186 G, WP 3187 F and WP 3231 B and Vigna ssp. sesquipedalis 1043.
The samples 1041, 1042, 1043 and 1040 came from the Botanical Garden, Basel, Switzerland. The other samples have been given to us from the E. Westphal Collection, Wageningen, which number them with WP.

Most astonishing was the fact, that most of the Vigna samples did not show any agglutinating effect nor did they stimulate lymphocytes. The reason for such a finding is not clear, but rises much more interest again in the question of which biological functions the plant lectins have within the individual species.

Discussion

As it has been shown nicely in this book by Guldager (9), modern techniques, especially the two-dimensional electrophoresis method, has the great advantage of demonstrating the heterogeneity of lectins from individual sources. Accordingly, one can assume now, that the term 'lectin' is only justified with respect to origin and specificity, but one must be aware of the fact, that we are dealing with a family of different isolectins, including even 'incomplete', non-agglutinating forms and precursors. As it can be deduced from Guldager's work this may be of importance, when monoclonal, labelled antibodies against lectins are used in order to investigate the physiological role of lectins in the plant kingdom. Reasons for such a heterogeneity may be searched less in different geological distribution, than in nutritional, seasonal or growth changes, but mainly in genetic variations.

The questions, which arise from this work of Guldager, from others and also from our findings may be summarized as follows:

(i) What is the biological function of lectins? Do they have an 'essential' function with respect to the life of the plant when even some members of the same botanical family lack them?

(ii) Can lectins have a function within the plant, when sugars, with which they react, do not accur in these and other plants at all, as for instance N-acetyl-D-galactosamine?

(iii) Lectins are genetic markers of plants: How is their evolutionary relationship in different plant tribes? Is there any relationship in amino acid sequence to certain glycosidases? We feel that anti-idiotypic antibodies to different combining sites of lectins could give further information and also reveal relationship to lectins from bacteria and animals.

Also the 'medical future' of lectins has interesting aspects. Possibly, most promising is the action of toxins in hybrids synthetized of lectin and toxin as 'therapeutical' tumor markers.

Acknowledgments

All samples of seeds have been made available to us by the courtesy of Dr. Claude Farron, kustos of the Seed Collection, Botanical Institute of the University of Basel. We feel greatly indebted to him also for his kind advice and help, otherwise these experiments could not have been performed.

References

1. Klenk, E., Uhlenbruck, G, Z. Physiol.Chem. 311 227-233 (1958)
2. Uhlenbruck, G., Krüpe, M. Nature 199, 1289-1290 (1963)
3. Uhlenbruck, G. Z.Immun.-Forsch. 127, 9-17 (1964)
4. Uhlenbruck, G., Pardoe, G.I., Bird, G.W.G. Z.Immun.-Forsch. 138 423-433 (1969)
5. Eichmann, K., Uhlenbruck, G., Baldo, B.A. Immunochemistry 13, 1-6 (1976)
6. Uhlenbruck, G., Vaith, P., Karduck, D., Haupt, H., Müller, W.E.G. Protides of Biological Fluids, Pergamon Press, Oxford and New York, 27, 595-598 (1979)

7. Kania, J., Uhlenbruck, G., Janssen, E., Klein, P.J., Vierbuchem, M. Immunobiol. 157, 154-168 (1980)

8. Pereira, M.E.A., Kabat, E.A. CRC Critical Reviews in Immunology 1, 33-78 (1979)

9. Guldager, P. These Proceedings.

LECTIN BINDING PROTEINS FROM PLANTS

H. Rüdiger, R. Gansera, G. Gebauer and H. Schurz
Institut für Pharmazie und Lebensmittelchemie der Universität Würzburg, 8700 Würzburg, W. Germany

The most striking property of lectins is their ability to interact with other partners specifically. We expected that lectins within their natural environment, that means within the plant, might be associated to some cell component in a similar way in which they associate to foreign cells or foreign molecules in vitro. The most simple approach to test this expectation was to prepare an affinity adsorbent by immobilizing a lectin, then to pass a crude extract of that plant from which the lectin had been isolated over this adsorbent and to look if any material was retained.

We did our experiments mainly with glucose binding lectins from Leguminosae seeds. Such seeds are readily available and the lectin can easily be prepared using Sephadex as an affinity adsorbent.

The first experiments I want to present were made with the lectin of the garden pea (Pisum sativum). A seed extract is applied to a column of pea lectin Sepharose at pH 8 and the column washed free from non-binding material. Then a protein fraction can be released from the column either by lowering the pH or by adding glucose to the eluting buffer (Fig. 1). We call this material lectin binder (1). The lectin binder was characterized by polyacrylamide gel electrophoresis. In absence and in presence of sodium dodecyl sulfate it displays chiefly one band. Minor side bands may appear after the first affinity chromatography; they are easily removed by repeating this step. Thus the lectin binder consists of one type of subunits with a M_r of 49 500 which is somewhat larger than

Lectins - Biology, Biochemistry, Clinical Biochemistry, Vol. I

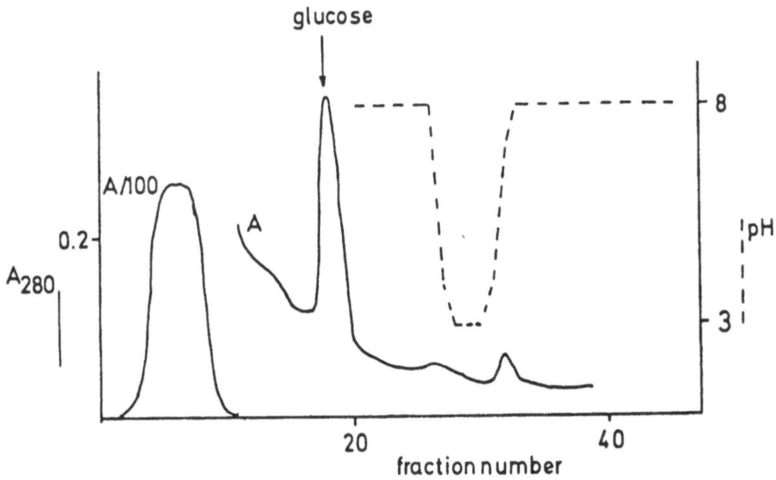

Fig. 1: Chromatography of a pea seed extract on a column of pea lectin Sepharose, taken from Ref. (1).

the whole lectin molecule. Presumably, the native lectin binder is a tetrameric protein. Gel filtration reveals a M_r of 219 000.

The pea lectin binder contains about 0.5% neutral sugar corresponding to 1 - 2 glucose equivalents per subunit. We believe this to be the binding site of the lectin since:

i. a glucose concentration as low as 5 mM is sufficient to desorb the lectin binder from the lectin column;
ii. the dextran gel Sephadex is able to compete with the lectin binder: the lectin - lectin binder complex is resolved on Sephadex leaving the lectin behind whereas the lectin binder runs through the column;
iii. Lectin adsorbed to Sephadex is not able to bind the lectin binder whereas lectin bound to a matrix either covalently (see Fig. 1) or by hydrophobic interaction combines with the lectin binder;
iv. Lectin and lectin binder at certain concentrations develop a slight turbidity which is abolished by glucose.

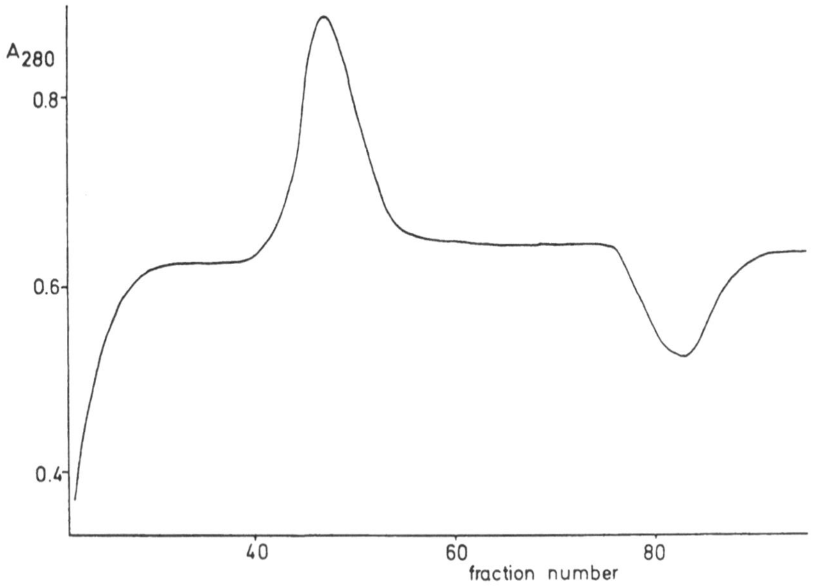

Fig. 2: Chromatography of the pea lectin binder on a column (100 x 0.9 cm) of Biogel P-300 equilibrated with pea lectin in 0.05 M Tris/HCl pH 8. Fractions of 0.55 ml/7.5 min were collected.

Moreover, interaction between both proteins can be demonstrated by gel filtration (Fig. 2). A column filled with an inert matrix is equilibrated with lectin of a certain concentration. Then a solution of the lectin binder which contains precisely the same concentration of lectin is applied to the column. As can be seen from Fig. 2, the lectin binder associated with lectin appears at a low elution volume whereas at a higher elution volume corresponding to the lectin a trough is formed. From the data of this experiment and assuming similar specific adsorptivities of both proteins, a stoichiometry of 4 lectins per 1 lectin binder may be deduced.

Concanavalin A (ConA) from the Jack bean (Canavalia ensiformis) is used most widely (2) because it is already known for several decades and because it is easily available in large amounts. Studying Canavalia, we were hampered by the fact

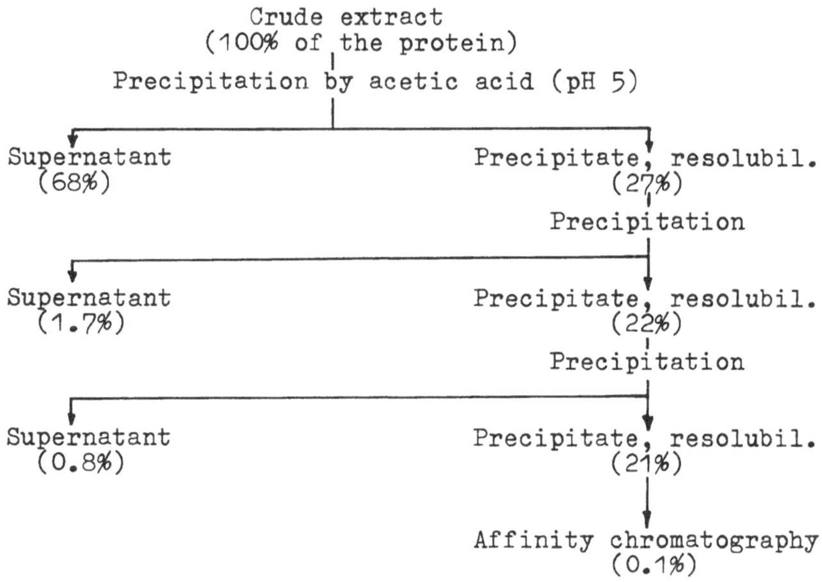

Scheme 1: Fractionation of a crude extract from 100 g Jack beans, taken from Ref. (1).

that a ConA column eagerly retains the lectin ConA itself instead of a lectin binder. As we know now, this is caused by the tendency of the fixed dimeric ConA to reconstitute forming a tetrameric immobilized protein above pH 6 (2). On lowering the pH which is our usual method of desorbing a lectin binder the tetramer is dissociated again and ConA appears in the effluent. Electrophoretical control, however, revealed that though most of the material migrated together with authentical ConA, a small amount of protein is different from it. This protein can be separated from ConA by acid precipitation at pH 5 (Scheme 1). The lectin remains in the supernatant whereas the lectin binding protein is precipitated. This procedure has to be repeated twice in order to get rid of all lectin which coprecipitates. From the precipitate of the last step, a lectin binding protein was isolated by affinity chromatography on ConA-Sepharose. The ConA binder thus isolated is remarkably homogeneous in different electro-

phoretical systems. It contains only one type of subunits of M_r = 35 000. If heating is omitted prior to SDS gel electrophoresis, dimers of M_r = 70 000 appear. Attempts to determine the size of the whole protein failed since it tends to aggregate and appears in the void volume of any gel type tested.
The lectin binders from pea and from Jack bean are not only different in their M_r but also in their behaviour towards their respective lectin. The pea lectin binder reacts with the pea lectin even in 2 M NaCl whereas the ConA - ConA binder interaction is abolished already at 5 mM NaCl. Sodium ions appear to act quite specifically since as high concentrations as 0.2 M (KCl, NH_4Cl) or 0.5 M (Tris hydrochloride) are required to exert the same effect with other salts. On the other hand, already 5 mM glucose solutions desorb the pea lectin binder from the immobilized pea lectin whereas the unphysiologically high concentration of 0.5 M glucose is required to dissociate the ConA - ConA binder complex.

At present, we can only speculate about a biological role of these proteins but in any instance it is tempting to assume a transient lectin - lectin binder complex which under physiological conditions is easily dissociated by minute variations of sugar/glycoside or salt concentrations.

We extended our studies to other plants. In each species which we tested we found a lectin binder by affinity chromatography. These results are summarized in Table 1. Each of the lectin binders listed there is homogeneous in electrophoresis except isoelectric focussing where sometimes weak side bands appear. Two types of subunits are found, one with a M_r of about 50 000 (Pisum sativum, Vicia sativa), another one of about 35 000 (Canavalia ensiformis, Vicia cracca, Vicia faba). We were not able to find a lectin binder directed against the second lectin of Vicia cracca which is blood group specific and binds to GalNAc (3). Lectin binders of the larger type are not desorbed by NaCl from their lectin

Plant	M_r Total	M_r Subunits	pI
Canavalia ensiformis	large 70 000 (SDS)	35 000	6.1, 5.4, 4.9
Pisum sativum	219 000	49 500	5.35, 5.7, 5.15
Vicia cracca (Glc)	n. d.	35 500	n. d.
Vicia faba	large	36 500	8.1
Vicia sativa	214 500	53 500	7.16, 6.88
Ricinus communis	n. d.	n. d.	n. d.

Table 1: Properties of lectin binders.
n. d.: not determined.

columns whereas the smaller type is. One result of Table 1 should be emphasized, that is the occurrence of a lectin binding protein in the castor bean (Ricinus communis). This result is remarkable for two reasons:

i. the lectin from castor beans differs in its sugar binding specificity from the lectins studied before;

ii. the castor bean belongs to the Euphorbiaceae which are far away from the Leguminosae in the botanical system.

Though we are convinced that lectin binding proteins of this kind are common in the plant kingdom it is premature to generalize too much at this state of our experiments.

If interactions between proteins are studied, one is of course confronted to the question of specificity. The lectin binders are quite specific for their respective lectin within the plant as the isolation procedure indicates which yields only one protein from one plant. This remarkably contrasts with the multiplicity of proteins which are usually found if human or animal serum or cell membranes are subjected to lectin affinity chromatography. Cross reactions between the lectin from one plant and the lectin binder from a second one may however be expected since it is known that lectins from different plants are very similar to each other

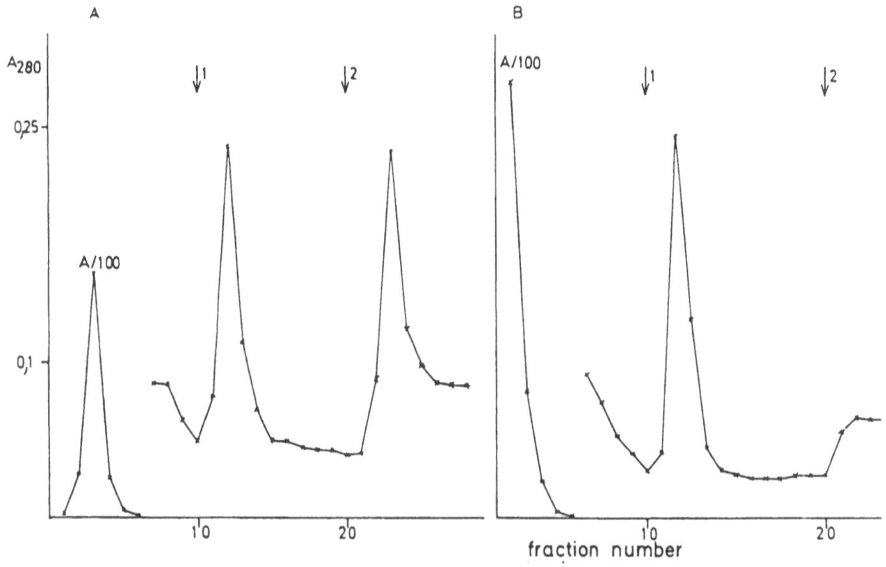

Fig. 3: Cross reaction of lectins with lectin binders. Columns (4 x 0.7 cm) were filled with ConA-Sepharose (A) or Vicia faba lectin Sepharose (B). Then 30 mg of an extract, purified by acid precipitation (see Scheme 1) from Vicia faba (A) or Canavalia ensiformis seeds (B) were applied. Elution by 0.05 M Tris/HCl pH 8, from (1) with 1 M NaCl, from (2) with 1 M glucose additionally. Fractions of 1 ml/10 min were collected.

in their amino acid sequence and, in some instances, even in their immunological behaviour (4). Indeed, when we tested the lectin binders from Canavalia ensiformis and from Vicia faba, we found that they cross react (Fig. 3). This reaction, of course, has no physiological significance.

A question more intriguing than cross reactions is: What is the fate of lectins and lectin binders during germination and growth of the plant? At present, we are not able to present a simple and common explanation because those three plants we investigated as far (Canavalia ensiformis, Vicia faba, Pisum sativum) behaved differently in this respect. If Canavalia plants are grown, the lectin slowly disappears from the plant and can only be detected in the cotelydones

age (days)	part of the plant	% of total protein lectin	lectin binder
10	cotelydones	0.30	0.18
	above ground	2.20	0.80
	subterraneous	0.00	0.00
40	above ground	3.00	0.10
	subterraneous	0.00	0.00

Table 2: Distribution of Vicia faba lectin and lectin binder during germination and growth.

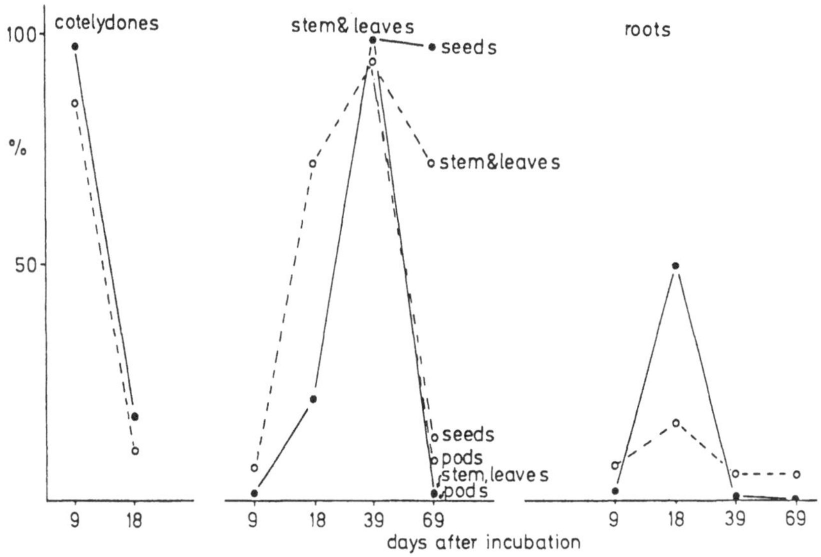

Fig. 4: Distribution of pea lectin (——) and lectin binder (----) during germination and growth. Ordinate: % of the respective protein.

of the young plant and after 40 days in its subterraneous part (1). One might discuss this finding in connection with the hypothesis that lectins are engaged in fixing Rhizobia to the Leguminosae roots (5). Such speculations, however, are premature at present since lectin and lectin binder of Vicia faba distribute differently during growth (Table 2). As in the case of Canavalia (1), this lectin disappears soon from the cotelydones but remains in the parts above ground

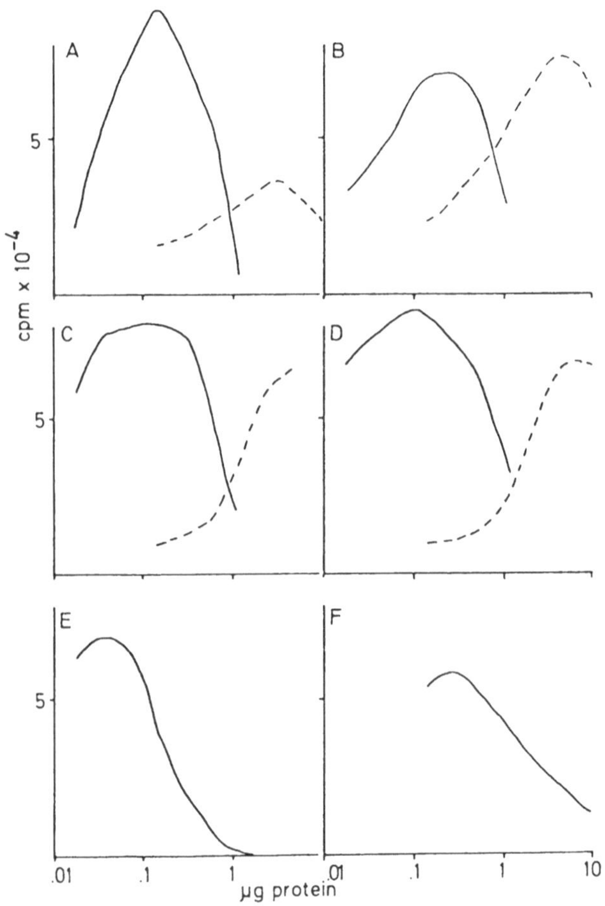

Fig. 5: Mitogenicity of lectins (——) and lectin binders (----) towards spleen lymphocytes of mice (Balb/c). Assay according to (6).
A: Canavalia ensiformis; B: Vicia faba; C: Vicia sativa; D: Pisum sativum; E: Vicia cracca (glucose binding); F: Vicia cracca (GalNAc binding).

as does the lectin binder. A more thorough investigation was made for the pea proteins (Fig. 4). In this case, the cotelydones soon lose both proteins. Their content, however, rises in the parts above ground and culminates in the appearance of the lectin in the seeds and of the lectin binder in stem and leaves. As a quite important phenomenon which de -

serves further elaboration we regard the transient appearance of both proteins in the roots between the second and third week of germination.

Many lectins are known to be mitogenic towards lymphocytes. The mitogenic effect of a protein is related to its concentration by a maximum curve. Fig. 5 shows the mitogenic effect of several lectins and lectin binders on a normal lymphocyte population from mice spleen. As may be seen, all lectins are mitogens with optimal effects in about the same concentration range as ConA, even the blood group specific lectin from Vicia cracca which is the only lectin in this experiment which does not bind to glucose and related sugars The lectin binders are mitogens, too, though their optimal concentration is shifted by 1 - 2 orders of magnitude to higher values.
Lectin binders, their properties and their behaviour during growth may contribute to the understanding of what lectins are. Moreover, as the last results indicated, they may turn out to be as valuable tools for cell biology as the lectins are.

Acknowledgements: The support of this work by the Deutsche Forschungsgemeinschaft and the Fonds der Chemischen Industrie is gratefully acknowledged.

References

1. Gansera, R., Schurz, H., Rüdiger, H.: Hoppe-Seyler's Z. Physiol. Chem. 360, 1579 - 1585 (1979).
2. Bittiger, H., Schnebli, H. P. (eds.): Concanavalin A as a Tool, John Wiley & Sons, London-New York-Sidney 1976.
3. Rüdiger, H.: European J. Biochem. 72, 317 - 322 (1977).
4. Baumann, C., Rüdiger, H., Strosberg, A. D.: This Meeting.
5. Rüdiger, H.: This Meeting.
6. Schechter, B., Lis, H., Lotan, R., Sharon, N.: European J. Immunol. 6, 145 - 149 (1976).

INTERACTIONS BETWEEN SEED PROTEINS AND THEIR RESPECTIVE LECTINS

P. Rougé and C. Chatelain
Laboratoire de Botanique et Biologie cellulaire, Faculté des Sciences Pharmaceutiques, Université Paul Sabatier
35 chemin des Maraîchers, 31 062 Toulouse Cedex, France

Although lectins have been reported in plants and particularly in seeds for almost one hundred years, their role is still obscure. The affinity of these proteins for sugars and glycoconjugates has led plant biologists to speculate that glycoprotein-lectin interactions could be of biological significance in seeds. Numerous reports on lectins are concerned with interactions between lectins and glycoproteins of animal origin, either free or membrane-bound. In contrast, much less data are available concerning plants and the purpose of this paper is to review seed protein-lectin interactions from the three points of view of methodology, data and prospects.

Concerning the methodology, four main techniques are presently used, namely :

- precipitation in agarose gel
- affinity electrophoresis with lectins in agarose gel
- labelling of proteins with labelled lectins
- affinity chromatography on immobilized lectins

1. Precipitation in agarose gel corresponds to the OUCHTERLONY's technique in which lectins and seed proteins respectively correspond to antibodies and antigens. Although specific, this technique may suffer from a major limitation due to the fact that not all the lectin-protein complexes precipitate. A prerequisite for the formation of such precipitates is that seed glycoproteins have at least two or more binding sites per molecule.

Lectins - Biology, Biochemistry, Clinical Biochemistry, Vol. I

2. The second technique, recently reviewed by BØG-HANSEN (1), is based on interactions between seed proteins and lectins during electrophoresis in agarose gel. This very promissing technique however requires antibodies whose specificities have been carefully controlled.
3. The third technique consists of the labelling of seed proteins by FITC- or ^{125}I-labelled lectins, once the seed proteins have been separated by gel electrophoresis. This is a very sensitive technique, the specificity of which can be easily checked by hapten inhibition with selected sugars. Controls including such inhibition tests are necessary to avoid possible confusing data due to non specific binding of free fluorochrome which may be associated with FITC-labelled lectins.
4. The last technique is concerned with affinity chromatography on immobilized lectins. This technique has proved very powerful and is now largely used because it allows the isolation of seed proteins interacting with lectins. However, some misleading results could result from non specific lectin-protein interaction, especially at low ionic strength.
One must conclude that a proper combination of these techniques is necessary to approach lectin-protein interactions in seeds for each of them can lead to somewhat different results : for instance, FITC-labelled pea lectins have a strong affinity to storage proteins (legumin and vicilin), whereas no precipitation reaction is found between seed proteins and lectins in agarose gel. Affinity electrophoresis or chromatography provide intermediate results since storage proteins are slightly retained.

The following exemples illustrate the benefits and limits of the data obtained with these techniques :
1. Precipitation reactions, for instance, have been found to occur only between Con A and seed proteins extracted from other species of legumes while other lectins including Con A don't react with their corresponding seed proteins (2)(3)(4)(Fig. 1).
2. Affinity electrophoresis with lectins indicate that pea seed storage proteins interact with their respective lectins (5)(GULDAGER, P., these proceedings).
3. By using ^{125}I-labelled lectins, BASHA (6) demonstrated that lectins

from several legume seeds (Jack bean, soya, peanut and kidney bean) do react with their corresponding seed proteins. We obtained identical data (7) with pea, lentil, broad bean, Jack bean and vetch (*Vicia sativa*) seed proteins and the corresponding lectins which had been labelled with FITC. In this case, the storage proteins were heavily labelled (Fig. 2).

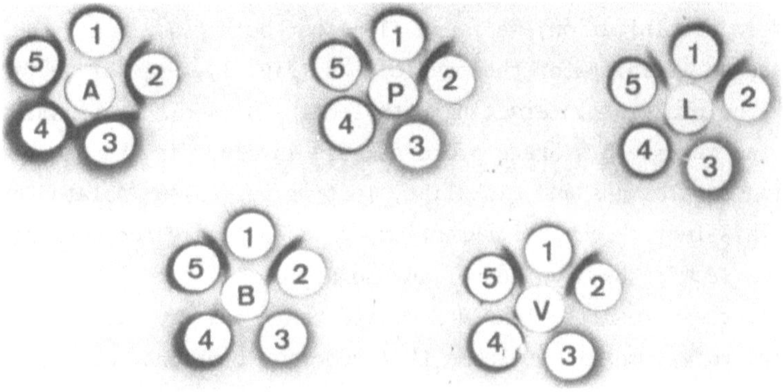

Fig. 1. Precipitation reaction in agarose gel between various lectins (Con A (A), pea (P), lentil (L), broad bean (B) and *Vicia sativa* (V) lectins) and the Tris soluble (0.05 M Tris, 0.15 M NaCl pH 7.6) proteins from the seeds of Jack bean (1), pea (2), lentil (3), broad bean (4) and *Vicia sativa* (5). In addition to the precipitation reaction between Con A (well A) and the other legumes seed proteins, a precipitation reaction occurs between the well 1 and the wells 2 and 5 due to the fact that the Tris soluble proteins from Jack bean seeds (well 1) contain Con A.

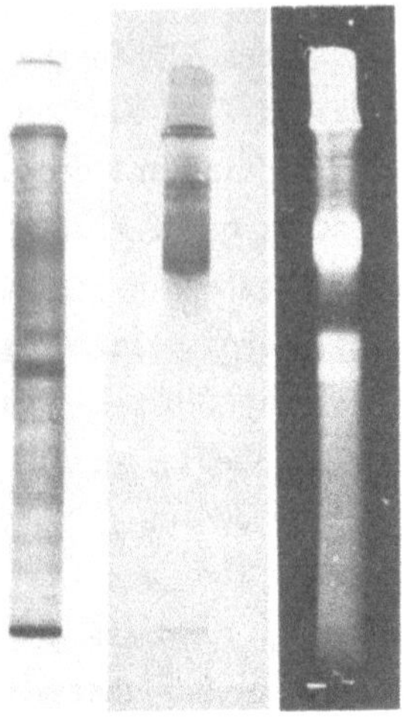

Fig. 2. Studies on interactions between pea seed proteins and pea lectins using FITC-labelled lectins.
A - Pea seed Tris soluble proteins (0.05 M Tris, 0.15 M NaCl pH 7.6).
B - Pea legumin + vicilin preparation (obtained according to (20)).
C - Pea seed Tris soluble proteins incubated with FITC-pea lectins.
Polyacrylamide gel electrophoresis and protein staining were performed according to (21) and (22).

4. By using the last technique, that is affinity chromatography, a *Lectin binder I* (LBI) was isolated by RÜDIGER's group (8)(9)(RÜDIGER, H., GANSERA, R., GEBAUER, G., SCHURZ, H., these proceedings) from different seeds including six legume species (Jack bean, pea, broad bean, vetches (*Vicia cracca, Vicia sativa*), kidney bean) and castor bean. In contrast to lectins these LBI precipitate at low pH (pH 5). Using the same technique but without acid precipitation, we found (10) that both pea albumins and globulins have an affinity for the pea lectins (Fig. 3). Further characterization of the proteins interacting with lectins showed that legumin and vicilin, the main seed storage proteins, are involved in the phenomenon. In addition to albumins and globulins, lectins were also isolated by this process. This lectin-lectin interaction could be totaly non specific since pea lectins seem to be devoided of any sugar.
Altogether, these results clearly indicate that seed protein-lectin interactions are very complex. It seems that numerous proteins could be involved in these interactions and that additional work is needed to have a clear picture of these interactions.

Despite these limitations, the available data allow interesting speculations on the possible role of these interactions in seeds. One may hypothesize, according to BASHA (6) that protein-lectin affinities could be involved in cross-linking the storage proteins (legumin and vicilin) into aggregates within the protein bodies. However, this hypothesis should be regarded with caution since only a small amount of seed proteins react with lectins and because there remains some incertainty about the general association of lectins with protein bodies. A close association has been demonstrated in the case of kidney bean (11)(12)(13) and castor bean (14) (15) whereas lectins are found associated with cytoplasmic components in Jack bean (16), in pea (17) and even in kidney bean (18). In *Datura stramonium* seeds, lectins seem to be equally distributed inside and outside protein bodies (19). Furthermore, BASHA's hypothesis does not take into account the fact that albumins interact with lectins, for instance in pea, although albumins are not present within protein bodies. Finally, one must remember that *in vitro* interactions don't necesseraly mimic *in vivo* interactions.

Obviously, a better understanding of seed protein-lectin interactions requires additional experimentation in two areas :

1. the distribution of lectins within the seed cells
2. the confrontation of data obtained with different techniques, both on a given species and on several ones, concerning the location and the characteriz ation of the various seed proteins involved in such interactions.

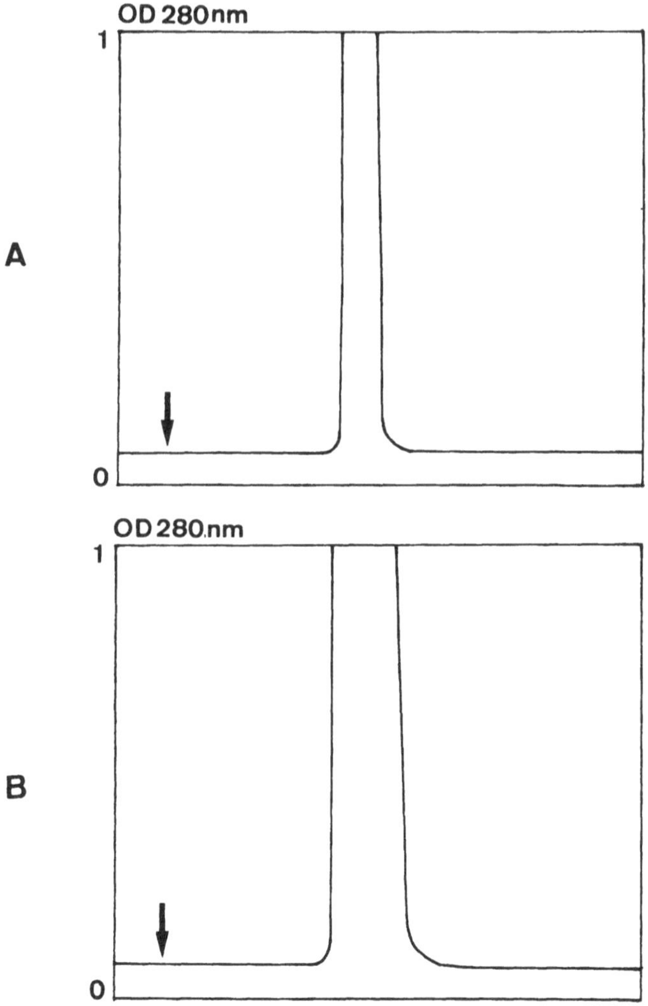

Fig. 3. Filtration of pea albumins (A) and globulins (B) through a column of pea lectins coupled to CNBr-Sepharose 4B. Albumin and globulin fractions interacting with lectins are eluted by adding 0.1 M glucose (arrows) to the elution buffer (0.05 M Tris, 0.015 M NaCl pH 7.6).

References

1. BØG-HANSEN, T.C. : In J.M. EGLY (ed.) : Third International Symposium on Affinity Chromatography and Molecular Interaction, Inserm Symposia Series, Paris, vol. 86, 399- 416 (1979).
2. JERMYN, M.A. : Aust. J. Plant Physiol. 2, 533-542 (1975).
3. GLEESON, P.A., JERMYN, M.A. : Aust. J. Plant Physiol. 4, 25-37 (1977).
4. ROUGE, P., PERE, M. : Bull. Soc. bot. Fr. 126, 387-398 (1979).
5. GULDAGER, P., BØG-HANSEN, T.C. : In H. PEETERS (ed.) : Protides of the Biological Fluids, 27th Colloquium, Bruxelles, Pergamon Press, Oxford, 401-404 (1979).
6. BASHA, S.M.M. : Plant Physiol. 63, suppl. 5, Abst. 324, 59 (1979).
7. ROUGE, P., CAUSSE, H. : Bull. Soc. bot. Fr. (in press).
8. GANSERA, R., SCHURZ, H., RÜDIGER, H. : Hoppe-Seyler's Z. Physiol. Chem. 360, 1579-1585 (1979).
9. GEBAUER, G., SCHILTZ, E., SCHIMPL, A., RÜDIGER, H. : Hoppe-Seyler's Z. Physiol. Chem. 360, 1727-1735 (1979).
10. ROUGE, P., CHATELAIN, C. (in preparation).
11. BARKER, R.D.J., DERBYSHIRE, E., YARWOOD, A., BOULTER, D. : Phytochem. 15, 751-757 (1976).
12. BOLLINI, R., CHRISPEELS, M.J. : Planta 142, 291-298 (1978).
13. PUSZTAI, A., CROY, R.R.D., WATT, W.B. : New Phytol. 79, 61-71 (1977).
14. TULLY, R.E., BEEVERS, H. : Plant Physiol. 58, 710-716 (1976).
15. YOULE, R.J., HUANG, A.H.C. : Plant Physiol. 58, 703-709 (1976).
16. CLARKE, A.E., KNOX, R.B., JERMYN, M.A. : J. Cell. Sci. 19, 157-167 (1975)
17. ROUGE, P. : Biologie des hémagglutinines chez le Pois (*Pisum sativum L.*), Thèse Doct. ès.Sc., Toulouse (1977).
18. MIALONIER, G., PRIVAT, J.P., MONSIGNY, M., KAHLEM, G., DURAND, R. : Physiol. vég. 11, 519-537 (1973).
19. KILPATRICK, D.C., YEOMAN, M.M., GOULD, A.R. : Biochem. J. 184, 215-219 (1979).
20. JACKSON, P., BOULTER, D., THURMAN, D.A. : New Phytol. 68, 25-33 (1969).
21. DAVIS, B.J. : Ann. N. Y. Acad. Sc. 121, 404-427 (1964).
22. CHRAMBACH, A., REISFELD, R.A., WYCKOFF, M., ZACCARI, J. : Analyt. Biochem. 20, 1-15 (1967).

SEED LECTINS FROM 4 LINES OF PEA

P. Guldager

Department of Genetics
The Royal Veterinary and Agricultural University
Bülowsvej 13 DK 1870 Copenhagen, Denmark

Introduction

Lectins isolated from seed of different species of leguminosae are known to bind specifically to various glycoproteins and glycocomplexes.

In pea we observed variation in lectin binding specificity in different lines (Guldager and Bøg-Hansen, 1979). Lectins from a specific line were shown to interact with a number of human serum proteins (IgG, HDL, α_1-antitrypsin orosomucoid and others) and to bind to agarose, while lectins from standard lines did not. Uhlenbruck et al. (1980, these proceedings) observed variation in mitogenic capability of cowpea lectins isolated from different seed sources.

These findings stress the necessity of precise information on origin of seed material in order to define a particular lectin species.

The findings might give a specific possibility to elucidate the unresolved biological function of lectins. A high degree of structural similarity is known to exist between lectins from the leguminosae (Foriers et al. 1977). This indicates that only few mutations have accumulated during evolution which again normally correlates with a specific and essential function of the molecule. The excistence of different binding specificities of lectins in pea gives an opportunity to study the vari-

Lectins - Biology, Biochemistry, Clinical Biochemistry, Vol. I

ation of the molecular structure developed on the relatively homogeneous genetic background found in pea.

In the present study, lectins from 4 different pea lines have been investigated. The antigenic properties of lectins, iso - lated by affinity chromatography and by immunoaffinity chromatography are investigated by rocket immunoelectrophoresis. The subunit composition is investigated by one and two dimensional polyacrylamide electorphoresis.

Material and methods

The pea lines studies are: L1293 and L1129 (Weibullsholm,Sweden), J.I. 181 (John Innes Institute, Norwich, England) and K42 (Dark Skin Perfektion, Ohlsens Enke, Denmark).

Lectins are extracted and isolated in accordance with Guldager (1978) based on methods developed by Trowbridge (1974). Antibodies specific to lectin are immobilized on CNBr-Sepharose 4B column (Casey 1979) and used to bind specifically the lectins

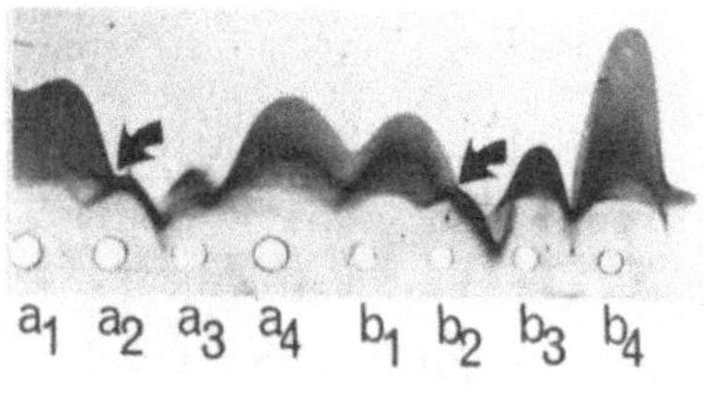

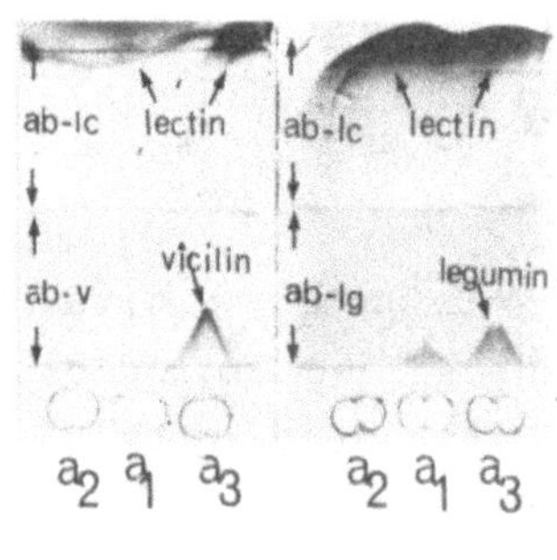

Fig. 1: Rocket immunoelectrophoresis of lectins. a_1-a_4: Lines JI 181, L1293, K42, and L1129, isolated by immunoaffinity chromatography. b_1-b_4: Lines JI 181, L1293, K42, and L1129, isolated by affinity chromatography (further explanation in text).

Fig. 2: Rocket immunoelectrophoresis with intermediate gel containing specific antibodies against vicilin (ab-v) and legumin (ab-lg) of lectins from lines L1293 (a_2), JI 181 (a_1) and K42 (a_3). ab-lc are antibodies specific against lectin.

in extracts from seeds. The purified lectins are subsequently eluted with KSCN buffer containing 0.15 M D-mannose. Immunoelectrophoretic methods were as described by Guldager (1978), based on methods by Axelsen et al. (1973). Antibodies are formed against lectin purified by affinity chromatography on Sephadex G 100. SDS and Urea PAGE is based on Laemmli (1970). Tris-glycine pH 8.8 is used as electrophoresis buffer with either a 10-15% (SDS PAGE) or a 2.5 - 16.8% (urea PAGE) gradient of polyacrylamid. For urea PAGE, urea is included in the gel (6 M) and in the electrophoresisbuffer (5 M). Two dimensional electrophoresis (1st dimension is isoelectric focusing and 2nd dimension is SDS PAGE)were performed in accordance with O'Farrel (1975).

Results and discussion

A. Native lectins and binding of vicilin and legumin. Results from rocket immunoelectrophoresis of lectins from 4 lines of pea employing antibodies specific against lectins from line K42 are illustrated in Fig. 1. All lectins seems to carry common antigenic sites and lectins from all lines give more than one precipitation arc. This could possibly be due to the exsistence of isolectins. The main lectin arcs of line L1293 lectin do not fuse with the main lectin arcs of JI 181, indicating structural differences of the lectin molecule from the two lines. Differences in arc size is presumably not line specific since the isolation procedure was not quantitative. Contamination with other seed proteins was observed only when the lectins were isolated by immunoaffinity chromatography. This contamination was further investigated by use of antibodies specific against vicilin and legumin. These proteins were both found as contaminents. From Fig. 2 it occurs that lectin from line K42 is contaminated with vicilin and with legumin. Lectins from line L1293 seems only to be slightly contaminated with vicilin. A possible explanation is that the seed proteins are bound, either by specific binding (Rougé and Chatelain, Rüdiger et al. 1980, theese proceedings) or by hydrophobic interaction (Ochoa et al. 1980, these proceedings) to lectin in the seed extract. This binding might be maintained when lectin binds to the immobilized lectin antibodies. When lectin is isolated by affinity chromatography no such contamination is detected, possibly because the lectin-seed protein binding is displaced by lectin-sephadex binding.

Extensive washing with D-mannose failed to remove protein from the antibody bound lectin, but washing with D-mannose and 1% Triton X-100 removed protein of none lectin origin according to band pattern in SDS gel. This indicates the existence of hydrophobic interaction between lectin and seed proteins under these conditions.

Differences between lines with respect to contaminating proteins in the lectins isolated by immunoaffinity chromatography might be due to differences in the binding specificity of lectins and might be due to differences in the content of carbohydrate residues or in hydrophobic properties of line specific seed proteins. The main seed proteins of pea, vicilin and le-

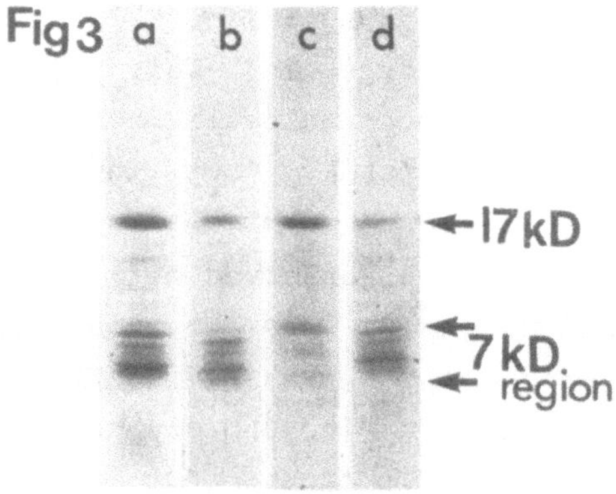

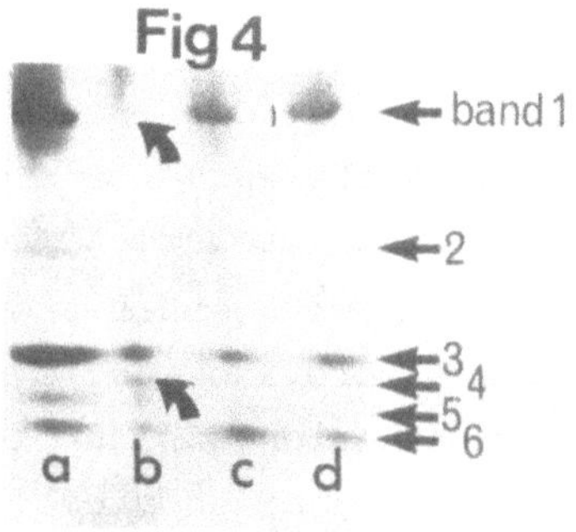

Fig. 3. SDS PAGE of lectins isolated by affinity chromatography. a: line JI 181 b: L1293 c: K42 d: L1129

Fig. 4. Urea PAGE, samples as Fig. 3 (cf. text).

gumin, are known to vary genetically in subunit composition between lines (Casey 1979, Guldager 1980).

B. Dissociated lectins. The band pattern for SDS dissociated lectins from the four pea lines is seen in Fig. 3. A major subunit is found at about 17 kD and several polypeptide bands are found in the 7 kD region. Lectins from the different lines exhibit differences in binding pattern in the 7 kD region. As contamination of the lectins, isolated by affinity chromatography, can not be detected, the differences are likely to reflect existence of different molecular forms of the minor subunits in pea lectin.

The isolated lectins are furher analysed in the urea PAGE system. Band 1 is missing in line L1293 (Fig. 4b). When band 1 is run in second dimension (SDS PAGE) the molecular weight is found to be 17 kD (not shown). Band 1 might be an aggregate or a polymer form of the 17 kD subunit. Bands 2-6 could not be detected after running the second dimension. As a 17 kD subunit is present in line L1293, one of the bands 2-6 is likely to be the 17 kD band and the other bands equivalent to the 7kD bands (arrows indicate differences).

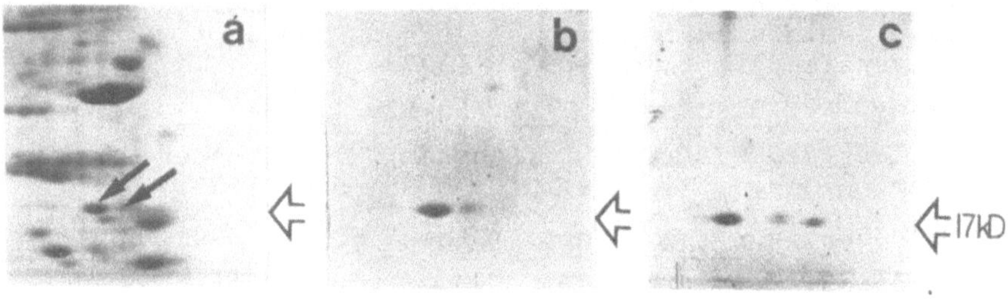

Fig. 5. Two dimesional electrophoresis, as described by O'Farrell (1975). Sections of the 2nd dimension (SDS PAGE).
a: total seed extract of line K42.
b: lectin from line K42 isolated by affinity chromatography.
c: lectin from line L1293 (cf. text).

The two-dimentional electrophoresis system (O'Farrell, 1975) revealed that dissociation of lectin from line L1293 deviates from the other investigated lines. Fig. 5 shows sections of the second dimension gel (SDS PAGE) of total seed extract from line K42 (Fig. 5a), of lectin from K42 (Fig. 5b), and of lectin from line L1293 (Fig. 5c). With lectin from lines K42, JI 181, and L1129, the 17 kD band resolves as two spots, where the minor deviates 0.3 pH unit from the major spot. In line L1293 the 17 kD band resolves as three spots with the pI of two minor deviating 0.6 and 0.9 pH unit from the major spot.

The observed differencies in molecular weight charge and pI between dissociation products of the different lines are possibly related to the differences in binding specificity. It is so far unresolved if the observed variation in the pattern of dissociation products of lectins results from differences in primary gene products or from post translational (but genetically determined) modifications.

References

1. Axelsen, N.H., Krøll, J., Weeke, B.: A manual of quantitative immunoelectrophoresis, methods and applications, Oslo, Universitetsforlaget pp 169 (1973).
2. Casey, R.: Biochem. J. 177 509-520 (1979).
3. Foriers, A. Wuilmart, C., Sharon, N., Strosberg, A.D.: Biochem. Biophys. Res. Commun. 75 980-986 (1977).
4. Guldager, P.: Theor. Appl. Genet. 53 241-250 (1978).
5. Guldager, P., Bøg-Hansen, T.C.: Protides of biological fluids, 27 colloquium 1979, 401-404 (1979).
6. Guldager, P., Buch, L., Henriksen, K.: Abhdlg. Akad. Wiss. DDR, Abt. Math., Naturwiss., Tech (1980, in press).
7. Laemmli, U.K.: Nature 227 680-685 (1970).
8. O'Farrell, P.H.: J. Biol. Chem. 250 4007-4021 (1975).
9. Trowbridge, I.S.: J. Biol. Chem. 249 6004-6012 (1974).

ACTIVATION OF ACID PHOSPHATASE FROM POA PRATENSIS BY BINDING TO LECTINS[§]

I. Lorenc-Kubis+ and T.C.Bøg-Hansen

The Protein Laboratory, University of Copenhagen, Denmark.
+)Permanent adress: Department of Molecular Biochemistry, Institute of Biochemistry, University of Wroclaw, Poland.

Lectins are known to interact with a variety of enzymes, as reviewed by Dulaney (13), see also (5,6,11,14,19,21-24,27-29,33, 35). The interaction takes place with the carbohydrate moiety of the enzymes, since the interaction is inhibitable by the usual carbohydrate inhibitors. Generally it could be expected that a binding to the carbohydrate part of an enzyme would not affect directly the enzymic activity (6,15,17,31) but would increase the stability in analogy to binding to an inert support or to binding of antibodies. Indeed, some examples of increased stability with lectin-binding have been reported (1,30) as well as inhibition of the catalytic properties after binding to lectin (1,4,5,21,25,28,29,35) and we wondered if generally lectin-binding could affect negatively as well as positively the enzymic properties of glycoprotein enzymes.

---

§) Presented at the Second Lectin Meeting in Copenhagen, 1979.

Lectins - Biology, Biochemistry, Clinical Biochemistry, Vol. I

The present paper describes the activating effect of conA and lentil lectin on the activity of acid phosphatase from gramineae seeds (grass seeds).

Materials and Methods

Ground seeds were extracted with 0.1 M acetate buffer, pH 5.1 at 0°C (details to be reported elsewhere). The acid phosphatase activity was measured spectrophotometrically by liberation of p-nitrophenol from p-nitrophenyl phosphate at 37 C at pH 5.1 (34). Affinity chromatography was performed with con A-Sepharose (Pharmacia Fine Chemicals, Uppsala) from where the bound proteins were liberated with 10% alpha-methyl-D-glucopyranoside (pH 5.6). Quantitative immunoelectrophoresis (crossed immunoelectrophoresis, fused rocket immunoelectrophoresis and lectin affinity immunoelectrophoresis) was performed at pH 8.6 in thin layers of agarose essentially as described in (3) cf. also other papers in these Proceedings. Antibodies were purified from rabbit antisera raised against protein extracts of Poa pratensis seeds (to be described elsewhere). After electro-immunoprecipitation, the acid phosphatase activity was revealed by the histochemical staining method used for acid phosphatase (incubation at 37°C in 0.1 M acetate, pH 5.1 with 0.1% sodium alpha-naphthyl phosphate and 0.04% Fast Blue B).

Results

During our attempts to characterize proteins and enzymes from the grass Poa pratensis by means of quantitative immunoelectrophoresis we found that the acid phosphatase activity in plates with conA was much more pronounced than in plates without conA. By visual evaluation we estimated the activity increase to be considerable. In Fig. 1 is shown crossed immunoelectrophoresis of an exctract from Poa pratensis. After electrophoresis the acid phosphatase activity of the two immuneprecipitated enzymes was revealed by histochemical staining. Fig. 1a is the control pattern without lectin incorporated, the normal crossed immunoelectrophoresis. In Fig. 1b and c is shown the increased staining with increasing amounts of free conA in the first dimension gel. The same increase of staining intensity was observed when conA was included in an intermediate gel, in the antibody-containing gel, or included as immobilized conA-Se-

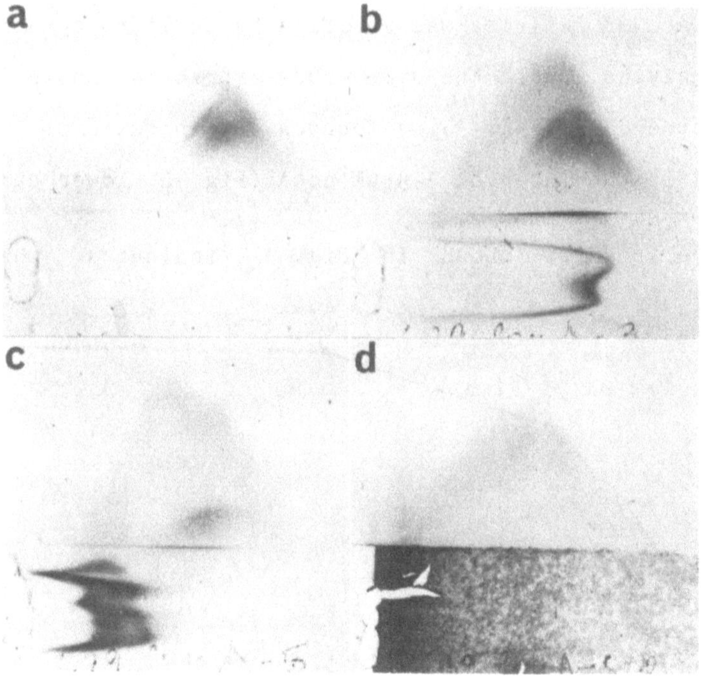

Fig. 1. Crossed immunoelectrophoresis of Poa pratensis acid phosphatases.
a: The control pattern.
b: ConA in the first dimension gel (5 µg/cm2).
c: ConA in the first dimension gel (25 µg/cm2).
d: ConA-Sepharose in the first dimension gel (150 µg/cm2).
A crude extract af proteins (180 µg) from Poa pratensis seeds was electrophoresed for 80 min in the first dimension at 10 V/cm (anode to the right), then into antibody-containing gel for 18 h at 2 V/cm (17.6 µl purified immunoglobulin fraction per cm2, anode at the top). Plates to be compared were always stained together at identical conditions.

pharose (Fig. 1d). We could distinguish between the two acid phosphatases with respect to their binding to conA. One of them (acid phosphatase-1) had an intermediate affinity to conA while the other (acid phosphatase-2) seemed to have a negligible affinity or to be heterogeneous with respect to conA-binding. Control experiments secured that the antibodies show all acid phosphatases precipitated by conA.

The increase of activity was verified by in vitro spectrophotometic analyses. With the same crude extract as shown in immunoelectrophoresis (Fig. 1) we found a 10% increase of activity with conA in the interval 1-9 µM conA (Fig. 2, lower curve).

Since the results shown in Fig. 1 indicated that the

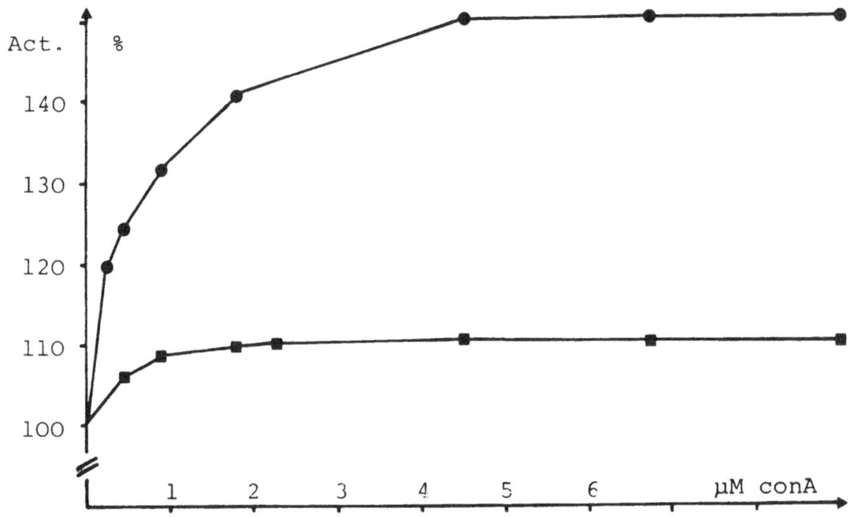

Fig. 2. Spectrophotometric measurements of activation of acid phosphatase activity with increasing amounts of conA. A sample of enzyme (1-20 ug protein) was preincubated with conA at pH 5.1 for 20 min at 25°C. The enzymatic hydrolysis of p-nitrophenyl phosphate was carried out for 10 min at 37°C.
Whole extract: ■----■. ConA-binding fraction: ●----●.

conA-binding acid phosphatase was the major component exhibit- and similar results were obtained but with 5-10 min preincubation we observed 5-10 % more activation. Precipitate formation was never observed.

Further analyses showed that the Km value for the conA-binding acid phosphatase fraction was 10-3 M without conA and 5 x 10-5 M with 9 µM conA. The normal pH optimum of 5.3 was broadened to the interval 5.0 to 5.4 with 9 µM conA. The pH stability as well as the thermal stability was greatly increased.

In control experiments with the specific inhibitor alpha-methyl-D-mannopyranoside we could correlate the degree of activa-

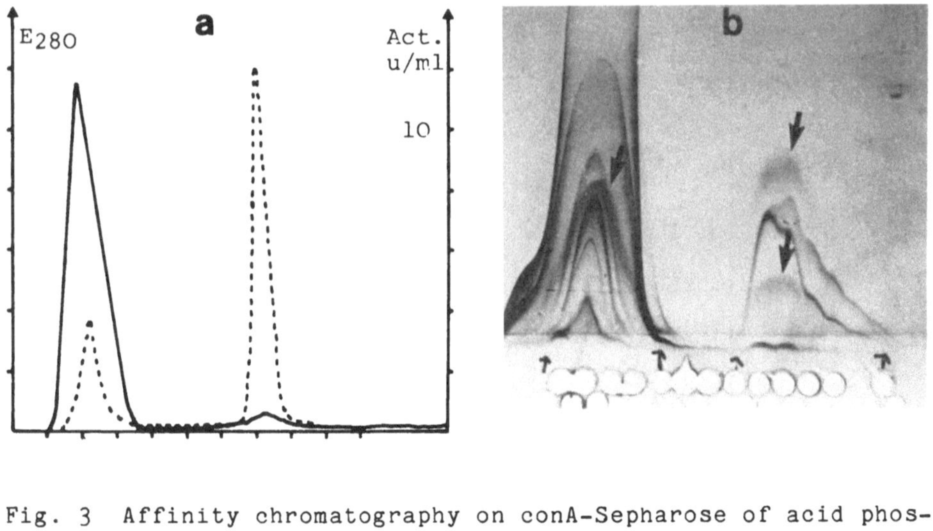

Fig. 3 Affinity chromatography on conA-Sepharose of acid phosphatases.
a. Elution profiles with UV (——) and spectrophotometric analysis (- - -).
b. Elution profile with fused rocket immunoelectrophoresis (stained for acid phosphatases (arrows) and proteins).
The bound proteins were liberated with 10 % alpha-methyl-D-mannopyranoside.

ing activity increase with conA, we purified this component by affinity chromatography on conA-Sepharose. Fig. 3 shows the elution profile of a typical separation, analyzed by UV and activity measurements (Fig. 3a) and fused rocket immunoelectrophoresis (Fig. 3b). When we analyzed the conA-binding fraction with respect to activation with conA in immunoplates, analogously to Fig. 1, we found that the fraction was heterogeneous and exhibited a similar activity increase as the original extract. However, in the in vitro spectrophotometric assay, the increase was considerably greater (Fig. 2, upper curve). In the interval 5-9 µM conA the activity increase was 50%. ConA was mixed before and after addition of substrate to the enzyme,

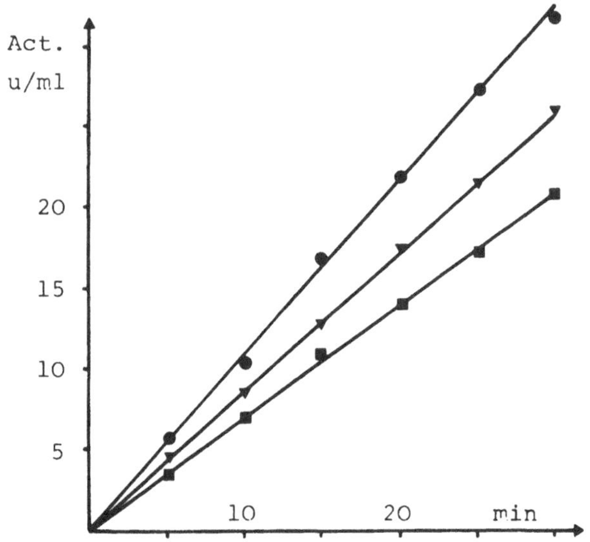

Fig. 4. Effect of free conA and lentil lectin on the activity of acid phosphatase purified by affinity chromatography on conA (cf. Fig. 3). Progress curve of p-nitrophenyl phosphate hydrolysis in presence of 9 µM lectin. Amount of enzyme protein in each assay was 1 µg.
Enzyme without lectin or with WGA: ■----■.
Enzyme with conA: ●----●.
Enzyme with lentil lectin: ▼----▼.

Other experiments revealed that the increase of the activity could not be a result of the increased stability of the enzymes with conA since the progress curve was linear and since the free enzyme was 100 % stable for the duration of the assay.

With lentil lectin a similar increase of the enzymatic activity was observed but we did not calculate the changes in Km (Fig. 4).

When the same immunoelectrophoretic and spectrophotometric analyses were performed with WGA, we did not observe any increase of acid phosphatase activity nor binding to immobilized WGA.

Discussion

Results with addition of lectins in crossed immunoelectrophoresis and in in vitro enzymatic assay show that acid phosphatases from seeds of the grass Poa pratensis are glycoproteins and that they are activated by reaction with the mannose specific lectins conA and lentil lectin, but not with the N-acetyl glucosamine specific lectin from wheat germ (WGA).

The occurrance of glycoprotein enzymes in plants is a well-established fact (2,8,10,12,20,26). We tend to take the present results as a pointer to the carbohydrate structure of the reacting enzymes, and we conclude that the carbohydrate of the acid phosphatases contains exposed mannose units but is devoid of exposed N-acetylglucosamine units.

The kinetic consequences of binding of a lectin to an enzyme vary. In some cases as for human placental alkaline phosphatase and phosphodiesterase (15) and for gamma-glutamyltransferase of rat kidney (31) there does not appear any change in the enzymatic properties. In other cases as for instance for beta-N-acetylhexosaminidase, acid phosphatase (5) and arylsulfatase (1) the formation of an enzyme-conA complex reduces the affinity of the enzyme for its substrate (the Km values of the conA-bound enzymes were two to three times higher than the Km of the free enzymes). Mainly this effect was explained in terms of steric hindrance (15). Oppositely, our results shows that Poa pratensis acid phosphatases increase the affinity to the substrate with lectin-binding (Km was diminished by a factor of 20 with conA). The difference in pH optimum with lectin was reported by Ahmad et al. (1). Generally a shift or a broadening of the pH optimum seems to be found with lectins. The activation was reversible with addition of specific sugar inhibitors of lectin-binding analogously to the reversal of inhibition of enzyme activity with conA for chick embryo prolyl hydroxylase (16).

The kinetic changes observed with lectins is in accord with the kinetic changes seen after immobilization to inert carriers and to antibodies, cf. (7). Therefore, presumably the underlying molecular event is totally unrelated to the removal by lectin affinity chromatography of factors responsible for the enzymic activity, as for instance the deactivation of lymphocyte adenyl cyclase by conA affinity chromatography by which an activating factor is removed (9). We interprete the activation as a result of some steric change of the active site, and consequently we conclude that the conA binding site is located near to or has an important structural effect on the active site. For yeast mitochondrial ATPase the same conclusions concerning the carbohydrate moiety were drawn on basis of inhibition of the enzymatic activity with conA (25). However, our experiments do not prove any direct participation of the carbohydrate moiety in the enzymatic function of grass seed acid phosphatases.

In conclusion we will suggest that the activity increase of enzymes interacting with lectins may indicate a regulatory function of lectins on the metabolism in plants. In this connection it must be noted that other authors have shown seed lectins in complex with glycoproteins (these proceedings). However, nothing is known about the Poa pratensis lectin, about its existence, its location, its specificity, its function. Presumably it would be related to other grass seed lectins, as for instance WGA, since related lectins occur in related species (these proceedings). The fact that we did not find any increase of Poa pratensis acid phosphatase activity with WGA, does not exclude a regulatory function of the hypothetical Poa pratensis lectin in vivo. In further studies we have shown that acid phosphatases and RNases from a number of plants are either activated or inhibited with lectins, especially with conA (these proceedings). Most thoughtprovoking is the finding that an enzyme (potato acid phosphatase) is activated by the homologous lectin (STA, potato lectin) (18).

Acknowledgments. The Danish Ministry of Education is thanked for a fellowship for ILK. The studies were supported by the Danish Medical Research Council (TCBH).

REFERENCES

1. Ahmad, A., Bishayee, S. and Bachhawat, B.K. Biochem.Bioph.Res.Commun. 53, 730-736 (1974)
2. Andreu, J.M., Warth, R., and Munoz, E. FEBS Lett. 86, 1-5 (1978)
3. Axelsen, N. (ed.) Scand.J.Immunol.Suppl. 10 (1981) in press
4. Bhattacharya, P., Simet, I. and Basu, S. Proc.Natl.Acad.Sci. 76, 2218-2221 (1979)
5. Bishayee, S. and Bachhawat, B.K. Bioc.Biop.Acta 334, 378-388 (1974)
6. Bøg-Hansen, T.C. and Brogren, C.H. Scand.J.Immunol. 4. Suppl. 2, 135-139 (1975)
7. Bøg-Hansen, T.C., Brogren, C.H. and McMurrow, I. J.Inst.Brew. 80, 443-446 (1974)
8. Bøg-Hansen, T.C. and Brogren, C.H. Scand.J.Immunol.Suppl. 10 (1981) in press
9. Bonnafous, J.C., Dornand, J. and Mani, J.C. FEBS-Lett. 99,152-156 (1979)
10. Carreira, J., Andreu, J.M., Nieto, M. and Munoz, E. Mol.Cell.Biochem. 1o, 67-75 (1976)
11. Deschodt-Lanckman, M., Robberecht, P., Camus, J.C. and Christophe, J. J.Cycl.Nucl.Res. 3, 177-187 (1977)
12. Dey, P.M. and Pridham, J.B. Biochem.J. 113, 49-55 (1969)
13. Dulaney, J.T. Mol.Cell.Biochem. 21, 43-63 (1978)
14. Farooqui, A.A. and Srivasta, P.N. FED PROC. 39, 2212 abstract (1980).
15. Gutensohn, W. H.-S. Z. Physiol.Chem. 359, 1599-1602 (1978)
16. Guzman, N.A., Berg, R.A. and Prokop, D.J. Bioc.Biop.Res.Commun. 73, 279-285 (1976)
17. Ishizaki, H. and Yasunobu, K.T. Bioc.Biop.Acta 611, 27-34 (1980)
18. Lorenc-Kubis, I., Morawiecka, B., Wieczorek, E., Wisniowska, J., Wierzba, E., Ferens, M. and Bøg-Hansen, T.C. These Proceedings
19. Mitranic, M.M., Strurges, J.M., Moscarello, M.A. Can. J. Biochem. 57, 1008-1013 (1979).
20. Morita, Y. and Kameda, K. Bull.Agr.Chem.Soc.Jap., 23, 28 (1959)

21. Olsnes, S., Refsnes, K. and Pihl, A.
Nature 249, 627-631 (1974)
22. Olsnes, S., Saltvedt, E. and Pihl, A.
J.Biol.Chem. 249, 803-810 (1974)
23. Riordan, J.R., Slavik, M. and Kartner, N.
J.Biol.Chem. 252, 5449-5455 (1977)
24. Saltved, E.
Bioc.Biop.Acta 451, 536-548 (1976)
25. Satav, J.G., R.F.Johnston, B.Monk and R.S.Criddle
Arch.Bioc.Biop. 199, 110-116 (1980)
26. Scocca, J. and Lee, Y.C.
J.Biol.Chem. 244, 4852-4863 (1969)
27. Shaw, L. and Petersen-Acher, L.
Clin.Biochem. 12, 256-260 (1979)
28. Slavik, M., Kratner, N. and Riodan, J.R.
Bioc.Biop.Res.Commun. 75, 342-349 (1977)
29. Stefanovic, V., Mandel, P. and Rosenberg, A.
Biochemistry 18, 357-361 (1979)
30. Sulkowski, E. and Laskowski, M.
Biochem.Biophys.Res.Commun. 57, 463-468 (1974)
31. Takahashi, S., Pollak, J. and Seifter, S.
Biochim.Bioph.Acta 371, 71-75 (1974)
32. Tanaka, H., Hiwada, K. and Kokubu, T.
Biochem.BRharm. 27, 815-816 (1978)
33. Young, M.E.M., Moscarello, M.A., and Riodan, J.R.
J.Biol.Chem. 251, 5860-5865 (1976)
34. Yoshida, H. and Tamiya, N.
J.Biochem. 69, 525-534 (1971)
35. Zygowicz, E.R., Sunderman, F.W., Horak, E. and Dooley, J.F.
Clin.Chem. 23, 2311-2323 (1977)

EFFECTS OF LECTINS ON ENZYMATIC PROPERTIES PLANT ACID PHOSPHATASES AND RIBONUCLEASES

I.Lorenc-Kubis, B.Morawiecka, E.Wieczorek,
J.Wisniowska, E.Wierzba, and M.Ferens
Institute of Biochemistry, University of Wroclaw, Poland

T.C.Bøg-Hansen
The Protein Laboratory, University of Copenhagen, Denmark

Many enzymes from animal tissues are glycoproteins which react with lectins. Especially acid hydrolases have been well characterized with respect to interaction with lectins (2,4,8,10, 12) cf. also references in the preceeding paper (7). Conversely for plant enzymes, information about the presence and structure of carbohydrate and interaction with lectins is rather scarce. Previously we have shown that some acid phosphatases (AcPases) and ribonucleases (RNases) from grass seeds and potato bulbs are glycoproteins (5,15). In the preceeding paper we have shown that the acid phosphatases in Poa pratensis seeds are activated with conA and lentil lectin (7).

We have purified further the acid phosphatase from Poa pratensis and the object of this work was to compare its reactions with lectins to those of other purified plant enzymes. We found a varied pattern of activation and inhibition of the pur-

Lectins - Biology, Biochemistry, Clinical Biochemistry, Vol. I

ified acid phosphatases and RNases from seeds of grasses, from potato tubers and from yeast after interaction of the pure enzymes with conA and potato lectin (STA, Solanum tuberosum agglutinin).

Materials and Methods

Extracts were prepared with 0.1 M acetate buffer, pH 4.7-5.1 from ground seeds of Poa pratensis and Dactylis glomerata (common grasses), from germs of Secale cereale (rye), from tubers of Solanum tuberosum (6) and from cells of the yeast Rhodotorula glutinis (14). Proteins were precipitated with ethanol or ammonium sulphate and the pellet was dissolved in acetate, pH 5.1. The proteins in the resulting supernatants were separated by gel filtration on Bio-Gels P-100 and P-150, Sephadex G-75 and Sephacryl S-200, by affinity chromatography on conA-Sepharose and ion-exchange on CM-cellulose. Fig. 1 shows a flow-sheet for purification of the enzymes. Active fractions were pooled, and the isolated enzymes were analyzed by disc electrophoresis on polyacrylamide gels at pH 4.5 and 8.5 according to (3 and 9). Acid phosphatase activity was found by histochemical staining of the gels to correspond to the main band of each preparation.

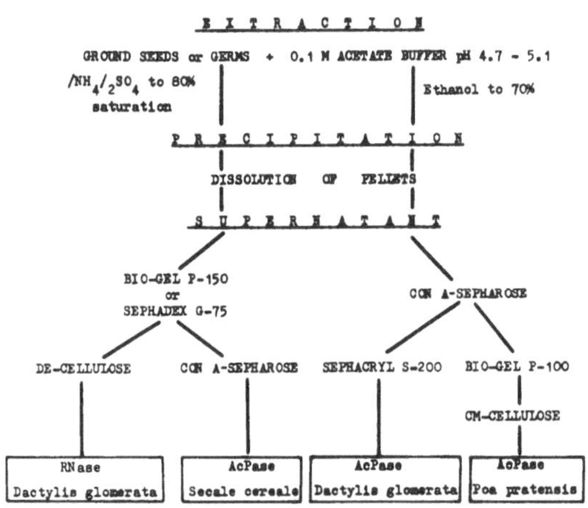

Fig. 1. Flow sheet for purification of acid phosphatase (AcPase) and ribonuclease (RNase) from plant materials.

Acid phosphatase activity was measured as release of p-nitrophenol from p-nitrophenyl phosphate at 37°C, pH 5.0 (17). RNase activity was measured as hydrolysis of yeast RNA (0.8%) according to (1).

Results

We studied the changes in the enzymatic activity of purified acid phosphatases and RNases by spectrophotometric analyses after interaction with conA and STA, and our special interest was devoted to the following parameters: Enzyme-lectin ratio, time of complex-formation, thermostability, and pH optimum. Table 1 shows the changes in activity observed at standard conditions (incubation for 30 min at 37°C at an enzyme-lectin ratio of 1 to 10). In all cases we could measure an effect of at least one of the lectins, conA or STA, on the enzymatic ac-

TABLE 1. Effect of lectins on the activity of some acid phosphatases and RNases

Source of enzyme	The relative activity of enzyme (%)	
	Con A	STA
ACID PHOSPHATASES:		
Poa pratensis seeds	160	-
Dactylis glomerata seeds	50	-
Secale cereale germs	140	150
Solanum tuberosum bulbs	100	130
RIBONUCLEASES:		
Poa pratensis seeds	65	-
Dactylis glomerata seeds	50	-

tivity, either as activation or as inhibition. Our earlier observations on activation of acid phosphatases from Poa pratensis with conA were verified. The more purified enzymes used in the present studies were activated to a higher degree than the less purified enzymes used earlier (7). Analogously the acid phosphatase activity from Secale cereale was increased with both conA and STA. However Solanum tuberosum acid phosphatase was activated with only STA, and Dactylis glomerata acid phosphatase was inhibited. Also RNase from the same seeds

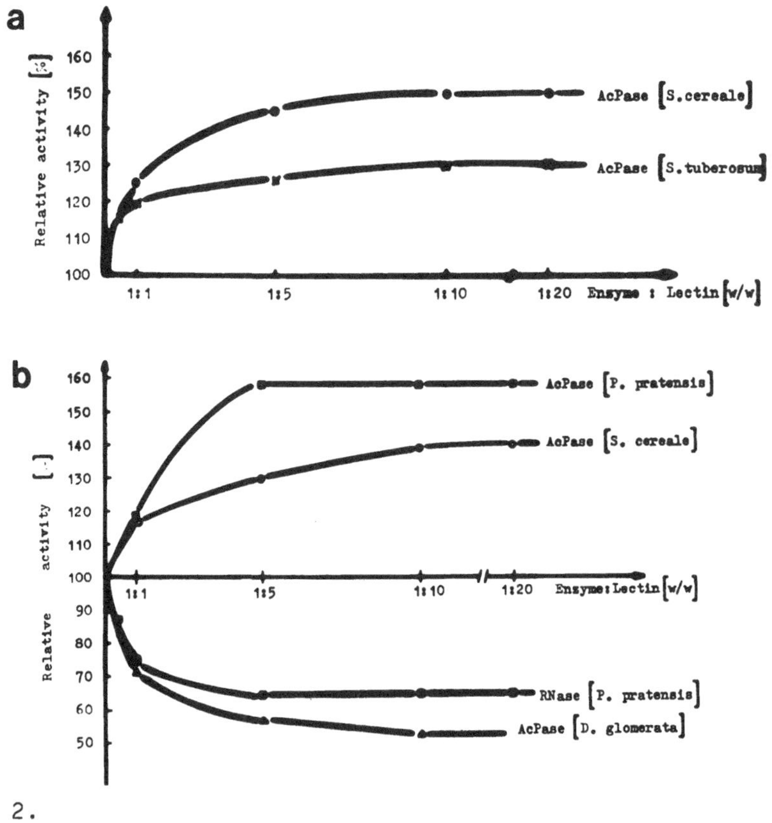

Fig. 2.
Effects of lectins on the in vitro activity of some enzymes.
a. Effect of conA at various concentrations.
b. Effect of STA at various concentrations.
Cf. text for details.

and from Poa pratensis were inhibited to about half the original activity. Acid phosphatase activity from Solanum tuberosum and Rhodotorula glutinis was unchanged in presence of conA.

The activation and inhibition as a function of the amount of lectin added to the glycoprotein enzymes is shown in Fig. 2. The largest effects were observed at enzyme-lectin ratios greater than 1 to 5. Precipitation was never observed during the incubation or assay.

When the incubation with lectin prior to the enzymatic incubation was varied with respect to time, we observed that a plateau was reached in from 10 to 20 min. A peculiar pattern of activation was exhibited by acid phosphatase from Secale cereale with a higher degree of activation after 5 min of incubation than after 10 min or more (cf. Fig. 3).

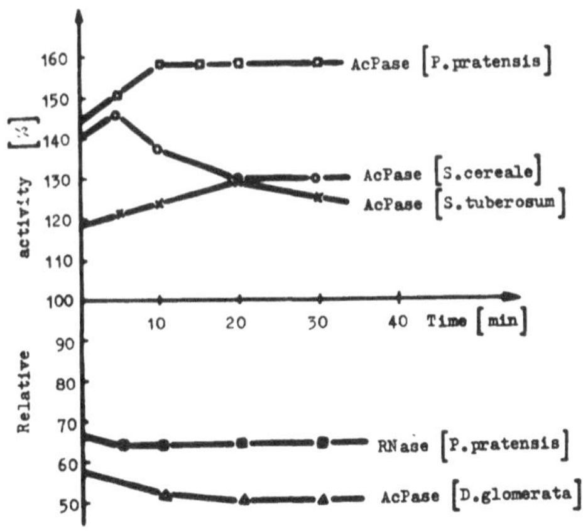

Fig. 3. Effect of length of preincubation of enzymes with conA. The relative activity of an enzyme is shown for a lectin-enzyme ratio of 10:1.

The stability of Poa pratensis RNase was tested by preincubation with conA at various temperatures (Fig. 4). At 60 C the thermal inactivation of the enzyme without conA was of the same degree as the lectin-induced inhibition at 37 C. At higher temperatures it was greater, meaning that conA protected the enzyme from heat denaturation.

Also we observed that some of the investigated enzymes changed their pH optimum in the presence of conA. Thus the pH optimum for acid phosphatases from S. cereale, P. pratensis and D. glomerata was shifted towards more acid values and were broadened. Fig. 5 shows a typical curve.

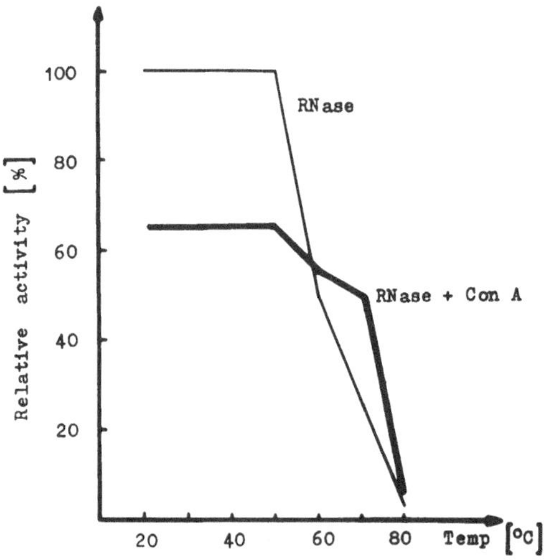

Fig. 4. Effect of conA on the thermostability of RNase from Poa pratensis seeds. Solid line gives the RNase activity at a lectin-enzyme ratio of 10:1.

Discussion

The results presented here demonstrate or verify the glycoprotein nature of some plant acid phosphatases and RNases. Characteristically these enzymes are either activated or inhibited by binding to lectin. Especially we studied the binding with conA, and amongst the enzymes, which we studied, we observed only two enzymes (acid phosphatase from potato and yeast) not affected by conA. However, acid phosphatase from potato was activated by the potato lectin, STA.

Largely inhibition has been reported for glycoprotein enzymes, cf. for instance the inhibition of human renin (13), serum (18) and milk fat globule membrane (11) 5'-nucleotidase, alkaline phosphatase (18) and microsomal electron transport enzymes (16). Earlier we found that a grass acid phosphatase changed its properties with conA (7). The present studies generalize

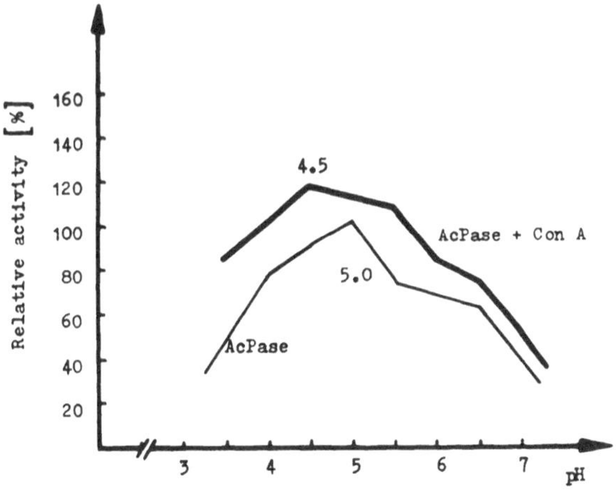

Fig. 5. Effect of conA on the optimum of acid phosphatase from Secale cereale germs. Solid line gives the acid phosphatase activity at a lectin-enzyme ratio of 10:1.

the activation of acid phosphatases to occur with the acid phosphatase from another grass (Secale cereale) and also with another lectin (STA).

We find it noteworthy that acid phosphatase and RNase from the same plant are inhibited in one case (for D. glomerata) while from another plant one enzyme is activated and the other enzyme is inhibited (for P. pratensis). Also we find that it is noteworthy that an enzyme may be effected by one lectin but not by another lectin (as for acid phosphatase from S. tuberosum). In our work we did not find any enzyme which was activated by one lectin and inactivated by another lectin.

Stabilization through binding to lectin is an effect analogous to binding to inert carriers or to antibodies as discussed in the previous paper (7). Similar results have been described for the conA-effect on the heat stability of ATPase from liver cell membranes by Riordan et al. (10). We also notice that expression of the interaction with lectin is timedependent, cf. the experiment with variation of incubation time, Fig. 3. In the case of activation, presumably the reaction with lectin stabilizes the enzyme in a conformation which is optimal or brings the active site in a conformation optimal to the cleavage of the artificial substrate. In the case of inhibition, a possible explanation would be the steric hindrance exerted by a lectin bound close to the active site, or some conformational change of the active site, cf. papers cited in (7).

Our experiments could not disclose any hint to the influence on the hydrolysis of natural substrates in vivo.

Earlier we suggested that the activity change of enzymes interacting with lectins may indicate a regulatory function of lectins on the metabolism in plants. However, this was suggested on basis of experiments with heterologous enzyme and lectin. Therefore, we find it extremely interesting and thought provoking that an enzyme is activated by its homologous lectin, as we found for potato acid phosphatase and potato lectin (STA).

This finding may indicate a possible physiological role for lectins in plants as well as in other organisms as enzyme effectors.

REFERENCES

1. Anfinsen, C.B., Bedfield, R.R., Choate, W.L., Page, J. and Carrol, W.R. J.Biol.Chem. 207, 201-210 (1954)

2. Bishayee, S. and Bachhawat, B.K. Biochim.Biophys.Acta 334, 378-388 (1974)

3. Davis, B.J. Ann.N.Y.Acad.Sci. 121, 404-408 (1964)

4. Dulaney, J.T. Mol.Cell.Biochem. 21, 43-63 (1978)

5. Kruzel, M. XV Meeting Polish Biochem.Soc. Gdansk, Abstract R-11, p. 106 (1977)

6. Kruzel, M. and Morawiecka, B. 12 FEBS Meeting Dresden, Abstract 3737 (1978)

7. Lorenc-Kubis, I and Bøg-Hansen, T.C. These Proceedings

8. Mintranic, M., Strugess, J.M. and Moscarello, M.A. Can.J.Biochem. 57, 1008-1013 (1979)

9. Reisfeld, R.A., Lewis, U.J. and Williams, D.E.
Nature 195, 291-283 (1962)

10. Riordan, J.R., Slavik, M. and Kartner, N.
J.Biol.Chem. 252, 5449-5455 (1977)

11. Snow, L.D., Doss, R.C. and Carraway, K.L.
Bioc.Biop.Acta 611, 333-341 (1980)

12. Stefanovic, V., Mandel, P. and Rosenberg, A.
Biochemistry 18, 357-361 (1979)

13. Tanaka, H., Hiwada, K. and Kokubu, T.
Biochem. Pharm. 27, 815-816 (1978)

14. Watorek, W., Morawiecka., and Korczak, B.
Acta Biochim.Polon. 24, 153-162 (1977)

15. Wieczorek, E., Lorenc-Kubis, I. and Morawiecka, B.
Acta Soc.Bot.Polon. 46, 481-488 (1977)

16. Winqvist, L., Eriksson, C. and Dallner, G.
J.Cell Sci. 39, 101-116 (1979)

17. Yoshida, H., and Tamiya, N.
J.Biochem. 69, 525-534 (1971)

18. Zygowicz, E.K., Sunderman, F.W. and Horak, E.
Clin.Chem. 23, 2311-2323 (1977)

ISOLATION OF LECTINS WITH DIFFERENT SPECIFICITIES USING IMMOBILIZED IMMUNOGLOBULIN

H. Franz and P. Ziska
Staatliches Institut für Immunpräparate und Nährmedien,
D-1120 Berlin, Klement-Gottwald-Allee 317/321

Introduction

Affinity chromatography is widely used for isolation and large scale production of lectins. Useful are either carbohydrate products as Sephadex or Sepharose or matrix bound carbohydrates or glycoproteins (glycoconjugates). The application of carrier fixed ovomucoid (1), mucin (2), fetuin (3) and thyroglobulin (4) for isolation of lectins with different specificities has been described. In this paper a report is given about the use of immobilized immunoglobulins, especially Sepharose bound IgG (Ig-Sepharose).

Materials and methods

1. Ig-Sepharose has been prepared by coupling human immunoglobulin (Ig), mainly IgG to CNBr-activated Sepharose 4 B (20 mg Ig/ml gel).

2. Common procedure of isolation of lectins using Ig-Sepharose

   50 gr of ground seeds or other plant material were stirred with 250 ml phosphate buffered saline pH 7,0 for 3 h at room temperatur, then filtered and solid ammonium sulphate (60 gr/100 ml) was added to the clarified extract. The precipitate was allowed to settle overnight, collected by centrifugation and resuspended in a minimum volume of saline. The solution was applied directly to a column (volume 60 ml) of Ig-Sepharose. The column was washed protein-

Lectins - Biology, Biochemistry, Clinical Biochemistry, Vol. I

free and the lectin was eluted with saline containing 0,15 M of the corresponding carbohydrate or with glycine/HCl buffer pH 2,8. The fractions containing the lectin were pooled, dialyzed against distilled water and freez-dried. For the preparation of the Helix pomatia lectin dried proteins glands were used as raw material.

3. Preparation of mistletoe lectins

The lectins were isolated from ground mistletoe plants without fruits grown on Robinia pseudoaccacia. Lectin I was isolated using acid treated agarose as carrier (5). The non-adsorbed material was applied to a column of Ig-Sepharose (2,6 x 30 cm) and the column was washed with phosphate-buffered saline, pH 7,0. Lectin II was eluted with 0,2 M D-galactose in saline. The fractions containing the hemagglutinating proteins were pooled, dialyzed against distilled water and freeze-dried. After the lectin II had been displaced from the column, lectin III was eluted with 0,2 M glycine/HCl buffer pH 2,6. The fractions with the active material were pooled and neutralized with sodium carbonate. The neutral solution was put on a column (1,5 x 30 cm) of Sepharose-N-(ε-aminocaproyl)-β-D-galactosamin. The column was washed and the lectin III was eluted with glycin/HCl buffer pH 2,8. The hemagglutinating protein fractions were pooled, dialyzed against distilled water and freeze-dried.

4. Preparation of lectin from Lepidium sativum (garden cress)

20 gr ground seeds from Lepidium sativum was defatted with petrol ether (6) and then stirred with 200 ml phosphate buffered saline pH 7,0 for 3 h at room temperatur. After centrifugation and filtration the clear solution was applied directly to a column of immunoglobulin-Sepharose (20 x 1,5 cm). The column was washed protein-free and the lectin was eluted with 1 M KSCN

in saline. The fractions containing the hemagglutinating protein were pooled, dialyzed against distilled water and freeze-dried. The solution that passed the column without any adsorption contains a further lectin. This lectin reacts with a glycopeptide isolated from whey which was bound to CNBr activated Sepharose. It was eluted from there using glycine-HCl buffer pH 2,8.

Results

1. Capacity of IgG-Sepharose columns for different lectins

Table 1 shows the amounts of well known lectins after affinity chromatography on Ig-Sepharose.

Table 1 Isolation of lectins by affinity chromatography using Ig-Sepharose (60 ml gel)

Species	Amount of rawmaterial	obtained lectin	specificity
Glycine max (soy bean)	50 gr	28 mg	D-GalNAc
Ulex europaeus I	50 gr	8 mg	L-Fuc
Triticum vulgaris	50 gr	16 mg	D-Glc-NAC
Dolichos biflorus	50 gr	18 mg	D-GalNAc
Lens culinaris	50 gr	50 mg	D-Man
Evonymus europaeus	50 gr	40 mg	
Arachis hypogaea	50 gr	22 mg	D-Gal
Helix pomatia	2 gr (dried protein glands)	9 mg	D-GalNAc

2. Mistletoe lectins

Using a combination of Ig-Sepharose we have been able to isolate three lectins from crude extracts of mistletoe. Isolation, properties and chemical modification of the D-galactosyl-specific lectin I have been published recently (7). Only lectin I is bound by partially-hydrolized Sepharose. After separation of the lectin I the extract contains two lectins, which are adsorbed by the carbohydrate moieties of the immobilized immunoglobulins. The lectin II can be eluted from this column by 0,2 M D-galactose solution. The lectin III cannot be eluted with 0,5 M D-galactose solution but with 0,2 M glycine/ HCl buffer pH 2,6. For further purification the solution of lectin III is applied to a column of Sepharose-N-(ε-aminocapryol)-ß-D-galactosamine and then eluted with glycine/HCl buffer.

Lectin I and II each give a single band by disc electrophoresis and lectin III shows a major band and three faint bands. After reduction with 2-mercaptoethanol each of the lectins shows two bands in polyacrylamide electrophoresis in presence of 0,1% sodium dodecylsulphate. Disc electrophoresis of mixtures of the lectin I and II, I and III and I, II and III showed that they are not identical. Thus, lectin I consists of two identical noncovalently-bound subunits.

Lectin II and III and also the subunits of lectin I contain two chains linked by a disulphide bridge. The lectins differ in their mol.wts. All three lectins agglutinated human erythrocytes with similar potency, but when tested against erythrocytes from other species marked differences in potency were observed. The sugar specificities of the three lectins were different (7). Agglutination of human erythrocytes by lectin I was strongly inhibited by D-galactose and α- and ß-methyl-D-galactosides. Lectin II was inhibited both by N-acetyl-D-galactosamine and the D-galactosides whereas lectin III was inhibited strongly only by N-acetyl-D-galactosamine. All the six chains of the three lectins are different. However lectins II and III crossreact with anti-lectin

I-antibody in radial immunodiffusion. Lectins I, II and III are stained by the PAS-reagent.

Lectin I is a hemagglutinating and toxic lectin. No mitogenic effect could be demonstrated upon guinea pig lymph node cells because the cells were killed within 6 hours by lectin I at a concentration of 25-80 /ug/ml per 10^6 cells (8). Lectin I inhibits protein synthesis in a lysate of rabbit reticulocytes with an ID_{50} (concentration giving 50% inhibition) of 2,6 /ug/ml. This effect is enhanced (ID_{50} 0,21 /ug/ml) if the lectin is first reduced with 2-mercaptoethanol. The lectin also inhibits protein synthesis by BL8L cells in culture. The ID_{50} is 7 ng/ml and the potency decreased after reduction of the lectin (9).

Discussion

The carbohydrate moiety of human immunoglobulins consists of the following sugars: D-mannose, D-galactose, L-fucose, N-acetyl-D-glucosamine and N-acetyl-neuraminic acid. Therefore Ig-Sepharose represents a multivalent material for the affinity chromatography of numerous lectins. Table 1 gives a choice only. Although immunoglobulins do not contain N-acetyl-D-galactosamin lectins known as N-acetyl-galactosamine-specific (like soy bean lectin) were bound by Ig-Sepharose, too. The reason may be the fact that the term "specificity" means the biologically occuring sugar which is bound with the highest affinity. Smaller affinity exists to other sugars. Thus, Ig-Sepharose can be recommended as a nearly universal material for affinity chromatography of lectins. Concerning mistletoe we have described the isolation of three different lectins (7). All the three mistletoe lectins are bound by Ig-Sepharose, but only lectin I binds to partially hydrolized Sepharose. Moreover, lectin II and lectin III are bound on Ig-Sepharose with different strength. This fact allows a fractionated elution using galactose and glycin/HCl buffer respectively.

Using Ig-Sepharose it was also possible to isolate a lectin from Lepidium sativum seeds. Extracts from Lepidium sativum seeds have been tested in 1960 by Adámková and Tobiska (10). The authors could not find any hemagglutinating activity because they used whole blood instead of washed erythrocytes. Serum glycoproteins are strong inhibitors of hemagglutination by the Lepidium sativum lectin.

Finally it may be mentioned that on principle glutaraldehyde crosslinked (11) and heat aggregated immunoglobulins can also be used with nearly the same effect as Ig-Sepharose.

References

1. Tsuda, M.: J. Biochem. 86, 1451-1461 (1979)
2. Pereira, M.E.A., Kabat, E.A.: J. Cell Biol. 82, 185-194 (1979)
3. Sela, B.A., Wang, J.L., Edelman, G.M.: J. biol. Chem. 250, 7535-7538 (1975)
4. Felsted, R.L., Leavitt, R.D., Bachur, N.R.: Biochim. biophys. Acta 405, 72-81 (1975)
5. Ziska, P., Franz, H., Kindt, A.: Experientia 34, 123-124 (1978)
6. Franz, H., Ziska, P.: 4. Arbeitstagung "Lektine", Berlin 1980
7. Franz, H., Ziska, P., Kindt, A.: in press
8. Scherbaum, I., Drössler, K., Ziska, P., Franz, H.: Allergie und Immunologie 24, 208-211 (1978)
9. Stirpe, F., Legg, R.F., Onyon, L.J., Ziska, P., Franz, H.: Biochem. J., in press
10. Adámková, B., Tobiska, J.: Z. Immunforsch. 120, 90-95 (1960)
11. Franz, H., Haustein, B., Luther, P., Kuropka, U., Kindt, A.: Acta biol. med. germ. 36, 113-117 (1977)

Part II

Methods Based on Reactions of Lectins

LECTIN BINDING MEASURED BY FLUORESCEIN FLUORESCENCE POLARIZATION (FFP) TECHNIQUES.

P. Balding, P. A. Light, A. W. Preece
Oncology Research Unit, Dept. of Radiotherapy, University of Bristol U.K.

Introduction

The measurement of fluorescein fluorescence polarization (abbreviated to FFP) and related events during the process known as fluorochromasia provides an interesting opportunity to observe the effect of a lectin, or other external stimulus on the integrity of the cell membrane, the cytoplasmic environment and the activity of intracellular hydrolytic enzymes.

The fluorochromatic reaction, first described by Rotman and Papermaster (1) involves the intracellular hydrolysis of fluorescein diacetate - FDA - to which the cell membrane is permeable, forming free fluorescein. Since the membrane is less perieable to the reaction product, the net result is an intracellular accumulation of fluorescein, increasing with the duration of the reaction.

When cells, for example lymphocytes, which are actively hydrolysing FDA to fluorescein are suspended in substrate in the cuvette of a spectrofluorimeter, the emission characteristics and rate of FDA hydrolysis can be measured (2,3,4). For my experiments a Perkin-Elmer MPF 4 spectrofluorimeter was used with excitation wavelength of 470nm, measuring the emission at 510nm (slit widths 20nm and 10nm respectively). If the excitation light beam is vertically polarized, the emitted light is also polarized but depolarization will occur because of the movement, for example rotation due to Brownian motion, of the intracellular fluorescein molecules. If the vertical and horizontal components of the emitted light are recorded by introducing a polarizing filter in the emission light path, the degree of polarization can be calculated from the simple formula developed by Perrin (5) :

Lectins - Biology, Biochemistry, Clinical Biochemistry, Vol. I

$$p = \frac{Iv - GIh}{Iv + GIh}$$

where Iv is the vertical component and Ih the horizontal component of intracellular fluorescence; G is the spectrofluorimeter grading factor compensating for unequal transmission of polarized light across the emission monochromator (6).

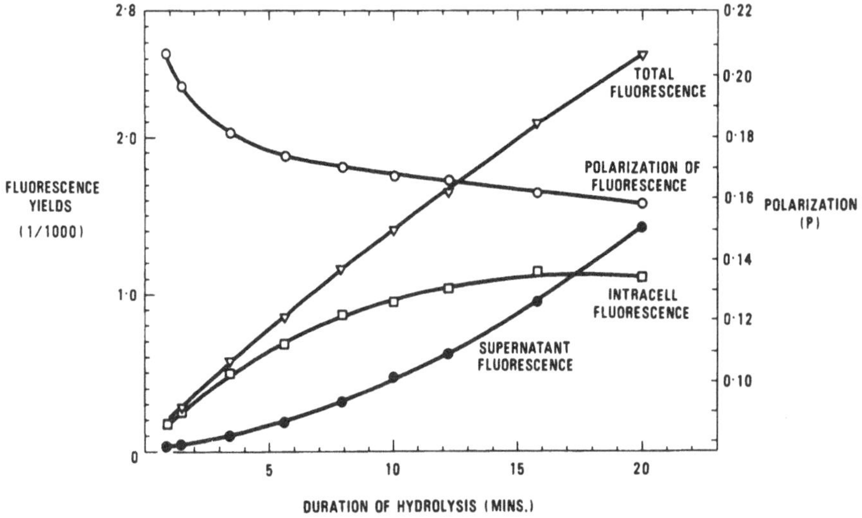

Figure 1 Fluorochromasia in human lymphocytes and the relationship between polarization of fluorescence and duration of FDA hydrolysis at 27°C (Fluorescence yield from 10^{-9} M NaF = 1.25×10^3).

However, during the course of the reaction some leakage of fluorescein out of the cells does occur. To measure, and compensate for this extra-cellular fluorescence, the cells in suspension are periodically filtered off using Millipore filters under controlled suction, and the components of supernatant fluorescence are recorded. Several filtration steps are necessary for accurate results (4). After a duration of about 20 minutes the measurement is terminated and the polarization (p) value, anistropy, rate of hydrolysis, rate of intracellular accumulation and rate of fluorescein leakage out of the cells can be calculated from the four components VT (vertical total), HT (horizontal total), VS (vertical supernatant) and HS (horizontal supernatant) obtained by manual extrapolation of chart recorder traces (2,3,4) since Iv = VT-VS and Ih = HT-HS. For the results reported here the data was collected using a purpose built analogue to digital converter and analysed by computer (7). Four equations, one for each of the four components of fluorescence intensity were calculated, and the various parameters described above were obtained by entering the required duration of hydrolysis in the equations. In addition plots of the change of p with time and intracellular fluorescence level, and the increasing fluorescence intensities may be produced as illustrated for human lymphocytes in Figures 1 and 2.

Similar results have been obtained for other lymphoid cell types including peripheral blood lymphocytes from other mammalian sources, murine thymus and spleen lymphocytes, and murine NK/LY/R lymphoma cells (4,7,8,9,10,11). However, only cells which hydrolyse FDA can be used. A second drawback is the need to measure the extra-cellular fluorescence components which can introduce a degree of error into the results. In general though, with properly controlled conditions, reproducibility is good as shown in Figure 3 where p value for the same lymphocyte suspension was measured on three successive occasions.

Results

When lymphocytes, or in some cases other lymphoid cells are pre-incubated with lectins for as short a time as 30 min. at 37°C, changes in several parameters can be measured during fluorochromasia (4), although generally only changes in p value have been reported, using PHA (10,12,13,14,15,16). There have also been reports describing p value measurements after PHA incubation using a fluorescence activated cell sorter, or cytofluorograph (11,17) but rarely have these instruments been used to measure FDA hydrolysis rates (18) or changes in cell leakage.

As the following results will show, the measurement of these parameters may be very interesting.

Experiments using a number of lectins have been carried out. *Ulex*

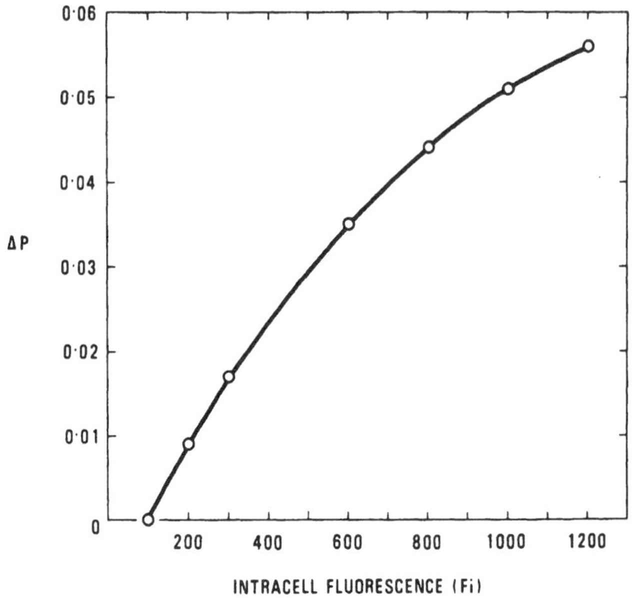

Figure 2 Relationship between intracellular fluorescence levels and changes in polarization values for human lynphocytes.

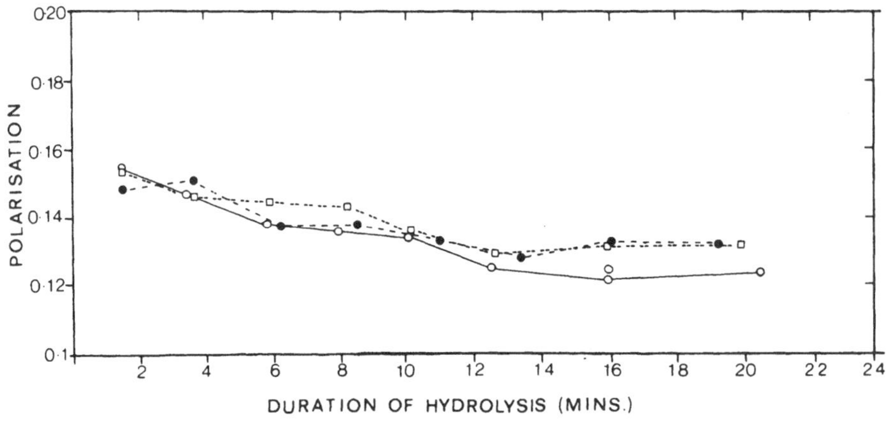

Figure 3 Reproducibility of p value measurement on human lymphocytes (identical conditions used for each measurement).

europaeus anti-H had no apparent effect on the p values of human lymphocytes; Lentil lectin (from Lens culinaris) appeared to have a marked effect on murine thymocytes, but the results are of doubtful validity since the lectin preparation used (obtained from Sigma Ltd.) showed hydrolytic activity. As an illustration of the potential application of the technique, some results using Poke Weed Mitogen (PWM) Wheat Germ agglutinin (WGA) and PHA with thymus and spleen lymphocytes from a BALB/C mouse will be presented. The cell suspensions were prepared in the usual way from homogenised tissue: red cells in the spleen but not thymus suspensions were lysed by treatment with tris buffered 0.83% ammonium chloride. Cells were finally suspended in phosphate buffered saline (Dulbecco complete A + B) to give a final concentration of 2.8 x 10^5 cells per ml. of FDA substrate in the cuvette.

After prior incubation with PWM for 30 min. at 37°C, thymocytes from an 8 week old normal mouse showed a concentration dependent increase in FDA hydrolysis rate of up to 50%, as shown in Figure 4. However, with

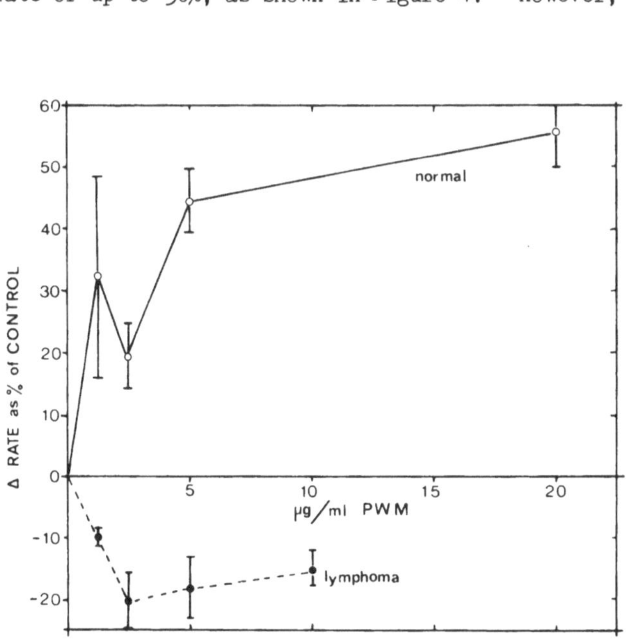

Figure 4. Effect of PWM concentration on the mean rate of FDA hydrolysis for thymocytes from normal and lymphoma bearing mice (±S.E.)

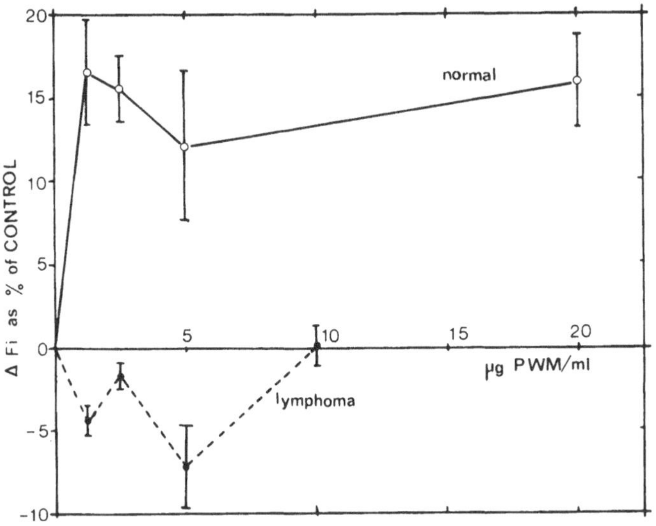

Figure 5 Effect of PWM concentration on intracellular fluorescence (Fi) levels in murine thymocytes (± S.E.)

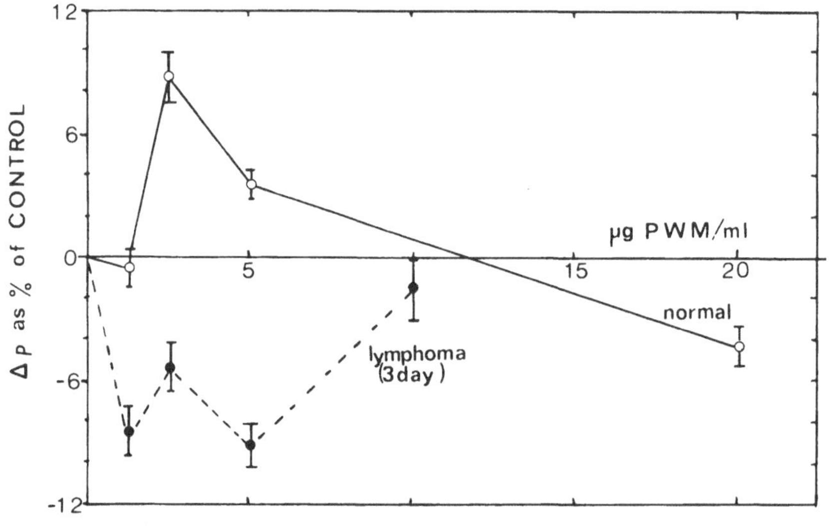

Figure 6 Effect of PWM concentration on p values at equivalent Fi (± S.E.)

thymocytes prepared from a mouse which had been inoculated intraperitoneally 3 days previously with an NK/LY/R ascites lymphoma, the response to PWM was the complete converse, i.e. a decrease in rate of up to 20%. Similarly, as shown in Figure 5 an increase in the intracellular fluorescein concentration in normal thymocytes was observed compared to a decrease in intracellular fluorescein level in thymocytes from the tumour bearing animal. The intracellular fluorescein concentration, or Fi, was determined at equivalent total fluorescence levels, and therefore corresponds to a significant change in the rate of leakage of fluorescein out of the cell. The full significance of these observations is not known at present, but must correspond to an alteration of some intracellular processes, since an increase in the rate of hydrolysis induced, for example, by a change in temperature of the reaction is usually accompanied by an increase in the leakage (1) rather than a decrease in leakage as found after PWM incubation of the cells.

It has been shown, both by our own previously reported results (4,8,9) and by others (10,11,15,19) that the polarization value is inversely related to the intracellular fluorescein concentration. It is therefore essential to compare p values at equivalent intracellular fluorescein concentrations, rather than at equivalent times of FDA hydrolysis, as has been reported (3,20). The changes in p value as shown in Figure 6 are calculated at equivalent intracellular fluorescein concentration. Thymocytes from the normal animal showed an increase in p value at low PWM concentration, although a decrease in p value was observed at the higher concentration of 20 µg/ml. Thymocytes from the tumour bearing animal, however, showed a decrease in p value.

The effects of incubating normal thymocytes with PHA were similar to those observed using PWM as shown in Figure 7. Thymocytes showed an increase in p value at low concentration but a slight decrease at higher concentration. In contrast, however, spleen lymphocytes showed a dose dependent decrease in p value. Results at equivalent intracellular fluorescein concentration are shown. When the thymocytes were prepared from the tumour-bearing animal there was a decrease in p value of up to 13% as shown in Figure 8.

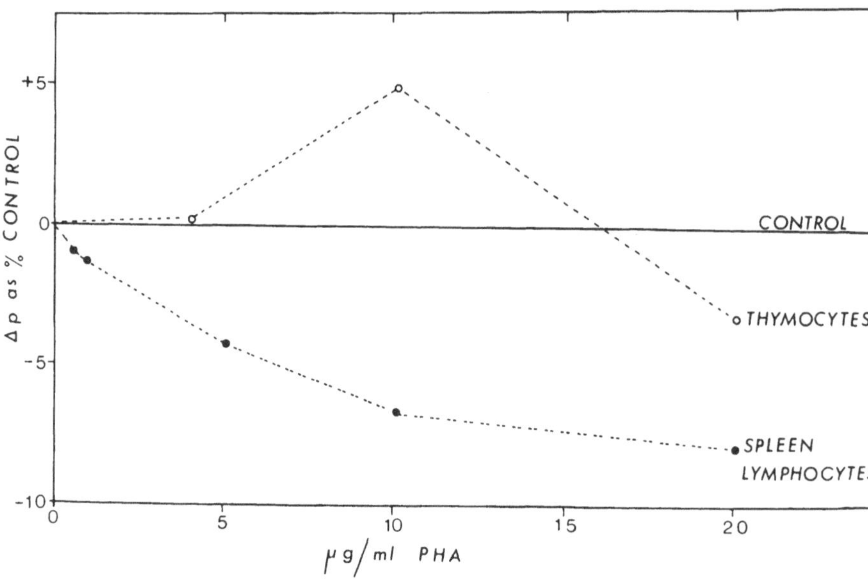

Figure 7 Effect of PHA concentration on p values of normal murine thymocytes and spleen lymphocytes.

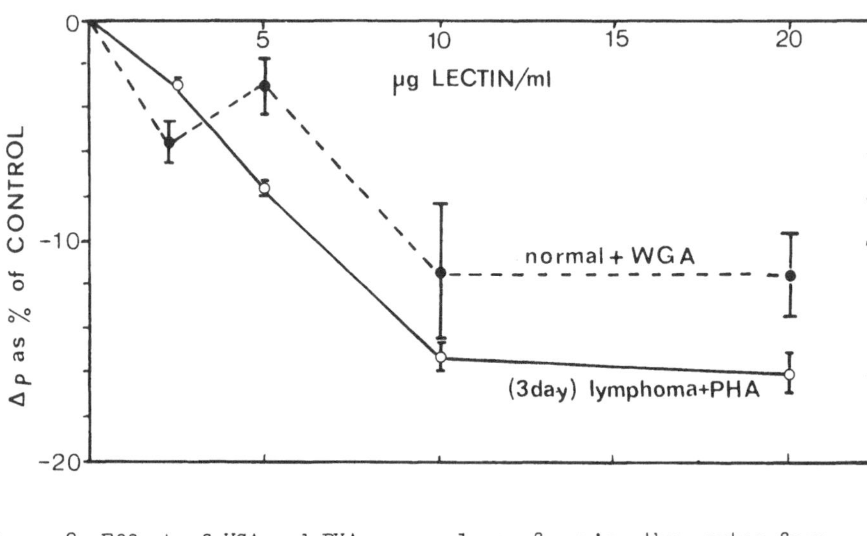

Figure 8 Effect of WGA and PHA on p values of murine thymocytes from normal and tumour bearing animals (± S.E.)

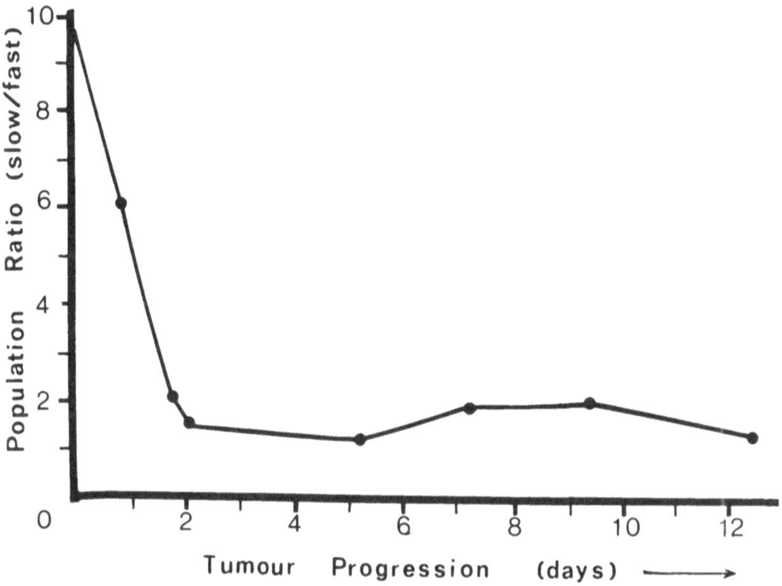

Figure 9 Effect of tumour progression on thymocyte population ratio (slow/fast)

Thymocytes do not respond to all lectins in the same way. Also shown in Figure 8 is the response of normal thymocytes to prior incubation with Wheat germ agglutinin, again for only 30 mins. at 37°C. The dose response curve is parallel to that of the thymocytes from the tumour-bearing animal with PHA. Thymocytes from the tumour-bearer showed a very much reduced response or a slight increase in p value with WGA.

To help understand these results, it is instructive to consider some changes in thymus cell populations during the growth of the lymphoma (21), illustrated in Figure 9. The electrophoretic mobility of individual cells in the suspension have been measured and as shown for the control animal there is a fast/slow population ratio of about 1 : 9 i.e. about 10% fast cells. These values correspond to the mature/immature cell ratio (21,22). As the tumour progresses the ratio of fast to slow cells increases, reaching about 1:2 after 2 to 3 days, at which point fluorescence was measured. The fast population has been identified as consisting of immunocompetent thymocytes which are released to the periphery. It is possibly this population which is responding to PHA and PWM, but not to WGA. It is therefore possible that it is the immunocompetent T cell population which is responding to PHA in the spleen lymphocyte sample.

The division of thymocytes into fast and slow populations as shown here depends on the cell surface charge which in turn is dependent upon the sialic acid content of the cell membrane (23). The response of thymocytes to different lectins may be related therefore to differences in cell surface glyosylation. It should be interesting to study the response of thymocytes to peanut lectin (Arachis hypogaea) during tumour progression: however, the early response of lymphocytes to lectin incubation as reported here suggests that an element of caution is necessary when interpreting results e.g. in transformation studies, when using lymphocyte subsets isolated using peanut lectin (24,25,26).

Results reported elsewhere (7,27) show that thymocytes from a PPD-sensitised animal respond to in vitro PPD incubation, showing a reduction in p value, an increase in hydrolysis rate and decrease in leakage of fluorescein out of the cells. Thymocytes from a non-sensitised control animal gave the opposite responses as shown by PWM. There was close correspondence between these results and those obtained in migration inhibition studies (28), again suggesting a relationship of response to immunocompetence. The technique of FFP may therefore be serving as an indication of the early events in mitogenesis and in cell mediated immunity.

References

1. Rotman, B. and Papermaster, B.W. : Proc. Nat. Acad. Science, 55, 134-,41 (1966)
2. Cercek, L and Cercek, B. : Int. J. Radiat. Biol.,21, 445-453 (1972)
3. Cercek, L., Cercek, B and Ockey, C..H. : Biophysik, 10, 187-194 (1973)
4. Balding, P., Light, P.A. and Preece, A.W. : Br. J. Cancer, 41, 73-85 (1980)
5. Perrin, F. : Ann. Phys., 10 (serie XII), 169 (1929)

6. Chen, R.F. and Bowman, R. C. : Science, 147, 729-732 (1965)

7. Balding, P., Grant, R.B., Light, P.A. and Preece, A.W. : J. Immunol. Methods, (submitted for publication, 1980).

8. Preece, A. W., Light, P.A. and Balding, P. Br. J. Cancer, 38, 197 (1978)

9. Balding, P., Light, P.A. and Preece, A.W. : Protides of the Biological Fluids, ed. H. Peeters, Pergamon Press, Oxford, 27, 435-438 (1979)

10. Mitchell, H., Wood, P., Pentycross, C.R., Abel, E and Bagshawe, K.D., Br. J. Cancer, 41, 772-777 (1980)

11. Udkoff, R. and Norman, A.J: Histochem. Cytochem. 27, 49-55 (1979)

12. Cercek, L., Cercek, B and Franklin, C.I.V. Br. J. Cancer, 29 345-352 (1974)

13. Cercek, L. and Cercek, B. : Eur. J. Cancer, 13, 903-915 (1977)

14. Pritchard, J.A.V., Sutherland, W.H., Seaman, J.E. and 6 others: Lancet, ii, 1275 (1978)

15, Hashimoto, Y., Takaku, F. and Yamanaka, T. : Br. J. Cancer 40, 156-160 (1979)

16. Dickinson, J. P., Dyson J.E.D., Smith, J.J. and Cowley, N. : Proc. 3rd. Int. Congr. Soc. Detection Prevention of Cancer. New York, Marcel Dekker (1976)

17. Stewart, S., Pritchard, K.I., Meakin, J.W. and Price, G.B., Clin. Immunol. Immunopathol. 13, 171 (1979)

18. Watson, J. Br. J. Cancer, 37, 39 (1978)

19. Epstein, M., Norman, A., Pinkel, D. and Udkoff, R. : J. Histochem. Cytochem. 25, 821-826 (1977)

20. Cercek. L., Cercek, B and Ockey, C.H. : Biophys. J. 23, 395-405 (1978)

21. Jenkins, R. : Immunology, 29, 893-902 (1975)

22. Dumont, F. : Immunolgy, 26, 1051 (1973)

23. Hoessli, D., Bron, C. and Pink, R.L. : Nature, 283, 576-578 (1980)

24. Reisner, Y., Linker-Israeli, M. and Sharon N. Cell Immunol. 25, 129-134 (1976)

25. London, J., Berrih, S. and Bach, J.F. : J. Immunol. 121, 438 (1978)

26. Berrih, S., Bonavida, B. and London, J : Protides of the Biological fluids, ed. H. Peeters, Pergamon Press, Oxford, 27 551-555 (1979)

27. Balding, P., Hughes, L., Light, P.A. and Preece, A.W. : 4th. Int. Congr. Immunol. Paris, abstract 13.8.01 (1980)

28. Hughes L., Preece, A.W. and Light, P.A. : Immunol.Lett. 1, 269-273 (1980)

Acknowledgements.
This work was supported by a grant from the Cancer Research Campaign.

STUDIES ON LECTIN-CARBOHYDRATE INTERACTIONS USING DIFFERENTIAL SCANNING CALORIMETRY.

B. Mattiasson and C. Borrebaeck
Department of Pure and Applied Biochemistry, Chemical Center, University of Lund, S-220 07 LUND, SWEDEN

Introduction

The technique of differential scanning calorimetry (DSC) has only recently been applied in studies on biological macromolecules (1-3). Endo- and/or exothermic processes which take place when heating a sample can be followed with this calorimetric method. From such studies information concerning denaturation enthalpies as well as transition temperatures (T_{tr}) may be obtained. Only a few reports on the application of DSC to study macromolecular interactions have been reported (4-6).

In the present paper we describe the results of a DSC-study on the interaction between the lectin Concanavalin A from <u>Canavalia ensiformis</u> and various carbohydrates, including both free sugars and glycoproteins. The metal ion dependence of the lectin for binding carbohydrate was also investigated.

Materials and Methods

Concanavalin A was obtained from Miles-Yeda Ltd, Rehovot, Israel or from Sigma Chemical Co., St. Louis, Missouri, USA. Horse radish peroxidase was also obtained from Sigma. All hapten sugars came from BDH Chemicals, Poole, Great Britain, and the two low-molecular weight dextrans were generous gifts from Dr. Kirsti Granath, Uppsala, Sweden and Professor Per-Åke

Lectins - Biology, Biochemistry, Clinical Biochemistry, Vol. I

Albertsson, Lund, Sweden.
All other chemicals used were of analytical grade.

Preparation of demetallized Concanavalin A.

Demetallization was carried out according to a previously published method (6).

Thermal analysis.

The DSC-instrument used (Perkin-Elmer DSC-2) operates with one sample vial and one reference vial. Those vials, usually called pans, are made of metal (aluminum, stainless steel or gold) and are sealed after the sample or the reference has been applied. The two pans are then heated at a programmed rate so that the temperatures of the two pans are identical. When an endothermic transition takes place in the sample pan, the reference pan is heated to compensate for the heat evolved upon transition of the sample. The extra effect added to the reference is registered on a chart recorder.
The samples were dissolved in 0.1 M Tris-HCl buffer pH 7.0 (25 ^{o}C) which normally contained 1 mM $CaCl_2$ and 1 mM $MnCl_2$.
A 1 % (w/v) solution of Concanavalin A was used throughout the study to which an equal volume of a solution containing carbohydrate (50 mM), glycoprotein or buffer was added. The mixture was incubated at room temperature for 15 min before the DSC study started.

Thermograms were run from 320 K to 390 K. Heating rate 10 K/min and amplification 0.42 mJ/s. In the studies when the transition temperature was above 370 K pans of aluminum could not be used because of a reaction between aluminum and water disturbing the measurements. At these elevated temperatures either gold or stainless steel pans were used. The volume of the sample added to a pan was 15 µL for the aluminum and gold pans whereas the stainless steel pans needed 50 µL.

Transition temperatures were read from the intersection of the two lines that could be drawn from the ascending and descending parts of the peak. (Fig. 1)

Results and Discussion

A question of interest when studying a lectin-glycoprotein interaction is whether there is a secondary protein-protein interaction besides the protein-carbohydrate interaction. It has been shown earlier that very weak interferences between ligands and proteins may be detected as a change in the thermogram. Based on these observations it could be postulated that a secondary interaction would also influence the T_{tr} for the glycoprotein. Fig. 1 illustrates the thermograms for a) horse radish peroxidase, b) Concanavalin A and c) a mixture of the two proteins. It can be seen from this figure that the T_{tr} of Concanavalin A markedly changes when horse radish peroxidase is bound, whereas no significant influence on the T_{tr} of horse radish peroxidase can be observed.

On changing the molecular ratio between the lectin and the glycoprotein the T_{tr} for Con A changed with a maximum for the ratio peroxidase/Con A of 4, i.e. one peroxidase molecule for each subunit of Con A (Fig. 2). The shape of the curve resembles that of antibody-antigen precipitation curves. The assymetric shape in this case is caused by the fact that at higher ratios peroxidase/Con A there is still a stabilization of the lectin.

The T_{tr} of peroxidase on the contrary did not change when varying the peroxidase/Con A ratio from 1 to 12. At lower ratios (<1) however a slight increase (~ 1 K) was observed. (Fig. 3).

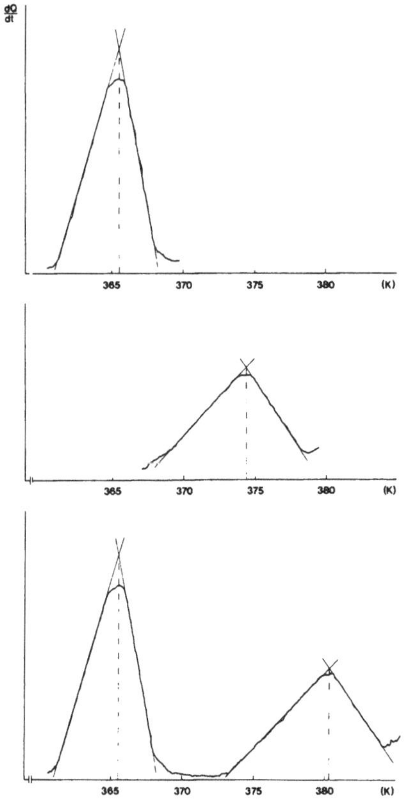

Fig. 1 Thermograms recorded for
a) horse radish peroxidase
b) Concanavalin A
c) a mixture of horse radish peroxidase and Concanavalin A (mol. ratio 4:1)

This fact was explained by the high degree of crosslinking of the enzyme.

The results described above point towards the fact that the carbohydrate moity of the glycoprotein interacts with the lectin without involvment of the protein part of the glycoprotein. The change in T_{tr} for Con A is thus caused by the binding of

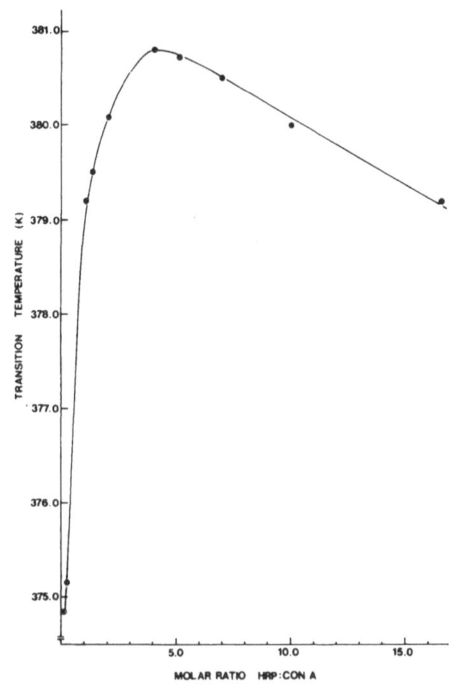

Fig. 2 The transition temperature for Concanavalin A associated with horse radish peroxidase as a function of the molar ratio of peroxidase:Concanavalin A. (from ref.6).

carbohydrate to the lectin, whereas that binding does not influence the tertiary structure of the protein part of peroxidase. These results are also in line with results from studies on enzyme activities for enzymes free in solution or when bound to lectins. It has been reported that the enzyme activity is unaffected or only marginally influenced by the binding of the lectin to the glycoenzyme (7,8). However, these observations do not seem to be general, since reports also exist where the enzyme activity is markedly influenced by the binding of a lectin molecule to the carbohydrate part of the glycoenzyme (this volume).

The transition temperature for native Concanavalin A was found to be 374.4 K. The accuracy in the measurements of all the transition temperatures was 0.1 K or better, whereas the other parameter that also could be read from the thermograms, the transition enthalpy, was not that reproducible. In the presence of hapten sugars (50 mM) a stabilization was observed as an

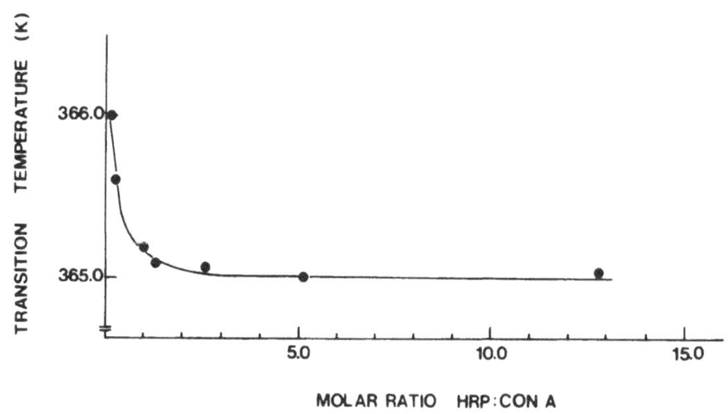

Fig. 3 The transition temperature for horse radish peroxidase as a function of the molar ratios of peroxidase:Concanavalin A.(from ref. 6).

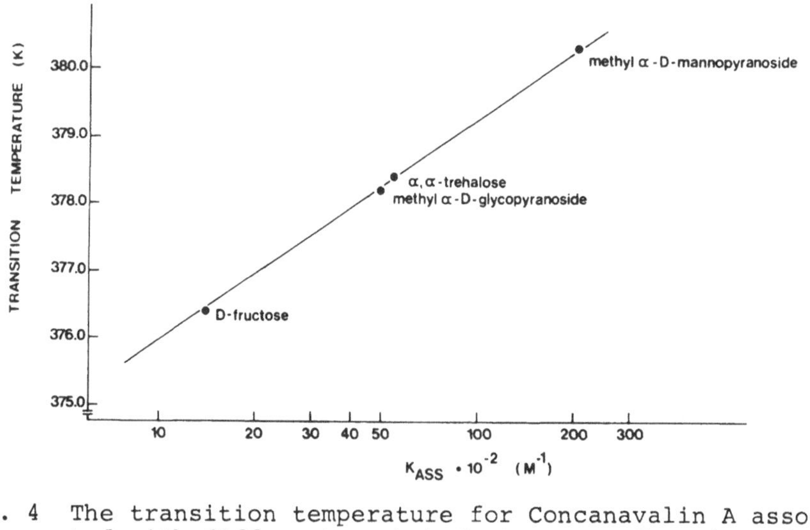

Fig. 4 The transition temperature for Concanavalin A associated with different carbohydrates as a function of their association constants. T_{tr} for native Concanavalin A is 374.4 K.(from ref. 6)

increase in transition temperature. A linear relationship was obtained when the K_{ass} values (obtained from literature) were plotted in a linlog diagram versus the T_{tr} for the Con A - carbohydrate complex. (Fig. 4)

Furthermore, since the binding of methyl-α-D-mannopyranoside changes the T_{tr} of Con A by 5.9 K this stabilizing effect was used as an indicator when studying the requirement for metal ions for carbohydrate binding to take place. Demetallized Con A had itself a T_{tr} of about 356 K (no sharp peak observed). When Mn^{2+} was added a T_{tr} of 370 K was observed (Fig. 5). How-

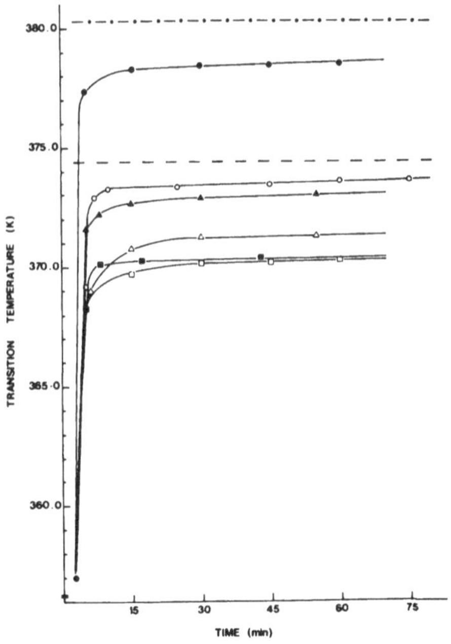

Fig. 5 The transition temperatures of different metal ion complexes of demetallized Concanavalin A as a function of the time of contact between the lectin and the ligand. Demetallized Concanavalin A was reconstituted with (□) Mn^{2+}; (■) Mn^{2+} and methyl-α-D-mannopyranoside; (△) Ca^{2+}; (▲) Ca^{2+} and methyl-α-D-mannopyranoside; (○) Ca^{2+} and Mn^{2+}; (●) Ca^{2+} and Mn^{2+} and methyl-α-D-mannopyranoside. Furthermore, the transition temperature of native Concanavalin A (---) and native Concanavalin A bound to methyl-α-D-mannopyranoside (-•-) are shown. (from ref. 6).

ever, no further stabilization was found on addition of methyl-α-D-mannopyranoside. In a parallel experiment with addition of Ca^{2+} to demetallized Con A a T_{tr} of 372 K was measured. In this case a slight further stabilization (1.7 K) was observed upon addition of methyl-α-D-mannopyranoside. The conclusion drawn from these last two experiments is that the lectin in the presence of Ca^{2+} but in the absence of Mn^{2+} still keeps some ability to bind carbohydrate whereas no such ability is observed in the reciprocal situation.

When both metal ions were added to the demetallized Con A a T_{tr} close to that of native Con A was observed. Likewise, binding of methyl-α-D-mannopyranoside resulted in a marked stabilization with a T_{tr} slightly lower than the corresponding value for native Con A with the hapten sugar bound.

In conclusion, the use of differential scanning calorimetry has proved a useful tool when studying macromolecular interactions. The model studies reported here clearly show the potential of calorimetric studies when investigating lectin-carbohydrate interactions.

References

1. Beck, K., Gill, S.J.,Downing, M.,J. Am. Chem. Soc. 87, 901-904 (1965).
2. Privalov, P.L., Khechinashvili, N.N., J. Mol. Biol. 86, 665-684 (1974).
3. Hägerdal, B., Martens, H., J. Food Sci. 41,933-937 (1976).
4. Donovan, J.W., Beardslee, R.A., J. Biol. Chem. 250, 1966-1971 (1975).
5. Donovan, J.W., Ross, K.D., Biochemistry 12, 512-517 (1973).
6. Borrebaeck, C., Mattiasson, B., Eur. J. Biochem. 107, 67-71 (1980).
7. Bishayee, S., Bachhawat, B.K., Biochim. Biophys. Acta 334, 378-388 (1974).
8. Mattiasson, B., J. Appl. Biochem., in press.

BIOCHEMICAL ANALYSIS BASED ON THE USE OF IMMOBILIZED LECTINS PLACED IN CONTINUOUS FLOW SYSTEMS.

B. Mattiasson
Department of Pure and Applied Biochemistry, Chemical Center, University of Lund, S-220 07 LUND, SWEDEN

Biological macromolecules, such as enzymes, antibodies and receptors have been increasingly applied in analysis. With the introduction of immobilization techniques, the basis was set for a new generation of analytical instruments. By using immobilized biochemical species in close proximity to a transducer, sensitive and versatile analytical units have been developed such as enzyme electrodes (1), immuno electrodes (2), apoenzyme electrode (3)and microbe and enzyme thermistors (4,5). An alternative approach is to place the immobilized enzyme in a small reactor and the transducer later on in the flow system (6,7). This has been applied when adapting conventional autoanalyzers to the use of immobilized preparations (8).

Very few reports are available on the connection of immobilized lectins to analytical applications. The aim of the present review is to present some different areas where immobilized lectins have been, or can be, used.

Lectin Electrode.

In theory it would be possible to use lectins in an analogous fashion to that of antibodies when setting up a binding assay; in the case of lectins for binding assays of carbohydrates.

The immobilized lectin was either in the form of a preparation bound to a nylon-net covering the sensitive surface of a

Lectins - Biology, Biochemistry, Clinical Biochemistry, Vol. I

polarographic oxygen electrode (9), or as a small precolumn (15-30 μL) placed in close proximity to the electrode (10). The competitive binding situation is set up by adding a fixed amount of labelled carbohydrate to the sample to be analyzed. The labelled carbohydrate used should preferentially be one of the naturally occuring glycoconjugates present in the form of catalytically active glycoproteins (i.e. enzymes). The mixture of the carbohydrate to be analyzed and the glycoenzyme is introduced as a pulse in the continuous flow. On passage of the immobilized lectin preparation competition for the binding sites will take place. After washing out all nonspecifically bound enzyme, the amount of bound enzyme is quantified by the addition of a pulse of substrate.

The procedure used involves non-equilibrium conditions, but at the same time isokinetic conditions, in the sense that the time of contact between the immobilized lectin and the molecular entities in the sample pulse is very reproducible, as are the pH, temperature etc.. Due to the isokinetic conditions very reproducible measurements can be performed (11). Fig. 1 shows calibration curves obtained for two different carbohydrates, methyl-α-D-mannopyranoside and methyl-α-D-glucopyranoside,competing with the glycoprotein horse radish peroxidase for the binding positions on a microcolumn of immobilized Concanavalin A.

The competitive binding assay as depicted here can be used to quantify different carbohydrates in pure solutions. In crude samples, however, it is not possible to differentiate between low and high molecular weight carbohydrates. To achieve such distinction a dialysis step has to be introduced. From the figure is seen that, at identical concentrations, the carbohydrates compete with horse radish peroxidase with varying success. On testing a broader selection of carbohydrates for their ability to compete with peroxidase in the binding assay, a relation such as that shown in fig. 2 was obtained. It can

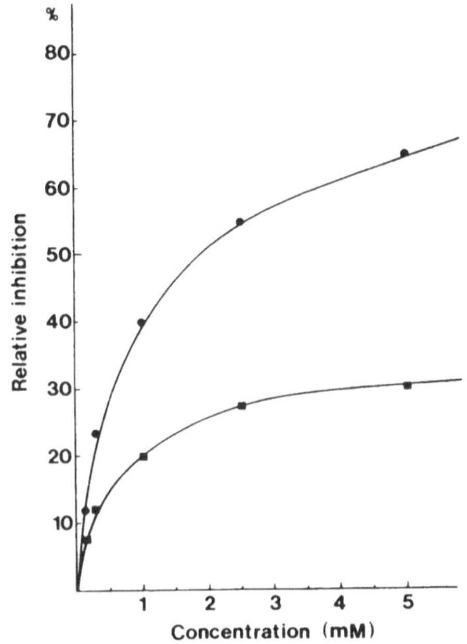

Fig. 1 Calibration curves for methyl-α-D-mannopyranoside (●) and methyl-α-D-glucopyranoside (■). The relative inhibition is given as a function of the concentration. (from ref. 12).

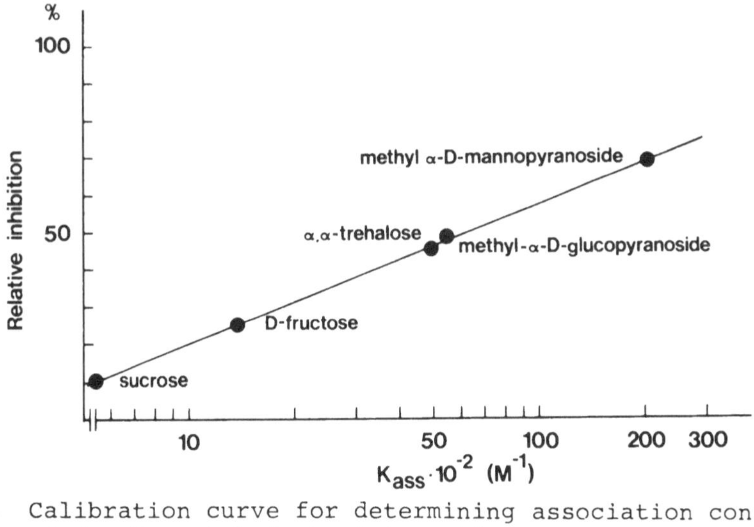

Fig. 2 Calibration curve for determining association constants of Con A-carbohydrate interaction.

thus be noticed that a simple analysis can be used to deduce the relative K_{ass} for the ligands interacting with the immobilized species.

Reversible Immobilization.

The procedure described for the lectin electrode involves reversible immobilization of the glycoenzyme to the immobilized lectin during a phase of the operating cycle. This feature can be utilized in another way in analysis. As stated elsewhere (13), reversible immobilization gives certain advantages and introduces a higher degree of flexibility into the system.See table 1.

Table 1.
Potential analytical applications of reversible immobilized biochemical structures.

Labile enzymes may be used.
Strong inhibitors may be quantified.
Cells and organells may be immobilized.

A common feature with all analytical systems based on immobilized enzymes is that they are designed to have a high operational stability, i.e. they should give the same response on repeated analyses over long periods of time. This stability is obtained at the expense of large quantities of enzymes - an excess of enzyme is added so that only a small portion is catalytically active - the other being inactive because of deficiency of substrate. Upon denaturation, earlier redundant enzyme molecules come into action. This stability-strategy can be avoided when reversible immobilization is applied since in this case a fixed small amount of enzyme is added each day with no excess being introduced.

The enzyme ascorbic acid oxidase (E.C. 1.10.3.3) was applied in an assay of ascorbic acid (14). Since the enzyme, is fairly expensive and also shows a rather low operational stability, it was an obvious candidate to be tried in an analytical system based on reversible immobilization. Small amounts (4.5 U) were introduced to the system each morning and analyses were then performed during the day. As the first step the following morning the old enzyme was washed off the column by a pulse of 0.2 M glycine-HCl pH 2.2 and after reconditioning in the appropriate buffer new enzyme was introduced.

Other cases where reversible immobilization is useful involve procedures where irreversible inhibitors are analyzed. Fig. 3 shows a calibration curve obtained for parathion using acetylcholine esterase reversibly bound to Concanavalin A (15).

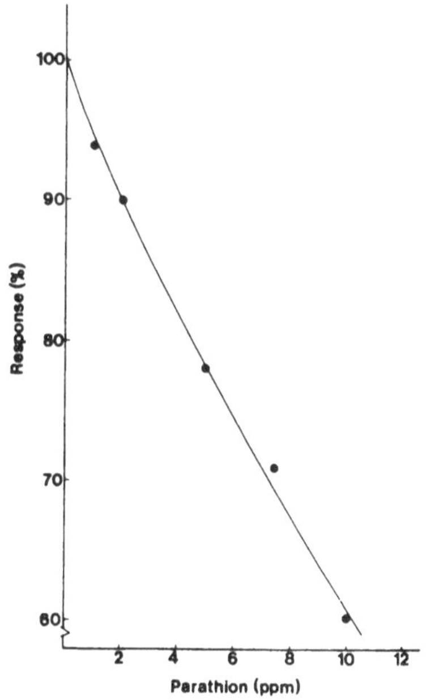

Fig. 3 Response of the enzyme thermistor containing 1 U of immobilized acetylcholine esteras to 1 min. pulses of 10 mM butyrylcholine after exposure of the bed to a 10 min. pulse of parathion containing buffer. (The response obtained when parathion conc. was zero was set at 100 %). (from ref. 15).

The unit must be reconditioned by addition of fresh enzyme after each assay.

Besides replacing old enzyme by fresh, it is also possible to change enzymes with a minimum of effort. Thus, as is schematically shown in fig. 4 it takes only 10 - 15 minutes to replace the enzyme in an analytical unit by another and to recondition the whole unit.

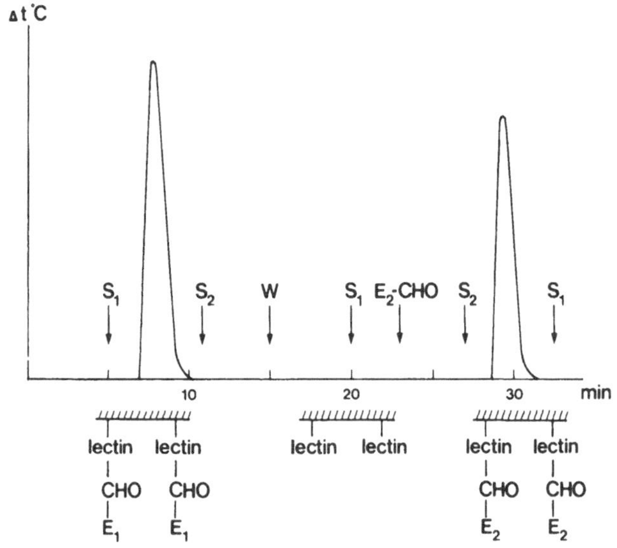

Fig. 4 Schematic presentation of an assay cycle using an enzyme thermistor. The arrows indicate changes in the perfusion medium. The cyle starts with glycoprotein (E_1-CHO) bound to the lectin-containing support material. At the arrows marked S_1 and S_2, substrate is introduced for enzymes E_1 and E_2 respectively. The heat signals obtained upon substrate pulses are represented by peaks. At the arrow W a pulse of0.2 M glycine-HCl pH 2.2, is introduced in order to split the complex and to wash the system. A new enzyme, E_2-CHO, is then introduced and substrate S_2 can be assayed.

Lectin-carbohydrate interactions may also be used for more complex biochemical structures. In a model study when glucose oxidase was bound to Con A and the enzyme was used in thermal analysis of glucose using an enzyme thermistor it was found that due to oxygen deficiency a very small linear range was obtained. If, however, red blood cells were added together with glucose oxidase when loading the thermistor column the situation was markedly changed (16). The concentration range where a linear response was obtained was enlarged and the amplitude of the signal was higher. These results were interpreted in terms of coimmobilization of the enzyme and an oxygen reservoir - the red blood cell. When, due to enzymic activity, the oxygen tension in the enzyme column decreased, oxyhemoglobin delivered oxygen so that the enzymic conversion could continue. After the pulse of substrate, the oxygen pressure of the perfusion buffer was normalized and the red blood cells were recharged with oxygen.

As the knowledge about lectins is rapidly increasing and the awareness of the existence of a great variety of lectins from different part of the biosphere is growing, one can predict that in the future a broad spectrum of lectins with many different specificities may be available for binding assays.

References

1. Guilbault, G.G., Methods Enzymol. 44,579-633 (1976).
2. Mattiasson, B., Nilsson, H., FEBS Lett. 78, 251-254 (1977).
3. Mattiasson, B., Nilsson, H., Olsson, B., J. Appl. Biochem. 1, 377-384 (1979).
4. Mattiasson, B., Larsson, P.-O., Mosbach, K., Nature 268, 519-520 (1977).
5. Danielsson, B., Mattiasson, B., Mosbach, K., Pure & Appl. Chem. 51, 1443-1457 (1979).
6. Borrebaeck, C., Mattiasson, B., Svensson, K., in "Enzyme Labelled Immunoassay of Hormons and Drugs" pp 15-27, (S.B. Pal ed.) Walter de Gruyter & Co., Berlin, 1978.
7. Ögren, L., Johansson, G., Anal. Chim. Acta 96, 1-11 (1978).

8. Hornby, W.E., Campbell, J., Inman, D.J., Morris, D.L., in "Enzyme Engineering" (E.K. Pye, L.B. Wingard,Jr.,eds) vol 2, pp 401-407, Plenum Press, New York, 1974.

9. Mattiasson, B., Nilsson, H., unpublished.

10. Borrebaeck, C., Mattiasson, B., in "Proc. from XXVII annual Colloquium Protides of the Biological Fluids", Brussels, April 30 - May 3 1979, (H. Peeters, ed.) Pergamon Press, Oxford, pp 607-610,(1979).

11. Mattiasson, B., Borrebaeck, C., in "Enzyme Labelled Immunoassay of Hormons and Drugs" pp 91-105, (S.B. Pal ed.) Walter de Gruyter & Co., Berlin, 1978.

12. Borrebaeck, C., Mattiasson, B., Anal. Biochem. in press.

13. Mattiasson, B., J. Appl. Biochem., in press.

14. Mattiasson, B., Danielsson, B., submitted.

15. Mattiasson, B., Rieke, E., Munneke, D., Mosbach, K., J. Solid-Phase Biochem. 4, 263-270 (1979).

16. Mattiasson, B., Borrebaeck, C., FEBS Lett. 85, 119-123 (1978).

LECTINS AS PROBES FOR MEMBRANE ASYMMETRY AND COMPARTMENTALIZATION OF SACCHARIDE MOIETIES IN CELLS

I. Virtanen, V.-P. Lehto and [1]P. Aula
Department of Pathology, University of Helsinki, Haartmanink. 3, and [1]Laboratory of Prenatal Genetics, University Central Hospital of Helsinki, SF-00290 Helsinki 29, Finland.

Introduction

Much interest has recently been devoted to the biosynthesis, metabolism and localization of membrane-bound glycoconjugates, glycoproteins and glycolipids, in cells (1-3). Lectins, plant or animal proteins that bind to specific sequence(s) in oligosaccharides have been widely used both as structural probes for oligosaccharides and in studies on the number, display, turnover and mobility of membrane glycosyl moieties (2,4,5). Here we will briefly summarize some recent results on the use of lectins in studies of cellular glycoconjugates.

A. Lectins as probes for membrane asymmetry

The binding of lectins to cells became evident already in classical agglutination experiments (cf. 2,4). Ferritin-coupled lectins as ultrastructural markers were first used by Nicolson and Singer in erythrocyte ghosts (6). These studies demonstrated lectin binding only at the noncytoplasmic surfaces of cell surface membranes from a variety of cells, but not at all to the cytoplasmic surfaces (6-8).
The binding of lectins to the noncytoplasmic surfaces of cells also revealed two of the main structural properties of

Lectins - Biology, Biochemistry, Clinical Biochemistry, Vol. I

cellular membranes: structural asymmetry of membranes in respect to their saccharide moieties and mobility of cell membrane components (cf. 5). Binding of lectins to cell surfaces was followed usually by a reorganization of the binding sites either into patches or caps (e.g. 1,5).

Later, however, several reports suggested lectin-binding sites also at the cytoplasmic surfaces of cellular endomembranes, e.g. nuclear and mitochondrial membranes (cf. for references 9-11). These results were usually made using lectin-agglutination methods or binding of radioactively labelled lectins. Using isolated rat liver cell endomembranes, plasma membranes, nuclear and endoplasmic reticulum membranes, we showed, on the other hand, that ferritin-coupled lectins bound only to the noncytoplasmic or cisternal surfaces of the endomembranes (9-11, Figs. 1,2). These studies were later confirmed by other groups (12,13).

Thus, it seems that lectins coupled to an ultrastructural marker are useful probes for cell membrane asymmetry. Appearance of specific membrane saccharide moieties on cell

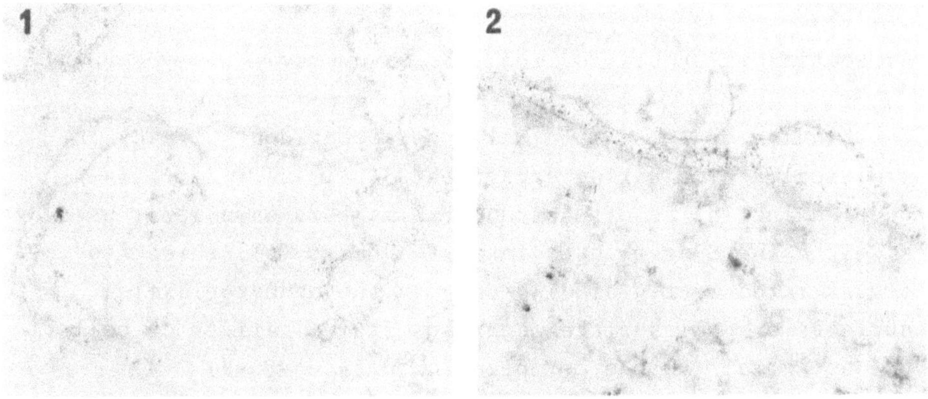

Figures 1,2. Binding of Concanavalin A-ferritin conjugate to rat liver cell bile canalicular membranes (1) and to nuclear membranes in isolated nuclei (2). Note the exclusive labelling of cytoplasmic surfaces of membrane sheets and of microvilli in Fig. 1 and the cisternal labelling in Fig. 2. Fig. 1, x 38000, Fig. 2, x 45000.

surfaces during embryonic development (14)or compartmentalization of saccharide moieties in different faces of structurally polarized cell membranes (11) have been recently elucidates using lectin techniques. In addition, several complementary methods have been developed to label cells and tissues for electron microscopy (cf. 15), e.g. peroxydase-gold-, mercury-, mannan-iron, hemocyanin- and radioactively-labelled lectins.

B. Lectins as probes for compartmentalization of saccharide moieties in cells

Both membrane and secretory glycoconjugates seem to share a common intracellular glycosylation and maturation pathway (3). The core-region saccharides of asparagine-linked oligosaccharides are probably linked to the polypeptide moiety already in the endoplasmic reticulum membranes whereas other, more terminal sugars and all o-glycosidically linked saccharides are added first in the Golgi apparatus (3). According to this concept, there should be a clear distinction between the saccharide moieties of cellular membranes, plasma membrane and Golgi apparatus membranes on one hand and the endoplasmic reticulum and nuclear membranes on the other hand, although several contrasting view have also been presented (e.g. 16).

Using fluorochrome-coupled lectins in cultured cells we could recently demonstrate that a strict compartmentalization apparently is present both in normal and malignant cells (17). The results showed that Concanavalin A and lens culinaris agglutinin, which bind to the mannose-rich cores of glycoproteins, stained a wide perinuclear region in all cells (Figs. 3,5), whereas other lectins, which bind to more terminal sugars in glycoproteins, stained a different juxtanuclear region in cells (Figs. 4,6). Both enzyme cytochemical

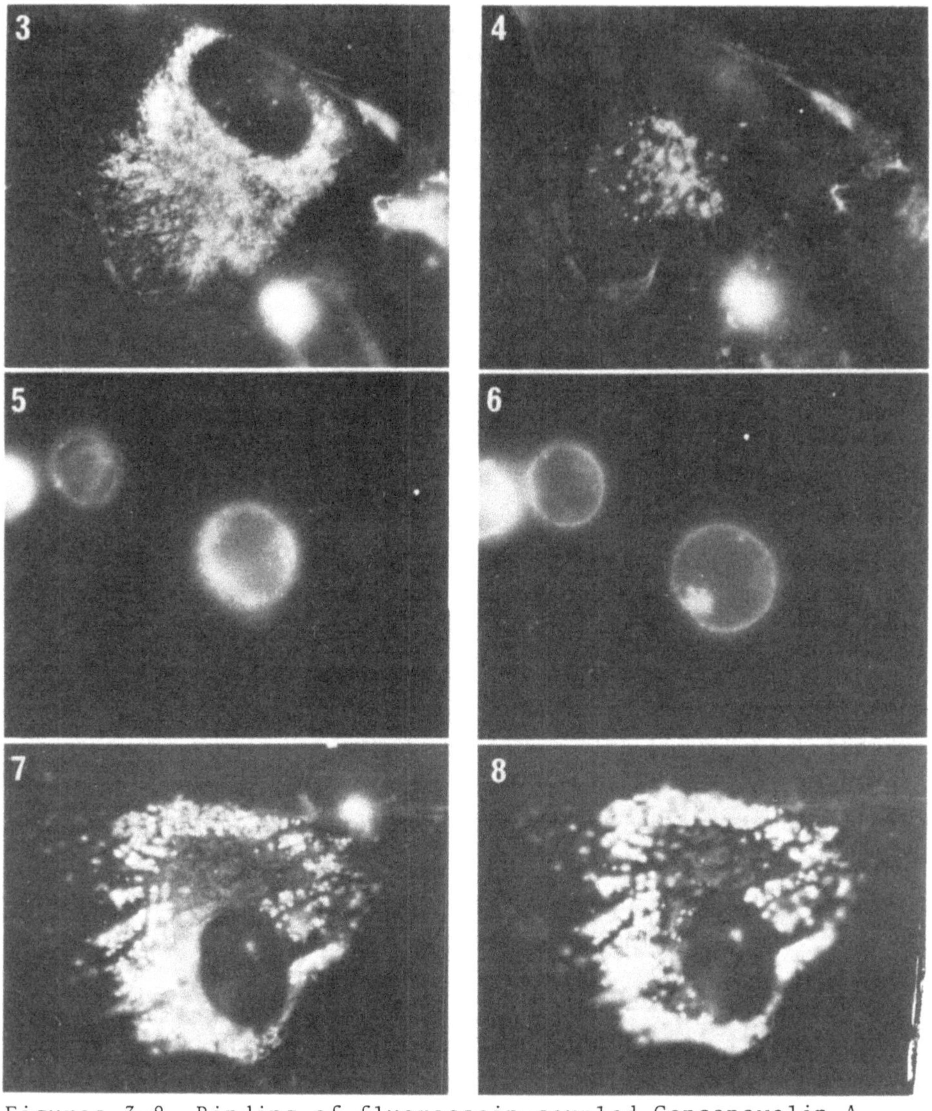

Figures 3-8. Binding of fluorescein-coupled Concanavalin A (3,5,7) and rhodamine-coupled wheat germ agglutinin (4,6,8) to paraformaldehyde-fixed and detergent-treated cultured huma fibroblasts (3,4), human monocytes (5,6) and I-cell disease fibroblasts (7,8). Note the wide perinuclear staining in 3,5, representing the region of endoplasmic reticulum and the different juxtanuclear staining in 4,8, representing the Golgi apparatus in these cells. In I-cell disease cells, a distinct labelling of the cytoplasmic storage granules is seen with both stainings.

methods (17) and inhibitors of cellular secretion, such as monensin (18), indicated that the latter staining represented the Golgi apparatus whereas the wide perinuclear staining apparently represented the region of endoplasmic reticulum in cells. These results gave further evidence for the strict compartmentalization of saccahride moieties in cellular membranes, the terminal saccharides of glycoconjugates being detectable only at the Golgi apparatus intracellularly. The early results of Hirano et al. (8) and recent findings of Yokoyama et al. (19) using peroxydase-coupled Ricinus agglutinin att the tissue level are in accordance with our results.

Intracellular lectin binding in cultured cells can apparently also be utilized for studing the defective cellular degradation of complex oligosaccharides in several lysosomal storage disorders. In these diseases accumulation of undegraded or partially catabolized oligosaccharides takes place in lysosomes both in vivo and in vitro. We have recently shown that in cells cultured from patients with a lysosomal storage disorder a characteristic lectin binding can be demonstrated: cells from I-cell disease (mucolipidosis II) bound all lectins used to typical storage granules, situated around the Golgi apparatus (Figs.7,8), whereas cells from patients with fucosidosis or mannosidosis oly bound fucose- or N-acetyl glucosamine specific lectins, respectively(20).

Conclusions

Recent results show that lectins can be utilized both at the light and electron microscopic level as probes for cellular saccharide moieties. These studies have revealed both the asymmetry of cellular membranes and the strict compartmentalization of saccharide moieties in endomembranes of cells. Lectin binding apparently provides wider possibilities for to study the dynamics of Golgi apparatus and metabolism of glycoconjugates both in normal and pathological conditions.

References

1. Sharon, N., Lis, H.: Methods in Membrane Biology 3, 147-200 (1974)
2. Gahmberg, C.G.: Cell Surface Reviews 3, 371-422 (1977).
3. Sturgess, J.M., Moscarello, M., Schachter, H.: Curr. Top. Membr. Transp. 11, 15-105 (1978).
4. Lotan, R., Nicolson, G.L.: Biochim. Biophys. Acta 559, 329-376 (1979).
5. Nicolson, G.L., Poste, G., Ji, T.H.: Cell Surface Reviews 3, 1-74 (1977).
6. Nicolson, G.L., Singer, S.J.: Proc. Nat. Acad. Sci. U.S.A. 68, 942-945 (1971).
7. Nicolson, G.L., Singer, S.J.: J. Cell Biol. 60, 236-248 (1974).
8. Hirano, H., Parkhouse, B., Nicolson, G.L., Lennox, E.S., Singer, S.J.: Proc. Nat. Acad. Sci. U.S.A. 69, 2945-2949 (1972).
9. Virtanen, I., Wartiovaara, J.: J. Cell Sci. 22, 335-344 (1976).
10. Virtanen, I.: Biochem. Biophys. Res. Commun. 78, 1411-1417 (1977).
11. Virtanen, I., Miettinen, A., Wartiovaara, J.: J. Cell Sci. 29, 287-296.
12. Feldherr, C.M., Richmond, P.A., Noonan, K.D.: Exp. Cell Res. 107, 439-444 (1977).
13. Rodriguez Boulan, E., Kreibich, G., Sabatini, D.D.: J. Cell Biol. 78, 874-893 (1978).
14. Maylie-Pfenninger, M.-F., Jamieson, J.D.: J. Cell Biol. 86, 93-103 (1980).
15. Nicolson, G.L.: Biological Electron Microscopy 2, 1-38 (1978).
16. Elhammer, A., Svensson, H., Autuori, F., Dallner, G.: J. Cell Biol. 67, 715-724 (1975).
17. Virtanen, I., Ekblom, P., Laurila, P.: J. Cell Biol. 85, 429-434 (1980).
18. Kääriäinen, L., Hashimoto, K., Saraste, J., Virtanen, I., Penttinen, K.: J. Cell Biol., in press (1980).
19. Yokoyama, M., Nishiyama, F., Kawai, N., Hirano, H.: Exp. Cell Res. 125, 47-53 (1980).
20. Virtanen, I., Ekblom, P., Laurila, P., Nordling, S., Raivio, K., Aula, P.: Pediatric Res., in press (1980).

STRUCTURAL BASIS FOR THE AFFINITY OF FOUR INSOLUBILIZED LECTINS, WITH A SPECIFICITY FOR α-D-MANNOSE, TOWARDS VARIOUS GLYCOPEPTIDES WITH THE N-GLYCOSYLAMINE LINKAGE AND RELATED OLIGOSACCHARIDES

H. Debray and J. Montreuil

Laboratoire de Chimie Biologique, Université des Sciences et Techniques de Lille I et Laboratoire Associé du CNRS n° 217, BP 36, 59655 Villeneuve d'Ascq Cédex and Institut de Recherches sur le Cancer de Lille, U. 124 INSERM (Pr. G. Biserte), BP 311, 59020 Lille Cédex, France

Introduction

Lectins are now widely used as tools to visualize the topography of saccharide moieties on cell surfaces (1), or to follow by lectin-affinity immunoelectrophoresis (2) variations in the glycosylation of plasma (3) or membrane glycoproteins (4). Insolubilized on various supports, lectins are also widely used to fractionate soluble or membrane glycoproteins of diverse origins (5) as well as glycopeptides obtained from soluble or membrane-bound glycoproteins (6).

Among the most used lectins are the so-called α-D-Mannose or Glucose binding lectins : Concanavalin A (Con A), *Lens culinaris* agglutinin (LCA), *Vicia faba* agglutinin (VFA) and *Pisum sativum* agglutinin (PSA) (7). However, it is now well known that it is not only a monosaccharide residue in a terminal non reducing position on a glycan that is recognized by a given lectin, but more complex saccharidic sequences (8).

In previous studies (9, 10), we used oligosaccharides or glycopeptides related to glycoproteins with the N-glycosylamine linkage (11) to precise the specificities of various lectins and

Lectins - Biology, Biochemistry, Clinical Biochemistry, Vol. I

in particular of the above α-D-Mannose binding lectins, by inhibition of agglutination of human red blood cells. These results were completed by studying the affinity of the insolubilized lectins towards these oligosaccharides or glycopeptides.

Materials and Methods

Lectins. All the lectins used in this study, except Concanavalin A which was purchased from Pharmindustrie, Clichy, France, have been purified by affinity chromatography.

Origin of glycopeptides and oligosaccharides. Oligosaccharides (N-9, N-5, N-6, S-6a) shown in fig. 1 were isolated by Dr. G. Strecker from the urines of patients with various lysosomal diseases (12). Glycopeptides isolated from human serotransferrin (GP-h-STF), from human lactotransferrin (GP-h-LTF$_1$) and from hen ovotransferrin (GP-ovoTF) were a gift from Dr. G. Spik. The glycopeptide of the oligomannosidic type, isolated from bovine lactotransferrin (GP-b-LTF) was a gift from Dr. A. Chéron.

All these structures shown in fig. 1 are related to glycoproteins with the N-glycosidic linkage and belong to 2 types of glycan structures shown in fig. 2 : the N-acetyllactosaminic and the oligomannosidic type (11).

Hemagglutination tests. Agglutination of human red blood cells group O^+ and hemagglutination inhibition by various oligosaccharides or glycopeptides were carried out according to Matsumoto and Osawa (13).

Results were expressed as the minimum concentration of sugar (mM) necessary to completely inhibit agglutination.

Insolubilization of lectins on Sepharose 4B. Lectins were coupled to Sepharose 4B activated according to the procedure of March *et al.* (14), at a concentration of 2 mg of lectin per ml of gel.

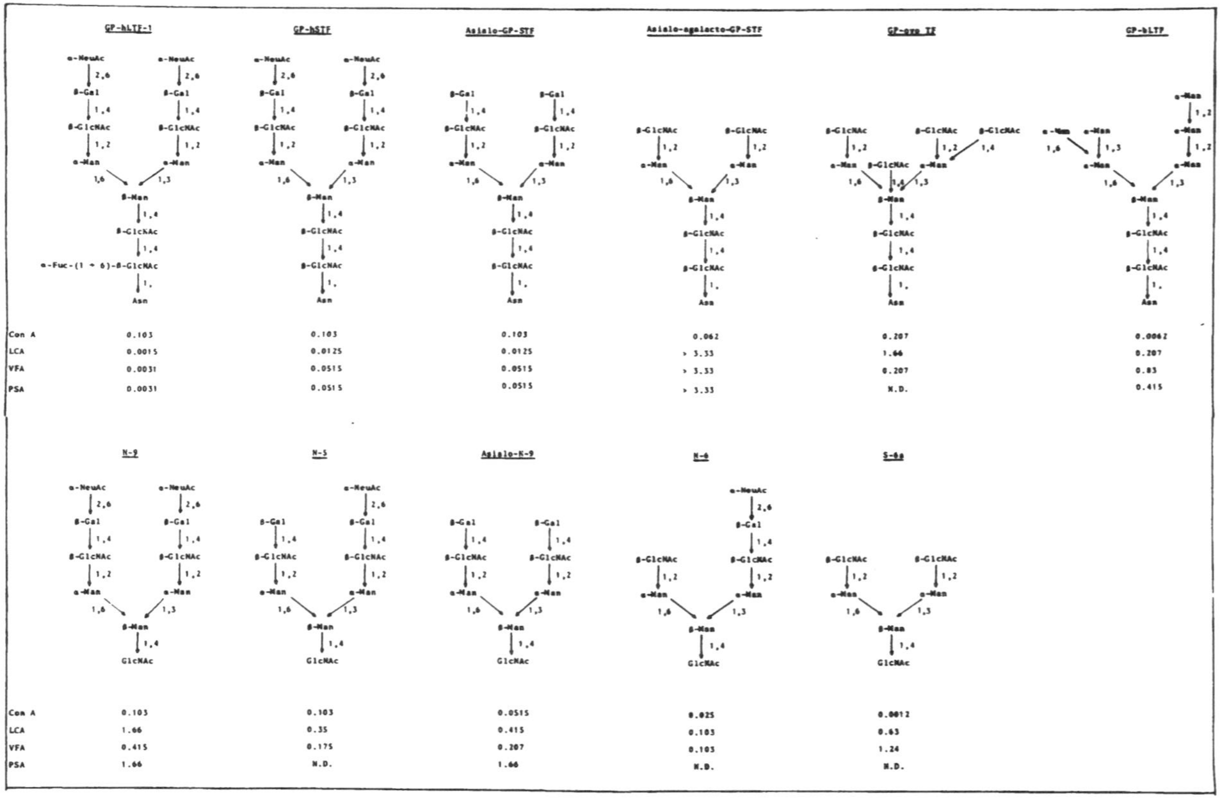

Fig. 1.- Saccharidic determinants recognized by Con A, LCA, VFA and PSA on glycans from glycoproteins with the N-glycosylamine linkage.

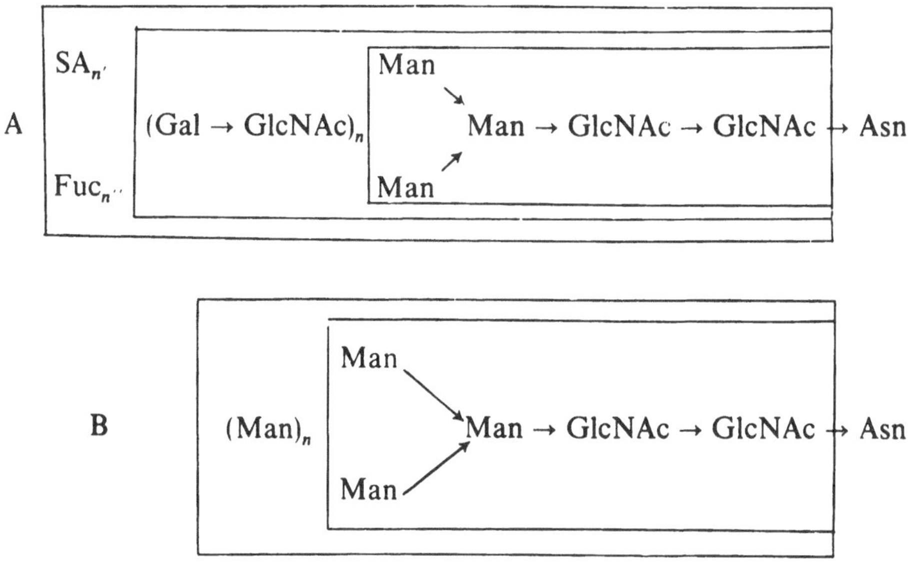

Fig. 2. General structural scheme of glycans from glycoproteins with the N-glycosylamine linkage. A) N-acetyllactosaminic type. B) Oligomannosidic type (according to Montreuil) (11).

Affinity chromatography of oligosaccharides or glycopeptides on lectin-Sepharose column. Glycopeptides, N-^{14}C acetylated according to Koide *et al.* (15), or related oligosaccharides, reduced on the terminal N-acetylglucosamine residue with tritiated sodium borohydride (16) were applied, after dissolution in phosphate buffered saline pH 7.4 (LCA, VFA, PSA) or in 0.05 M sodium acetate buffer pH 5.2 containing 0.1 M NaCl and $CaCl_2$, $MgCl_2$ (1 mM each) in the case of insolubilized Con A, to the column of lectin-Sepharose (10 x 1 cm) equilibrated with the same buffer. Elution was carried out first with buffer and then with buffer containing 0.15 M methyl-α-D-glucoside.

Fractions of 2 ml were collected and samples of 100 µl analysed for radioactivity.

Results and Discussion

Hemagglutination Inhibition Results

These results are summarized in fig. 1. The best inhibitors found for Con A are asialo-agalacto GP-h-STF and the related oligosaccharide S-6a and Con A presents a great affinity for the trimannosidic core substituted by two N-acetylglucosaminyl residues. It is noteworthy that this affinity, as measured by inhibition of agglutination, is not decreased by the addition of a β-1,4-linked N-acetylglucosamine on the β-linked mannose residue.

On the other hand, Con A also presents a great affinity for GP-b-LTF which possesses the complete trimannosidic core with α-1,3 and α-1,6-Mannose substituted by four additional α-linked Mannose residues (oligomannosidic structure). Substitution of core α-1,3-Mannose at position C-4, as in GP-ovoTF, decreases the affinity of Con A.

However, the best inhibitor found for the three others lectins LCA, VFA and PSA is glycopeptide $GP\text{-}h\text{-}LTF_1$ and α-1,6-linked Fucose seems to be a major determinant of the binding, for its removal with an α-L-Fucosidase from rat kydney gives glycopeptides 10-fold less inhibitory (GP-h-STF). However, if presence of an α-1,6-linked Fucose near the N-glycosidic linkage seems to be a necessary condition, its presence is not a sufficient one and glycopeptides possessing this fucose residue, but where terminal neuraminyllactosamine sequences are lacking are very poor inhibitors (10). This suggests the existence of an extended carbohydrate binding site for LCA, VFA and PSA and the methyl group of L-Fucose near the N-glycosidic linkage could interact with an hydrophobic area in or near this carbohydrate binding site. This hydrophobic area had been previously suggested for LCA, VFA and PSA by Allen *et al.*(17). Such an hydrophobic region, close to the carbohydrate-binding site had also been demonstrated for Con A (18).

Affinity Chromatography of Glycopeptides with the N-Glycosylamine Linkage and Related Oligosaccharides on Insolubilized Lectins

The hemagglutination inhibition values given in fig. 1 can be compared to the true association constants between various oligosaccharides or glycopeptides and a given lectin and from these inhibition values, the affinity of a saccharide for the insolubilized lectin could be predicted.

In fact, for insolubilized Con A, results obtained by this second method, confirms previous data (19, 23). The four elution profiles which can be obtained with insolubilized Con A are shown in fig. 3 :

- The non-retained profile (FNR) is obtained with tetra- or triantennary glycopeptides of the N-acetyllactosaminic type or with biantennary structures possessing an additional β-1,4-linked N-acetylglucosamine on the β-linked mannose residue. One can note that this last point is in contrast with the hemagglutination inhibition result where this β-1,4 substituted glycan possesses the same inhibitory power than the unsubstituted one (S-6a). This discrepancy is not explained.

- The retarded profile (FR) is obtained by elution with the starting buffer and reflects a weak affinity between Con A and the saccharide. This profile is obtained with biantennary structures as GP-h-STF (21, 22). However, with a higher concentration of Con A in the gel, these structures can be weakly retained and then eluted specifically with low concentration of α-methyl-D-glucoside or mannoside (20).

- The sharp elution profile (FE) with 0.15 M α-methyl-D-glucoside is obtained with oligomannosidic type glycans as GP-h-LTF (21, 22). However, with a higher concentration of Con A in the gel, a broad peak can be obtained (20) and some glycans cannot be eluted with α-methyl-D-mannoside but only with detergents (19).

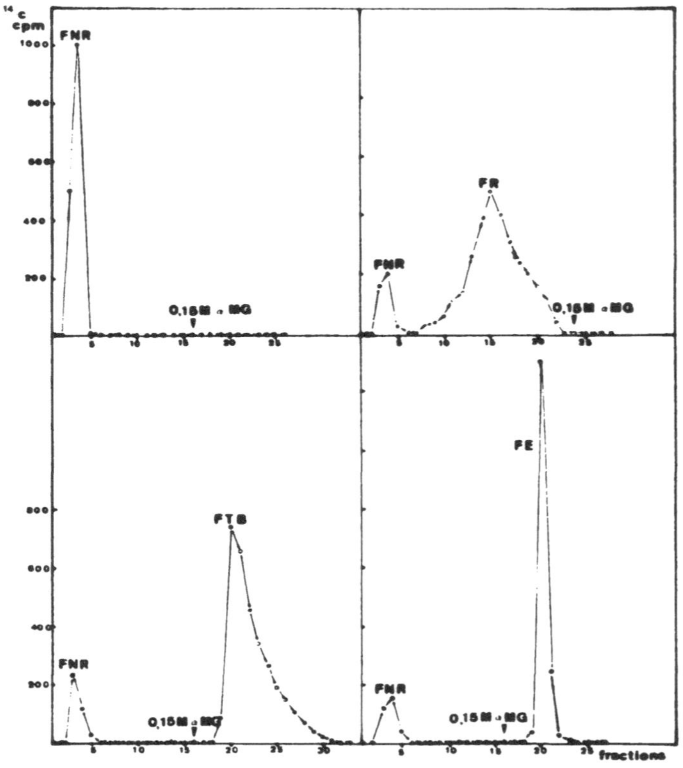

Fig. 3.- Con A-Sepharose elution profiles.

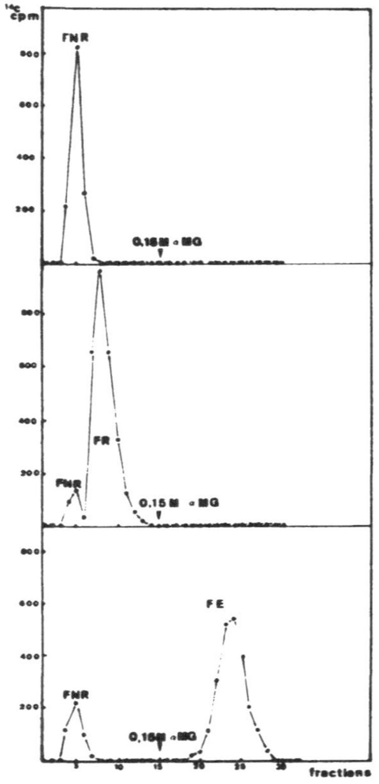

Fig. 4.- LCA, VFA or PSA-Sepharose elution profiles.

- The "tightly bound" profile (FTB) is obtained with biantennary structures such as S-6a or asialo-agalacto-GP-h-STF, where at least one N-acetylglucosamine β(1 → 2) α-Mannose sequence is very accessible to the lectin. Same results were described by Narasimhan *et al.* (22) and Baenziger and Fiete (23) and this broad, trailing profile expresses a very great affinity between the saccharide and the lectin.

With insolubilized LCA, VFA or PSA, three elution profiles shown in fig. 4 can be obtained :

- The non-retained profile (FNR) is obtained with tetratri or biantennary glycopeptides lacking the α-1,6-linked Fucose on the N-acetylglucosamine involved in the N-glycosylamine bond.

- The retarded profile (FR) is obtained for insolubilized VFA or PSA, with biantennary glycopeptides (GP-h-LTF) possessing this α-1,6-linked Fucose.

- This glycopeptide (GP-h-LTF) is tightly bound (FE) on insolubilized LCA and is only displaced with 0.15 M α-methyl-D-glucoside.

However, if the presence of this α-1,6 linked Fucose seems to be a necessary condition for the recognition by LCA or PSA and VFA, its presence is not a sufficient one, for the glycan, liberated from glycopeptide GP-h-LTF by hydrazinolysis and which still possesses this α-1,6-linked Fucose, is no longer retarded on VFA and PSA or retained on insolubilized LCA.

These interactions between glycopeptides such as GP-h-LTF and LCA can also involved charge effects and (or) some hydrophobic interactions between an hydrophobic area in or near the carbohydrate binding site of the lectin and aromatic residues of the peptidic part of the glycopeptides or glycoproteins.

Conclusion

Our results show that four lectins (Con A, LCA, VFA and PSA), considered to be identical in terms of α-D-Mannose or Glucose specificity, possess the ability to recognize fine differences in more complex carbohydrate structures.

While insolubilized Con A represents a very useful tool for the fractionation of both N-acetyllactosaminic and oligomannosidic type glycopeptides or related oligosaccharides, the three others lectins will also be very useful for the separation of N-acetyllactosaminic type glycopeptides on the condition that an α-L-Fucose residue substitutes in C-6 the N-acetylglucosamine residue involved in the N-glycosylamine bond.

Acknowledgements

This work was supported in part by the Centre National de la Recherche Scientifique (L.A. n° 217), by the Institut National de la Santé et de la Recherche Médicale (U. 124) and by the Délégation Générale à la Recherche Scientifique et Technique (Grant n° 79.7.0669).

References

1. Virtanen, I., Letho, V.P. : These proceedings.
2. Bøg-Hansen, T.C., Bjerrum, O.J., Ramlau, J. : Scand. J. Immunol. 4, suppl. 2, 141-147 (1975).
3. Kerckaert, J.P., Bayard, B. : These proceedings.
4. Bjerrum, O.J., Bøg-Hansen, T.C. : Biochim. Biophys. Acta 455, 66-89 (1976).
5. Lotan, R., Nicolson, G.L. : Biochim. Biophys. Acta 559, 329-376 (1979).
6. Finne, J., Krusius, T., Järnefelt, J., 27th International Congress of Pure and Applied Chemistry (ed. A. Varmavuori) Pergamon Press, Oxford-New York, pp. 147-159, 1980.

7. Goldstein, I.J., Hayes, C.E. : Adv. Carbohyd. Chem. Biochem. 35, 127-340 (1978).

8. Kornfeld, R., Ferris, C. : J. Biol. Chem. 250, 2614-2619 (1975).

9. Debray, H., Decout, D., Strecker, G., Montreuil, J., Monsigny, M. : Protides of Biological Fluids, 27th Colloquium, Brussels (ed. H. Peeters), pp. 451-454, Pergamon Press, Oxford, 1979.

10. Debray, H., Decout, D., Strecker, G., Spik, G., Montreuil, J. : Eur. J. Biochem. (submitted for publication).

11. Montreuil, J. : Adv. Carbohyd. Chem. Biochem. 37, 157-223 (1980).

12. Strecker, G., Montreuil, J. : Biochimie 61, 1199-1246 (1979).

13. Matsumoto, I., Osawa, T. : Arch. Biochem. Biophys. 140, 484-491 (1970).

14. March, S.C., Parikh, I., Cuatrecasas, P. : Anal. Biochem. 60, 149-152 (1974).

15. Koide, N., Muramatsu, T. : J. Biol. Chem. 249, 4897-4904 (1974).

16. Takasaki, S., Kobata, A. : J. Biochem.(Tokyo) 76, 783-789 (1974).

17. Allen, A.K., Desai, N.N., Neuberger, A. : Biochem. J. 155 127-135 (1976).

18. Loontiens, F.G., Van Wauwe, J.P., De Gussem, R., De Bruyne, C.K. : Carbohyd. Res. 30, 51-62 (1973).

19. Ogata, S., Muramatsu, T., Kobata, A. : J. Biochem. 78, 687-696 (1975).

20. Krusius, T., Finne, J., Rauvala, H. : FEBS-Lett. 71, 117-120 (1976).

21. Debray, H., Montreuil, J. : Biochimie 60, 697-704 (1978).

22. Narasimhan, S., Wilson, J.R., Martin, E., Schachter, H. : Can. J. Biochem. 57, 83-96 (1979).

23. Baenziger, J.U., Fiete, D. : J. Biol. Chem. 254, 2400-2407 (1979).

INTERACTIONS OF PHA WITH HUMAN NORMAL SERUM PROTEINS

G.A. Spengler and R.-M. Weber
University of Bern, Institute for Clinical Protein Research,
Tiefenauspital, CH-3004 Bern, Switzerland

Introduction

Although as early as 1888, Stillmark had described formation of a precipitate upon addition of castor bean lectin (ricin) to serum (10), and a large number of plant lectins had since been studied for hemagglutinating and mitogenic activities, it was only in 1960 that Nakamura and collaborators (5) reported on precipitin reactions of various bean extracts with human and animal sera in double diffusion tests. Furthermore, they concluded from electrophoretic experiments that phytohemagglutinin (PHA) from red kidney beans (Phaseolus vulgaris) chiefly reacted with α_2-globulins. In Ouchterlony tests, Beckman (1) found PHA to produce precipitation lines with α_2- and γ-globulins. Morse (4) was the first to use a solution of commercial PHA in immunoelectrophoretic examination of human serum. Her list of PHA-precipitable proteins included IgM, ß-lipoprotein (LDL), α_2-macroglobulin, α_1-acid glycoprotein, α_1-lipoprotein (HDL), β_2-glycoprotein, and some of the IgA myeloma proteins tested. Using a similar immunoelectrophoretic technique, Osunkoya and Williams (6) noticed marked differences among various species in the formation of precipitation lines with α-, ß- and γ-globulins in human and animal sera, but did not further identify the classes of PHA-precipitable proteins. A first report on subclass dependent precipitation of PHA with monoclonal IgA was published in 1972 (7). In the

This work was supported by Grants No. 3.1130.73, 3.506-1.75 and 3.080-0.76 from the Swiss National Science Foundation.

present paper, methods will be described which for analytical and preparative purposes utilize the selectivity of PHA interactions with a restricted number of proteins in human normal serum. Furthermore, some remarks will be made on the question of inhibition by carbohydrates of the PHA precipitin reaction.

Materials and Methods

PHA in our experiments was Bacto-Phytohemagglutinin P (Difco Laboratories, Detroit, Michigan) which is a nonhomogeneous preparation from red kidney beans and contains, aside from inactive material, two different lectins with distinct in vitro activities: PHA-L (leukoagglutinin) which induces lymphocyte stimulation and agglutinates white blood cells, and PHA-H (hemagglutinin) which causes red cell and mixed agglutinations, and also contributes to the induction of blast formation in lymphocyte cultures (11,12). As was evident from absorption experiments the protein precipitating activity of commercial PHA has also to be attributed to PHA-H.

	Absorbed with	Hemagglutination	Precipitation with		Lymphocyte stimulation (S.I.)	
			IgM	IgA2		
Saline control	-	-	-	-	-	(1.0)
PHA	-	+++	+++	+++	+++	(63.6)
PHA	RCM	-	-	-	++	(29.4)
PHA	IgM	-	-	-	++	(34.3)
PHA	IgA2	-	-	-	++	(32.8)

Table 1. Effects of various absorption procedures on the in vitro activity of PHA. - Abbreviations: S.I. = stimulation index. IgM = monoclonal pentameric IgMλ. IgA2 = monoclonal dimeric IgA2κ. RCM = red cell membranes (washed ghost cells from hemolyzed erythrocytes of blood groups A and O).

Except for polyvalent antiserum (from horse) to human serum proteins, in tests with PHA whole antisera were omitted in order to avoid crossreactions with PHA-precipitable animal proteins. Instead, preparations of specific 7S antibodies (IgG from immunized rabbit or goat) were used. These preparations either were made from whole antisera by gel filtration through Sephadex G 200 (Pharmacia Fine Chemicals AB, Uppsala, Sweden) or equivalent commercial products were used.

Immunoelectrophoresis on microslides was performed in agar gel following standard procedures. Usually the application pattern consisted of two troughs (2 mm x 40 mm) and a central sample hole (ø 2 mm). In most tests, precipitation lines were developed with antibody in one trough and with PHA solution (20 mg/ml, in 0.15 M NaCl) in the other. After immunodiffusion (20 hours at 4°C) the slides were left for 24 hours in a saline bath (0.15 M NaCl) for elution of nonprecipitated proteins, before they were dried and stained. Occasionally a line produced by PHA was noticeable after immunodiffusion but readily dissolved in the saline bath. Therefore, only those lines which were insoluble in 0.15 M NaCl and appeared on the stained slide were regarded to represent positive PHA precipitin reactions.

PHA selection electrophoresis (8) differed from conventional immunoelectrophoresis only insofar as the tests were performed in PHA-containing agar gel (2 mg PHA/ml of gel). Microslides were coated with 2.5 ml Special Agar Noble (Difco), 20 mg/ml in 0.1 M sodium barbital/HCl buffer, pH 8.2. In a second step, PHA was incorporated by diffusion into the agar gel. The troughs were filled (0.12 ml each) with PHA solution (20 mg/ml) and the slides were stored 24 hours at 4°C for equilibration of PHA in the gel. Selection slides were then ready for use. During the elctrophoretic run, PHA-reactive proteins formed a precipitate in the central part of the slide and were selected from PHA-nonprecipitable proteins which migrated freely and could subsequently be identified as separate precipitation lines from immunodiffusion with specific antibodies.

PHA affinity chromatography. For affinity chromatographic separation of PHA-reactive from PHA-nonreactive proteins, PHA was insolubilized by coupling to Sepharose (35 mg of PHA per 1 g of pre-conditioned CNBr-activated Sepharose 4B) following the manufacturer's instructions (Pharmacia). The gel was then packed in a column (200 mm x 20 mm), and non-covalently bound PHA was removed by repeated perfusion cycles, until the eluate was protein-free as controlled spectrophotometrically, at 280 nm. Each cycle consisted of a wash at pH 4 (0.1 M acetate buffer containing 1 M NaCl) followed by a wash at pH 8 (0.1 M Tris buffer containing 1 M NaCl). Finally the gel was equilibrated with 0.15 M NaCl, and the column was stored at 4°C. The average adsorbent gel volume was 15 ml per column. For affinity chromatographic protein separation 0.2 ml of serum diluted tenfold with saline was applied to the column, and was incubated within the gel for 2-3 hours at 4°C. Unbound serum proteins were then removed with 0.15 M NaCl, followed by elution of the PHA-reactive proteins with 0.1 M glycine/HCl buffer, pH 2.6, containing 0.15 M NaCl. For restoring a neutral pH the gel was eluted with 0.1 M Tris/HCl buffer, pH 7.5, containing 0.15 M NaCl, and was finally re-equilibrated with 0.15 M NaCl. Flow rates were kept at 1 ml per minute. Eluates were collected in 5 ml fractions, and protein concentrations were recorded at 280 nm. Acidic elution fractions were neutralized with concentrated Tris buffer.

Inhibition and reversibility of protein-binding to PHA had to be established in developing the PHA affinity chromatographic method. Hemagglutination by PHA is known to be inhibitable or reversible by the addition of N-acetyl-D-galactosamine (2). On the other hand, a potent inhibitor of PHA was isolated from human red cell membranes and was shown to be a branched

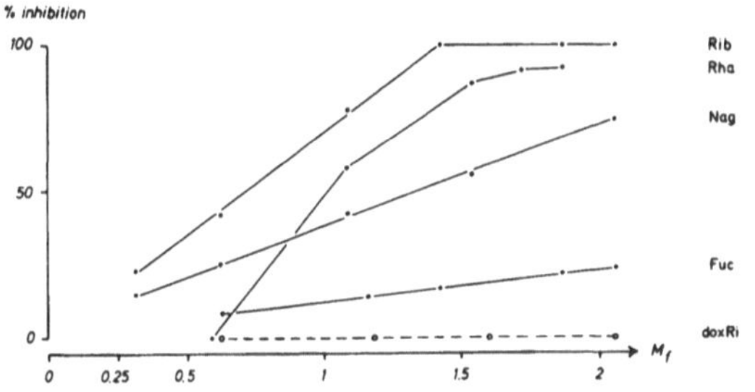

Figure 1. Carbohydrate inhibition of protein precipitation by PHA. Dose response curves from serial tests with increasing final molarities (M_f) of D-ribose (Rib), L-rhamnose (Rha), N-acetyl-D-galactosamine (Nag), D-fucose (Fuc), and 2-deoxy-D-ribose (doxRi). Reaction mixtures consisted of 0.5 ml PHA solution (20 mg/ml), 0.5 ml sugar solution, and 0.2 ml of a myeloma serum containing 1.2 mg monoclonal IgA2.

glycopeptide with the determinant sugar being ß-D-galactose and mannose in the core (3). In order to establish a sugar suitable for the elution of proteins bound to insolubilized PHA in affinity chromatography several carbohydrates were tested for their capacity to inhibit PHA precipitin reactions. At 0.1 M final concentrations no measurable inhibition was noticed with any of the sugars. At 0.833 M final concentrations, inhibition occurred with N-acetyl-D-galactosamine, D-ribose, L-rhamnose and, to a slight degree, also with D-fucose. D-glucose, 2-deoxy-D-glucose, α-methyl-D-glucopyranoside, N-acetyl-D-glucosamine, D-mannose, α-methyl-D-mannopyranoside, D-galactose, D-fructose, D-sorbose, 2-deoxy-D-ribose, L-arabinose, D-xylose, L-fucose, maltose, and α-lactose did not inhibit protein precipitation by PHA. In experiments with increasing concentrations of the inhibiting sugars complete inhibition of PHA precipitation could be achieved only with D-ribose at final molarities above 1.4 (see Fig. 1). High concentrations of L-rhamnose and N-acetyl-D-galactosamine markedly but incompletely inhibited the reaction. However, in equilibrium dialysis with tritiated D-ribose or N-acetyl-D-galactosamine no binding affinity of these sugars

Figure 2. Immunoelectrophoresis of human normal serum. Precipitation lines were developed with polyvalent antiserum (upper trough) and with PHA (20 mg/ml, in 0.15 M NaCl).

to PHA could be established. Therefore it was concluded that inhibition of protein precipitation by PHA in the presence of high concentrations of several sugars was a nonspecific phenomenon which was not applicable to PHA affinity chromatography. PHA-bound serum proteins were rather eluted with acid buffer because PHA-precipitates proved to readily and reversibly dissolve at pH 2.6.

Results

Immunoelectrophoresis. When PHA was used like an antiserum in immunoelectrophoretic tests with human normal serum, only few precipitation lines occurred which were usually faint and hazy (see Fig. 2). They corresponded to α_2-macroglobulin, LDL (apoprotein B) and IgM, as was confirmed in tests with purified preparations of these proteins. When α_1-acid glycoprotein was markedly increased (over 2 mg/ml) it was also visualized by PHA. No additional PHA precipitation lines were found with over 300 sera from normal donors and from patients with high levels of acute-phase proteins or polyclonal immunoglobulins or both.

PHA selection electrophoresis rendered little information in tests with human normal sera, and was therefore rarely used. The application of this method in examination of pathological sera will be discussed elsewhere (9).

Serum protein	Concentration before	Concentration after	Percentage of protein bound to insolubilized PHA
	passage through PHA-adsorbent column		
Prealbumin	29	29	
α_1-acid glycoprotein	64	44	31 %
α_1-antitrypsin	297	297	
Ceruloplasmin	28	28	
α_2HS-glycoprotein	69	69	
Haptoglobin	176	174	
α_2-macroglobulin	230	114	50 %
LDL (apoprotein B)	380	150	60 %
Hemopexin	86	86	
Transferrin	258	258	
C'3 (β_1A-globulin)	87	84	
C'4 (β_1E-globulin)	40	38	
β_2-glycoprotein I	25	25	
IgA	588	512	13 %
IgM	162	-	100 %
IgE (I.U./ml)	154	68	56 %
IgG	1078	1072	

Table 2. Selective binding of human serum proteins to insolubilized PHA in affinity chromatography. Except for IgE, concentrations of proteins (mg/100 ml) were determined by radial immunodiffusion.

PHA-affinity chromatography. In order to assess the total number of PHA-reactive protein classes, human normal serum was subjected to PHA-affinity chromatography, and the amounts of proteins bound to the insolubilized PHA were estimated (see Table 2). Only IgM which in normal serum is present in the regular pentameric form was entirely bound by PHA. The other PHA-reactive protein classes were composed at varying proportions of molecules which were bound by PHA, and molecules which were not. It was felt that this difference in avidity to PHA might reflect microheterogeneity among molecules of PHA-reactive protein classes, at least as far as their carbohydrate moieties are concerned. The same seems to be true for IgD because, although this immunoglobulin was not included in the affinity chromatographic experiment, coexistence of PHA-precipitable and PHA-nonreactive monoclonal IgD in myeloma sera was established by PHA selection electrophoresis. On the other hand, no α_1-lipoprotein was detectable among PHA-reactive serum proteins recovered from the PHA adsorbent column by acidic elution, in immunoelectrophoretic examination with specific antiserum to HDL (apoprotein A). The only unexpected finding in this experiment was the assessment of a minor proportion of normal serum IgA to be PHA-reactive because PHA had never been found to precipitate with IgA in immunoelectrophoresis with numerous sera containing high levels of polyclonal IgA. When PHA-reactive IgA isolated by affinity chromatography was examined on PHA selection electrophoresis, only a small amount of the IgA was precipitated. It was concluded that most PHA-reactive IgA molecules of polyclonal origin form only soluble complexes with PHA, and are thus bound to insolubilized PHA in affinity chromatography, but are not detectable by the formation of an immunoelectrophoretic precipitation line when exposed to the lectin. This conclusion was substantiated by findings in IgA myeloma (9).

Discussion and Conclusions

In vitro activities of a lectin are based on its carbohydrate affinity the specificity of which can usually be assessed by the strongly inhibiting effect of one selected simple sugar. However, this was not the case in tests with PHA although this lectin has been shown to cause hemagglutination through specific binding to a branched glycopeptide on human erythrocyte membranes (3). Neither D-galactose being the determining sugar on the outer branch nor mannose in the core of this receptor were found to be capable of inhibiting hemagglutination or precipitation by PHA whereas inhibition by high concentrations of D-ribose, L-rhamnose, or N-acetyl-D-galactosamine proved to be nonspecific and to not reflect selective affinity of PHA towards these sugars. In contrast to the structure of the red cell receptor the conformation of PHA-receptors on serum proteins has not yet been analyzed. The results of the PHA absorption tests indicate, however, that precipitin reactions with serum proteins are due to PHA-H rather than to PHA-L. It may further be concluded that the same selective binding affinity of PHA is involved in hemagglutination and in protein precipitation. Therefore it seems reasonable to assume structural similarity of PHA-receptors on red cell membranes and on PHA-precipitable serum proteins. Furthermore it is likely that different serum proteins carry PHA-receptors of the same type because absorption of PHA with one protein also removes its reactivity towards other classes of serum proteins. The fact that PHA precipitates only with a restricted number of serum proteins reflects the requirement for specific receptors to which PHA can bind. Microheterogeneity involving the carbohydrate structure of serum glycoproteins may be responsible for the observation that within a PHA-reactive class of serum proteins, PHA precipitates with part of the molecules and does not react with the others.

The mechanism of precipitin reactions between lectins and serum proteins closely resembles a specific antigen antibody reaction (8). On the basis of the above observations, methods have been developed which utilize the precipitating activity of PHA for the visualization of PHA-reactive proteins in immunoelectrophoresis, for the discrimination in PHA selection electrophoresis of PHA-precipitable from nonprecipitable molecules of the same serum protein class, and for separation by PHA-affinity chromatography of PHA-nonreactive from PHA-reactive glycoproteins, including those proteins which are not precipitable but form soluble complexes with PHA.

References

1. Beckman, L.: Nature 195, 582-583 (1962).
2. Borberg, H., Woodruff, J., Hirschhorn, R., Gesner, N., Miescher, P., Silber, R.: Science 154, 1019-1020 (1966).
3. Kornfeld, R., Kornfeld, S.: J. Biol. Chem. 245, 2536-2545 (1970).
4. Morse, J.H.: Immunology 14, 713-724 (1968).
5. Nakamura, S., Tanaka, K., Murakawa, S.: Nature 188, 144-145 (1960).
6. Osunkoya, B.O., Williams, A.I.O.: Clin. exp. Immunol. 8, 205-212 (1971).
7. Spengler, G.A., Weber, R.-M., Skvaril, F.: Schweiz. med. Wschr. 102, 1618-1619 (1972).
8. Spengler, G.A., Weber, R.-M.: J. Immunol. Methods 32, 71-84 (1980).
9. Spengler, G.A., Weber, R.-M.: In preparation
10. Stillmark, H.: Inaug. Diss., Dorpat 1888.
11. Weber, T., Nordman, C.T., Gräsbeck, R.: Scand. J. Haemat. 4, 77-80 (1967).
12. Yachnin, S., Svenson, R.H.: Immunology 22, 871-883 (1972).

PARAMETERS OF ELECTROPHORETIC DETERMINATION OF DISSOCIATION CONSTANTS BETWEEN GLYCOPROTEINS AND LECTINS

T.C.Bøg-Hansen, Pia Jensen, F.Hinnerfeldt and K.Takeo*

The Protein Laboratory, University of Copenhagen, Denmark
and *Department of Biochemistry, Yamaguchi University, Japan

Introduction

Though the last decade has been one of rapid developments of preparative affinity methods, the corresponding analytical methods seems to be much less developed. When we introduced our combination of biospecific interaction of lectins with glycoproteins during electrophoresis and electro-immunoprecipitation, the terms 'affinity electrophoresis' and 'affino-immunoelectrophoresis' were chosen to describe the analytical electrophoretic systems in which ligand-induced mobility shifts are the results of reactions of interacting components [4]. Other electrophoretic systems could be described as affinity electrophoresis also, as for instance counter immunoelectrophoresis, rocket immunoelectrophoresis (electro-immunoassay) and crossed immunoelectrophoresis (see for instance [17]). A short introduction to affinity electrophoresis was given by Horejsi and coworkers [14].

Lectins - Biology, Biochemistry, Clinical Biochemistry, Vol. I

Our two-dimensional lectin affinity electrophoresis was developed for identification and characterization of glycoproteins and for prediction of preparative separations [2-8]. Until now these methods have been used mainly qualitatively.

Other methods permitting quantitative studies of ligand-protein interaction require that the ligand is a low molecular component or require large amounts of purified components (cf. equilibrium dialysis, ultrafiltration, gel filtration or spectroscopic methods). Therefore, the on basis of earlier quantitative uses of one-dimensional affinity electrophoresis for determination of dissociation constants with immobile ligands [18-21] we generalized our analytical approach to an alternative for determination of dissiociation constants of interacting macromolecules [10]. As one of the characteristics of our approach is that crude mixtures of proteins may be assayed, cf. [5], it appeared important to illuminate the effects on the quantitative measurements of interfering contaminating substances. In the present work we investigated the influence of the amount of glycoprotein present and the effects of detergent on determination of the dissociation constant.

Materials and methods

Chemicals. Lectins were obtained from Pharmindustrie (IBF, France), Pharmacia Fine Chemicals (PFC, Uppsala), and from Sigma (USA). The gel was 1% agarose of intermediate or low electroendosmosis (Litex, Glostrup, Denmark) in TRIS-veronal buffer (73 and 24.5 mM), pH 8.6 with 0.36 mM calciumacetate. Antibodies were multivalent antibodies against mouse or human serum proteins [9].

Electrophoresis was performed in thin layers of agarose; the first dimension electrophoresis (10 V cm^{-1}) for 1 hr or more (specified); the second dimension electrophoresis (2 V/cm^{-1}) for 18 h. Antibodies and lectins were mixed with the melted (55°C) agarose gel and cast immediately after mixing. After electrophoresis the plates were pressed, washed with water, pressed, and stained for protein (Coomassie Brilliant Blue R 250) or for esterase enzyme activities with 1- and 2-naphthyl acetate and Fast Red TR salt [9]. The electrophoresis equipment was thermostated at 16°C (Holm Nielsen Analyseudstyr, Soborg, Denmark). Migration distances were determined as first dimension migration distances from the origin to the projection of the peak of the immunoprecipitate on the base line (the first dimension migration pathway). Generally migration was expressed relatively to that of a non-reacting protein (mouse or human albumin, mouse CxE-0, or human prealbumin or lipoprotein).

Calculation of dissociation constants. We derived the general equation for electrophoretic determination of dissociation constants [10]:

$$\frac{1}{Rmo - Rmi} = \frac{K}{Rmo - Rmc} \cdot \frac{1}{c} + \frac{1}{Rmo - Rmc} \qquad (1)$$

where K is the dissociation constant of the conA-glycoprotein complex, c is the concentration of conA, Rmo is the mobility of the glycoprotein without conA, Rmi is the mobility of the glycoprotein in presence of conA, and Rmc is the mobility of the conA-glycoprotein complex, taken in relation to an internal standard (Fig. 1). This equation and its derivation was subsequently published by Horejsi using other symbols [13]. This equation represents a straight line, which gives the dissociation constant K as well as the mobility of the conA-glyco protein complex, Rmc (cf. Fig. 2). The values of K and Rmc were calculated by the least squares method. The symbols used here are in agreement with our earlier work.

Results

Fig. 1 shows the pattern of mouse serum proteins in crossed immunoelectrophoresis. When we included a lectin as conA in the first dimension gel, the glycoproteins interacted with the lectin and gave a deviating pattern, where the shifts were dependent upon the concentration, specificity and charge of the lectin. In these experiments at pH 8.6 free conA had a very slow cathodic migration, and some typical reactions could be seen, notably formation of an affinity precipitate in the first dimension gel in addition to changed position and multi-peaked profiles. We believe that each change is indicative of the carbohydrate nature of the protein. With increasing amounts of free conA in the first dimension gel we observed a gradual retardation of several glycoproteins including alpha-1 carboxyl esterase (CxE, a genetic marker [11]) that was seen to give a multi-peaked precipitate with four distinct peaks: CxE-0 (non-

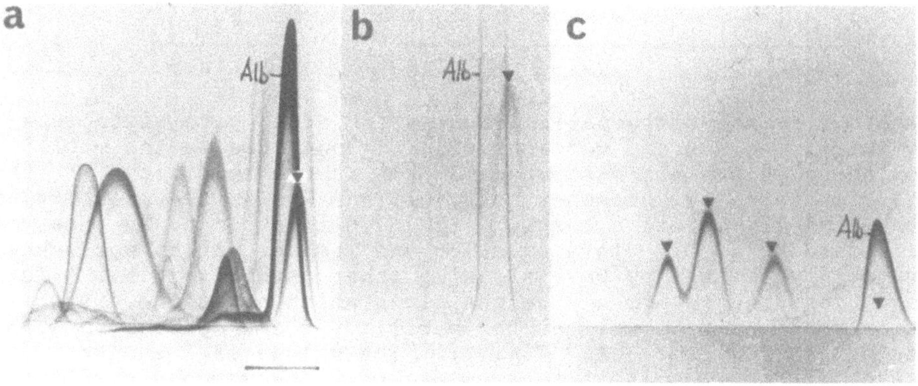

Fig. 1. Lectin affinity crossed immunoelectrophoresis.
a: The control pattern of mouse serum proteins (0.5 µl serum) analyzed with 1.8 µl antibody preparation per cm^2 (purified immunoglobulins from rabbits immunized with mouse serum proteins).
b: The pattern of a purified alpha-1 carboxyl esterase. Only anodal part of plate is shown. 0.2 µl of electrophoretically purified enzyme analyzed as in Fig. 1a exept with 10 times less antibody preparation.
c: The pattern of another alpha-1 carboxyl esterase preparation with conA in the first dimension gel (7.8×10^{-5} N IBF). First dimension electrophoresis 80 mm, antibody prepatation: 0.5 $\mu l/cm^2$.
The bar represents one cm. The arrow indicates CxE. Alb indicates albumin. (Immunoelectrophoresis was performed according to standard procedures. Details and identification of proteins given in [9].

binding), CxE-1 (weakest binding), CxE-2 (intermediate binding), and CxE-3 (strongly binding). We will refer to these forms as the microheterogeneity forms, because we assume that they represent molecular forms differing in their carbohydrate content and configuration.

By plotting the relative migration velocity (Rmo) as a function of the lectin concentration after equation (1) we calculated the dissociation constant for several glycoprotein constituents of the mouse serum including the microheterogeneous forms of the alpha-1 carboxyl esterase (see Fig. 2 and TABLE 1).

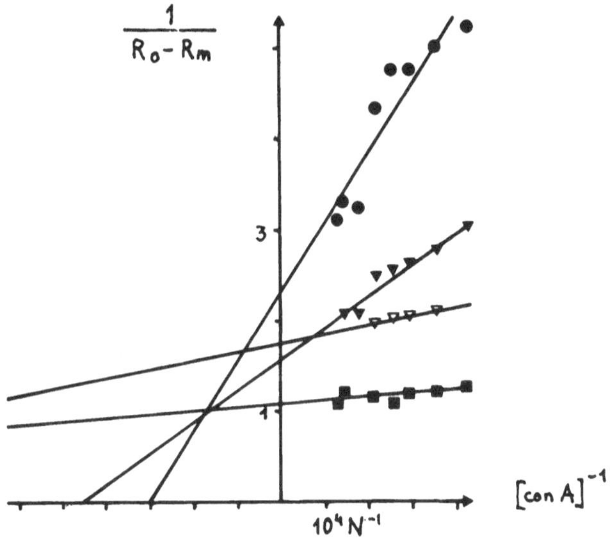

Fig. 2. Plot for calculation of the dissociation constants for conA complexes of the microheterogeneous forms of mouse serum alpha-1 carboxyl esterase.
The plots were drawn from data from the same set of experiments.
CxE-1: ●——●
CxE-2: ▼——▼
CxE-3: ▽——▽
CxE-4: ■——■

Effect of interfering glycoproteins. To test the influence of competing glycoproteins we measured relative mobilities and calculated dissociation constants at different glycoprotein-lectin ratios. For this purpose we chose to purify the mouse serum alpha-1 carboxyl esterase since it is a minor constituent of mouse serum which can be detected in minor amounts. When we stained with coomassie brilliant blue (the normal protein stain) the esterase was not visualized, but only by specific histochemical staining the enzyme could be detected. This means that it was present in our experiments in amounts less than 1 ng. For comparison there was more than 10 µg of lectin-binding glycoproteins in an experiment with total serum as in Fig. 1a.

TABLE 1. Dissociation constants for conA complexes of the microheterogeneous forms of mouse serum alpha-1 carboxylic esterase

	PFC	Sigma	IBF	IBF+)
CxE-0	∞	∞	∞	∞
CxE-1	1.6-3.2	1.7	1.1-1.5	2.2
CxE-2	1.1-2.7	1.3	1.0-1.5	1.6
CxE-3	0.9	–	0.8	1.0

Dissociation constants given as 10^{-5} M. Determined with normal first dimension gels (45 mm migration of human serum albumin).
+) Determined with long first dimension electrophoresis (80 mm)

We purified the conA-binding alpha-1-carboxyl esterase (CxE) from mouse serum by preparative electrophoresis in flat-bed agarose gels. Fig. 1b shows the result of a typical purification. Only few other proteins were present, and only one protein-staining glycoprotein was contaminating some of our preparations (absent in this preparations). Series of experiments were performed with and without added glycoproteins (up to 5 ul human serum, corresponding to more than 100 ug of glycoprotein). The same four molecular variants of the enzyme that could be seen in serum, were seen with the purified enzyme: CxE-0 (non-binding); CxE-1 (weakest binding); CxE-2 (intermediate binding) and CxE-3 (strongly binding). In addition to these components we observed a minor component which bound

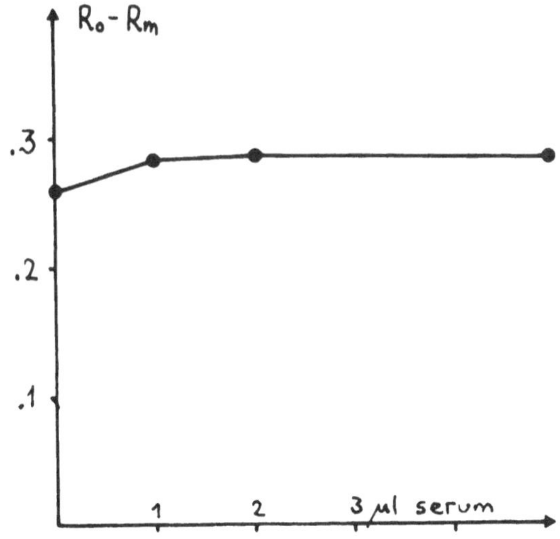

Fig. 3. Plot of the difference in relative mobility of CxE-1 as a function of added glycoprotein.
150 µg glycoprotein corresponds to 5 µl serum. The normally used amount of crude serum is 1 µl or less. The first dimension electrophoresis was performed for about 65 min corresponding to 45 mm migration of human serum albumin.

strongest to conA: CxE-4 (Fig. 2). The dissociation constants were calculated as before, cf. TABLE 1.

Fig. 3 shows how the difference in relative mobility changes for CxE-1 with increasing amounts of competing glycoproteins at an intermediate conA concentration. The change is in the order of 10% for the interval of 20-150 µg of inhibiting glycoproteins and is an expression of the inhibition. For comparison the analysis is normally performed with 1-2 µl serum corresponding to less than 50 ug competing glycoproteins. Fig. 4 shows the relative mobility (Rmi) of CxE-1 and CxE-2 with and without a large excess of glycoproteins at an intermediate concentration of conA. With a short first dimension electrophoresis a deviation of the relative mobility of less than 10% was

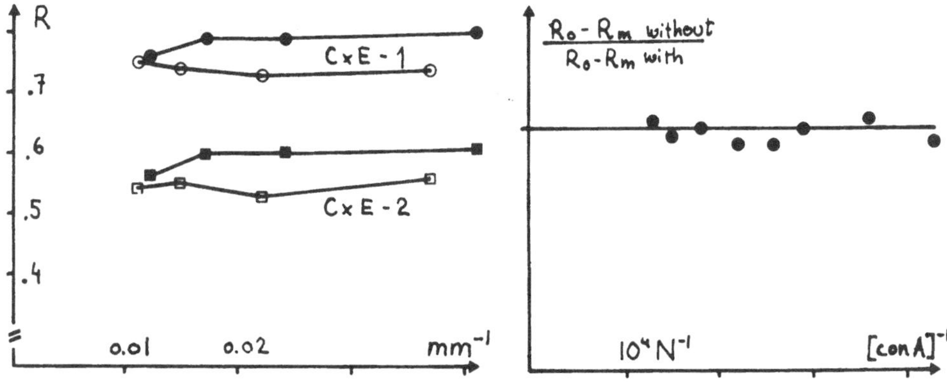

Fig. 4. Influence of added glycoprotein.
a: The relative mobility of CxE-1 and CxE-2 as a function of the length of the first dimension electrophoresis
with ●——●
and without ○——○
conA (IBF).
b: The ratio of the differences of the relative mobility with and without added glycoprotein as a function of the conA concentration.
First dimension electrophoresis was performed for 65 min corresponding to 45 mm migration of human serum albumin. ConA was from IBF.

measured. When the first dimension was performed for 2 h, corresponding to a migration of about 80 mm for albumin, the difference in Rmi was negligible. This is seen in Fig. 4b where the ratio of the difference between Rmo and Rmi without and with excess glycoprotein is plotted as a function of the lectin concentration. The difference in the calculated Km was never greater than 30%.

With the highest concentration of added glycoprotein, Cx3 could not be distinguished but as a shoulder on Cx2, even after a long electrophoresis in the first dimension.

The effect of detergents was studied to gain information of the function of the hydrophobic site of conA and its interaction with detergent micelles and lipoproteins. Preliminary experiments with the human serum alpha-1-lipoprotein-associated arylesterase (ArE) indicate that non-ionic detergent micelles may influence the binding of glycoproteins to conA. Therefore we tested the retardation pattern of serum glycoproteins with conA in presence of non-ionic detergents. We did not find any deviation of the retardation of the normal hydrophilic serum proteins with neither Berol nor Triton up to a concentration of 1%. This included experiments with purified CxE from mouse serum. However, when we measured the retardation of the alpha-1-lipoprotein-associated arylesterase in human serum we found that Berol inhibited strongly and Triton inhibited less strongly the complex formation with conA. Fig. 5 gives the plots for determination of the dissociation constant with and without detergents: Km without detergent: 1.58 x 10^{-5} M; Km with Triton X-100: 5 x 10^{-4} M; and Km with Berol: 10^{-4} M.

The electrophoretic conditions. Other parameters were varied to test the susceptibility of the electrophoretic system with the conclusion that the best conditions to choose are those which are optimal for electroimmuno-precipitation. This is a limitation to the general use of the method, as the variation of the dissociation constant as a function of pH and of temperature can be studied only within narrow limits. Routinely, electroimmuno-precipitation is performed in the pH range of 8.5 to 8.9 and in the temperature interval of 16 to 18 C; exceptionally, pH can be chosen as low as 5.5 and as high as 9.2, and the temperature may be from 10 to 30 C. Only small changes in the values of dissociation constants were observed in the narrow intervals. It was outside the scope of this work to test extreme conditions as well as lectins with other mobility properties. Immobilized lectin (Sepharose-bound conA) was found to influence the electrophoretic properties of the first dimension gel. In addition, control of the binding site concentration was difficult with immobilized lectin. Therefore free lectin was more convenient and satisfying.

Generally we kept the conA concentration in the first dimension gel below 6 x 10^{-5} N. At higher concentrations (7-8 x 10^{-5} N conA PFC and Sigma) we observed that the second dimension electrophoresis could not be completed without detrimental drying of the agarose gels.

Differences in lectin preparations. Our main concern was the difficulty to prepare reproducibly the lectin solution, the variation in content of active lectin, and the differences between various commercial lectins. For this study as for most of our previous studies we used frozen solutions of lectin, prepared as large batches and divided into aliquots convenient for one day's use. However, when we determined the day to day variation of the determination of dissociation constants we found a large variation. We interprete this variation as an

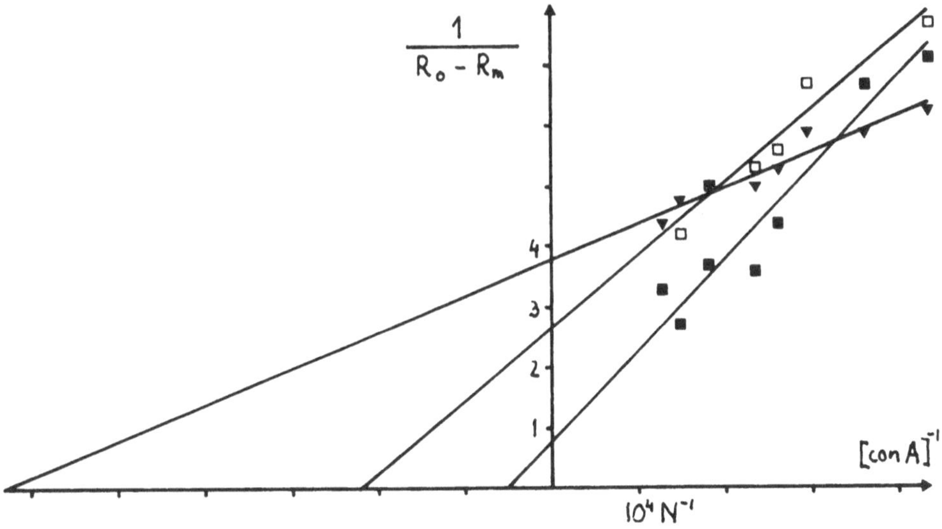

Fig. 5. Influence of non-ionic detergents on binding of conA to human serum alpha-1 lipoprotein-associated arylesterase.
Without detergent: ▼——▼
With Triton X-100 0.5% (w/w): □——□
With Berol 0.5% (w/w): ■——■
The electrophoretic experiments were performed with 1 µl of human serum. The first dimension was performed for 75 min corresponding to 85 mm migration of human serum albumin. In the control experiments without detergent in the first dimension gel, a 1 cm intermediate gel with 1% Triton X-100 was interposed between the first dimension gel and the antibody-containing gel to liberate the enzyme from its association with alpha-1 lipoprotein to give a distinct enzyme precipitate. ConA was from IBF.

expression of a variation in the effective concentration of lectin in the aliquots. In the future we will tend to keep the lectin in solution at 4°C to minimize the non-reproducible effects of freezing and thawing.

When we compared different commercial conA preparations we observed the same general pattern of microheterogeneous forms in most cases, though the effective concentration of conA seemed to be different. With conA Sigma we could not detect Cx3. Kerckaert and Bayard reported recently on the affinity differences between some commercially available immobilized conA preparations [15] and a genetic variation of conA might be expected in the light of the findings for other lectins, cf. other papers in these proceedings. In our experiments with conA we could not distinguish between variation in specificity, variation in affinity and effective concentration. It should be noted that the concentration of conA is calculated as the monomer concentration, the 'normality'. We did not estimate what effects would result from changes of the polymer-monomer ratio.

In experiments with orosomucoid (not shown) we could show that WGA PFC and WGA IBF gave a totally different pattern in our analyses. WGA PFC gave three distinct peaks whereas WGA IBF gave only one peak with the purified protein.

Discussion

The first theoretical considerations of affinity electrophoresis were published by Takeo and Nakamura in a work on enzymes

[18]. Implicitly a mobility of zero was presumed for the complex. Thus the theory behind the present work [10] represent a generalization of the principle of affinity electrophoresis. Earlier our combinations of electroimmuno-precipitation and affinity electrophoresis were used only for qualitative purposes [2,8]. In our experiments the electrophoresis is analogous to affinity chromatography; in effect, the gel with lectin could be compared to an affinity column in which the protein is bound and retarded. The general equation for calculation of dissociation constants by affinity electrophoresis, equation (2), permits the complexed macromolecules (here: the glycoprotein-lectin complex) to be mobile. If the complex has a lower mobility than the free protein, higher affinity means a higher degree of retardation; if the complex has a higher mobility than the free protein, higher affinity means increased mobility [10]. Qualitative differences were not found between analytical affinity electrophoresis and affinity chromatography [4,16]. We did not compare experimentally our present results to determination of dissociation constants by either inhibition precipitation, frontal analysis, elution analysis or any other method mainly because the need for pure components in large amounts.

Earlier we concluded on basis of qualitative experiments that the best conditions to choose for our systems with lectins are those which are optimal for electroimmuno-precipitation. This appears to be limiting for a general use of the method with antibodies [10].

Influence of competing glycoproteins. Our present experiments render probable that quantitative measurements can be performed facefully with our system of combined affinity interaction and electro-immunoprecipitation. We find that the influence of contaminating glycoproteins on the determination of dissociation constants for lectin-glycoprotein complexes is negligible as long as the amount of glycoproteins is less than 150 µg per analysis with conA. Therefore our system may be utilized for determination of such parameters for complex protein systems as for instance totally unpurified extracts or fluids.

Our earlier observations with the pattern of orosomucoid (alpha-1 acid glycoprotein, AGP, cf. [23]), where we observed the same three-peak precipitate in serum and in a purified preparation of orosomucoid [5], were verified using a purified enzyme from mouse serum. We concluded that the influence from other glycoproteins is negligible as long as an excess of ligand (conA) is present. In this connection we find it noteworthy that generally, the values obtained in one-dimensional affinity electrophoresis seem to coincide with those obtained in kinetic and other studies [19,22].

Reproducibility of the method. Our analytical approach is very reproducible what concerns the identification of each protein or microheterogeneity form. However, many factors influence the actual measurements of relative mobility. It appeared of great importance that the electrophoresis was performed neatly to give straight peaks in the antibody-containing gel. Even slightly slanted or skew peaks influenced greatly the possibility to measure accurately the first dimension mobility. This involved careful cutting and handling of the first dimension gels, careful mixing of antibodies into the antibody-containing gel, level casting of the antibody-containing gel, and careful pressing and staining of the plates after electrophoresis. If care was not taken we found that the error of determination of the mobility could be as great as 1-2 mm. Consequently the best determined dissociation constants are found for proteins of high relative mobility and of high affinity. Generally, we find the method suited for determination of dissociation constants in the order of 10^{-5} M.

It is a drawback that the electrophoretic method for determination of dissociation constants requires a number of analyses performed at well-determined ligand concentrations. On the other hand, even with the requirements mentioned above the electrophoreses are simple to perform. Most important is that only a minimum of material is required for the determination and the ligand need not be immobilized. Experimentally, a simpler way to express the relative affinity of a group of proteins to the same ligand is to calculate the retardation coefficients, $R = lo/lr - 1$, where lr and lo are the migration distances in the first dimension electrophoresis with and without ligand [5].

Hydrophobic proteins. Based on modification experiments, Kerckaert and Bayard assume that appearance of a component of orosomucoid is due to hydrophobic interaction or to a special conformation of the polypeptide backbone. We did not find any change of the reaction pattern of hydrophilic proteins, including orosomucoid, when non-ionic detergents were introduced in the electrophoretic system. However, for a hydrophobic protein we found that the complex formation with conA is inhibited when the hydrophobic protein is solubilized in non-ionic detergent. Consequently, binding constants cannot directly be compared when hydrophobic components are involved. This may mean that the choice of detergent is important for solubilization and affinity chromatographic purification of membrane glycoconjugates. The important parameter to consider may be the size of the detegent micelle.

Applicability of the method. In general, any macromolecular or particle-bound ligand or interacting substance may be included during electrophoresis to give a ligand-induced shift of the electrophoretic mobility. Even uncharged ligands may be used. A discussion of some generalizations of the present approach for description of receptor functions of membrane proteins were given recently [3] and Horejsi has given some theoretical considerations on the general quantitative use of affinity electrophoresis [12,13].

As a result of our present findings one of the main advantages of the analytical electrophoresis approach appears to be that is not necessary to purify a protein before its interaction with a binding macromolecule can be studied quantitatively.

Heterogeneity of glycoproteins. The extreme heterogeneity of serum glycoproteins in their reaction with lectins as observed before [4,8,10] cf. also other papers in these proceedings, may now be ascribed to the existence in serum and plasma of populations of microheterogeneous glycoproteins though the apparent heterogeneity has been assumed to be partly explained by a result of non-carbohydrate interaction as mentioned above [1]. We presume that each peak we see in our experiments with lectins included in the first dimension gel has a characteristic carbohydrate structure, possibly related to the function or metabolic turnover of the glycoprotein. Thus the ratio of the three distinct components of orosomucoid seems to be the same in normal individuals but shifted in pregnancy and during treatment with hormone-like drugs [23]. However, their binding properties to conA are unchanged.

One of the facinating aspects is the diagnostic potential of analysis of microheterogeneous glycoproteins as for instance alpha-fetoprotein (AFP), cf. other papers in these proceedings.

Acknowledgements. The studies were supported by the Danish Medical Research Council.

REFERENCES

1. Bayard, B. and Kerckaert, J.P. (1980) Biochem.Biophys.Res.Commun. 95, 777-784.
2. Bjerrum, O.J. and Bøg-Hansen, T.C. (1976) Biochim. Biophys. Acta 455, 66-89.
3. Bjerrum, O.J., Ramlau, J., Bock, E. and Bøg-Hansen, T.C. (1980) in S.J.Jacobs and P.Cuatrecasas (eds.) Techniques for Membrane Receptor Characterization and Purification, Chapman and Hall, London (in press).
4. Bøg-Hansen, T.C. (1973) Anal.Biochem. 56, 480-488.
5. Bøg-Hansen, T.C. (1979) in J.M.Egly (ed.) Proceedings of the Third International Symposium on Affinity Chromatography and Molecular Interactions, Inserm Symposia Series, Paris (in press).
6. Bøg-Hansen, T.C. (1980) in B.J.Radola (ed.) Electrophoresis '79, De Gruyter, Berlin, pp. 193-202.
7. Bøg-Hansen, T.C. (1981) Scand.J.Immunol. Suppl. 10 (in press).
8. Bøg-Hansen, T.C., Bjerrum, O.J. and Ramlau, J. (1975) Scand.J.Immunol. 4, Suppl. 2, 141-147.
9. Bøg-Hansen, T.C., Krog, H.H. and Back, U. (1978) FEBS Lett. 93, 86-90.
10. Bøg-Hansen, T.C. and Takeo, K. (1980) J.Electrophoresis 1, 67-71.
11. Brogren, C.H., Warren, H.S., Nielsen, E. and Rygaard, J. (1977) Proceedings of the Second International Wrokshop on Nuce Mice/ University of Tokyo Press, Tokyo and Gustav Fischer Verlag, Stuttgart, pp. 157-165.
12. Horejsi, V. (1979) in Egly, J.M. (ed) Affinity Chromatography and Molecular Interactions, Les colloque de l'INSERM, Vol. 86, INSERM, Paris, pp.391-398.
13. Horejsi, V. (1979) J.Chromatography 178, 1-13.
14. Horejsi, V., Ticha, M. and Kocourek, J. (1979) TIBS 4, 1, N6-N7.
15. Kerckaert, J.P. and Bayard, B. (1980) Biochem.Biophys.Res.Commun. 92, 95-101
16. Kerckaret, J.P., Bayard, B. and Biserte, G. (1979) Biochim.Biophys.Acta 576, 99-108.
17. Laurell, C.B. (1972) Electroimmuno assay. Scand.J.clin. Lab. Invest. 29, Suppl. 124.
18. Takeo, K. and Nakamura, S. (1972) Arch. Biochem. Biophys. 153, 1-7.
19. Takeo, K., Nitta, K. and Nakamura, S. (1974) Clin.Chim. Acta 57, 45-54.
20. Takeo, K., Kuwahara, A., Nakayama, H. and Nakamura, S. (1976) in H.Peeters (ed.) Protides of the Biological Fluids, Vol. 23, Pergamon Press, Oxford, pp. 645-649.
21. Takeo, K. and Kabat, A. (1978) J.Immunol. 121, 2305-2310.
22. Ticha, M., Horejsi, V. and Barthova, J. (1978) Biochim. Biophys. Acta 534, 58-63.
23. Wells, C., Cooper, E. and Bøg-Hansen, T.C. These proceedings.

AN AUTORADIOGRAPHICAL APPROACH TO THE STUDY OF BINDING OF LECTINS AND THEIR CORRESPONDING ANTIBODIES TO GLYCOPROTEINS IN CROSSED IMMUNOELECTROPHORESIS

O.J.Bjerrum, T.C.Bøg-Hansen, T.Plesner and M.Wilken
The Protein Laboratory, University of Copenhagen, and
Department of Clinical Chemistry, The Finsen Institute,
Copenhagen, Denmark.

Within the last decade lectins have become a valuable tool for identification and characterization of soluble and membrane-associated glycoproteins. When used in combination with electroimmunoprecipitation, lectin binding has been established on basis of secondary reactions i.e. changes of the involved immunoprecipitates, cf. the preceeding paper (5). Earlier, binding of radio-labelled conA has been used to enhance faint immunoprecipitates (7) and to identify glycoproteins after polyacrylamide-SDS gel electrophoresis, see for instance erythrocyte membrane proteins (15).

To investigate autoradiographical methods for detection of the primary binding of lectins to electroimmunoprecipitated glycoproteins, we have applied radioactively labelled lectins and radioactively labelled anti-lectin antibodies. From various biological sources we chose glycoproteins known to react with lectins.

Lectins - Biology, Biochemistry, Clinical Biochemistry, Vol. I

Materials and methods

Antigen preparations. Chronic lymphocytic leukemia lymphocytes were isolated from untreated patients with a typical disease pattern by gradient centrifugation according to (6,12). Solubilization of 10^9 cells/ml was performed as described in (14). The supernatant was stored at -20°C for subsequent electroimmunoprecipitation analysis. Erythrocyte membranes were isolated from outdated bankblood according to (8) and solubilized as described in (1). The supernatant fluid was used for immunoelectrophoresis within 2 h. Human serum was taken from frozen aliquots.

Antibodies. Antibodies to wheat germ lectin (WGA) were raised in rabbits according to (9). The immunoglobulin fraction of the pooled serum from 2 rabbits was isolated giving a final protein concentration of 30 g/l. Plasmin was inactivated by addition of 1% (v/v) aprotinin (Novo, Copenhagen). Well-defined rockets of wheat germ lectin was formed with 0.2-0.5 µl antibody preparation per cm gel. However, our antibody preparation was found also to react with glycophorin of human erythrocyte membrane, presumably due to presence in the commercial WGA of contaminating glycophorin derived from the purification procedure. Therefore the antibody preparation was absorbed twice with human erythrocyte membranes (4 mg ghost protein per ml). Anti-human lymphocyte antibodies were raised in rabbits as described in (13) and rabbit antibodies to human proteins of serum (code 100SF) and erythrocyte membrane were obtained from Dako-Immunoglobulins, Copenhagen.

Chemicals. Concanavalin A (conA), lentil lectin and wheat germ lectin (WGA) were from Pharmacia Fine Chemicals. $Na^{125}I$ with a specific activity of 8-16 Ci ^{125}I/mg I was obtained from Behringwerke or Amersham. Triton X-100 scintilation grade was from BDH Chemicals. Agarose type HSA was from Litex, Glostrup, Denmark.

Radioactive labelling. ConA was labelled with 0.1 mCi ^{125}I/mg protein at a protein concentration of 10 g/l with the lactoperoxidase method (1,8). After desalting, the protein concentration was 0.33 g/l. 5 µl lentil lectin (30 g/l) was labelled with 2 mCi ^{125}I giving rise to a specific activity of 2 x 10^4 cpm/ug lectin. 5 µl anti-WGA antibody was labelled in the same way giving rise to a specific activity of 3 x 10^4 cpm/µg immunoglobulin.

Autoradiography. The dried immunoplates were placed on X-1 Kodak film in X-ray casettes and the film exposed at room temperature for 1-22 d. For some experiments the plates were exposed at -90°C using a 'Lightning plus' intensifying screen, Dupont, which reduced exposure time to 1/3 (11). Two different exposure times of the film was used to ensure optimal exposure with minimal background staining.

Results

ConA-reactive serum proteins detected by electrophoresis with labelled lectin.

Crossed lectin affinity interaction immunoelectrophoresis (4) of serum with ^{125}I-conA was performed either by direct mixing of 2.5 μg lectin per μl serum before immunoelectrophoresis, or

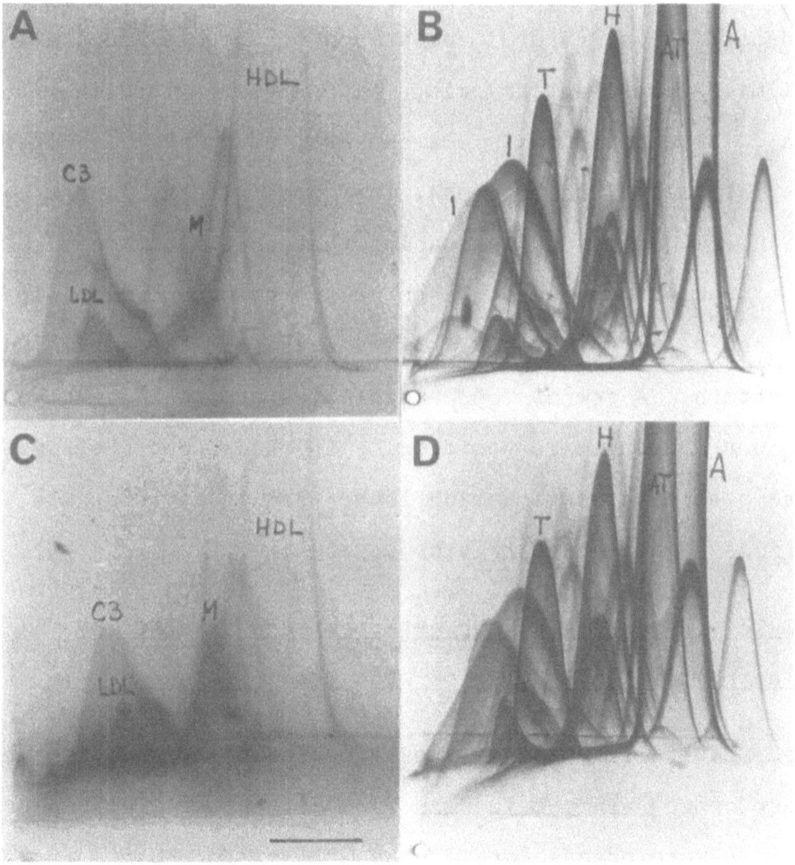

Fig. 1. Crossed radio-immunoelectrophoresis of human serum (2 μl) with ^{125}I-labelled conA in the gel.
A and B. ConA in the first dimension gel, 4 μg/cm^2.
B. Protein stained.
C and D. ConA in the intermediate gel, 4 μg/cm^2.
D. Protein stained.
A = albumin, AT = alpha-1 antitrypsin, C3 = complement C3, LDL, HDL = lipoproteins, H = haptoglobin, I = immunoglobulins, M = macroglobulin, T = transferrin.
The bar represents 1 cm, other details, see (3,4).

by incorporation of the radioactive lectin in the first dimension (66 µg/ml corresponding to 10 µg/cm^2), or in an intermediate gel (28 µg/ml corresponding to 4 µg/cm^2). electrophoresis the plates were pressed followed by soaking in 0.1 M NaCl for 15 min (repeated twice), dried and stained with Coomassie Brilliant Blue (16).

After autoradiography a number of conA-binding proteins could be identified by careful comparison of the stained plates and the films: alpha-1-antitrypsin, alpha-2-macroglobulin, haptoglobin, immunoglobulins, LDL and HDL, cf. Fig. 1. Generally with this small amount of conA, the precipitates appeared at their normal position, indicating that precipitation with conA did not occur in the testtube or in the gels. However, in all three cases we observed slight changes from the normal reference pattern. A few of the precipitates of the major glycoproteins were changed in position or in shape, but the number of shifted precipitates were much lower than those observed with 5-20 times higher concentration of conA.

Lentil lectin-reactive lymphocyte proteins detected by postelectrophoretic incubation with labelled lectin.

Crossed immunoelectrophoresis of solubilized lymphocyte protein was performed as described in (12,14). After electrophoresis the plates were pressed and washed 3 times for 30 min in 0.15 M NaCl. Postelectrophoretic incubation of the immunoplates with ^{125}I-lentil lectin was performed for 18 h in 2 l 0.05 M phosphate buffer pH 7.4 with 0.1 M NaCl, bovine albumin (1 g/l) and the radioactive lectin (75 µg/l). Then the pressing and washing procedure was repeated and the plates were dried before they were exposed (see below). Finally the plates were stained for protein (16). Fig. 2 shows the results. Comparison of the autoradiograph (Fig. 2A) with the protein-stained plate (Fig. 2B) reveals that selective labelling of some precipitates has been obtained. Thus a distinct precipitate due to contaminating human serum albumin (Fig. 2B) has not been tagged by the radioactive label (Fig. 2A). The radioactive precipitates have been shown to be membrane proteins which bind to lentil lectin-Sepharose in column affinity chromatographic experiments as well as in conventional crossed affinity immunoelectrophoresis at a lentil concentration of 50 µg/cm (17). The precipitate denoted by an open arrow (Fig. 2A) corresponds to the HLA-A,B,C histocompatibility antigen (14); the closed arrow denotes the HLA-DR histocompatibility antigen (17). The third

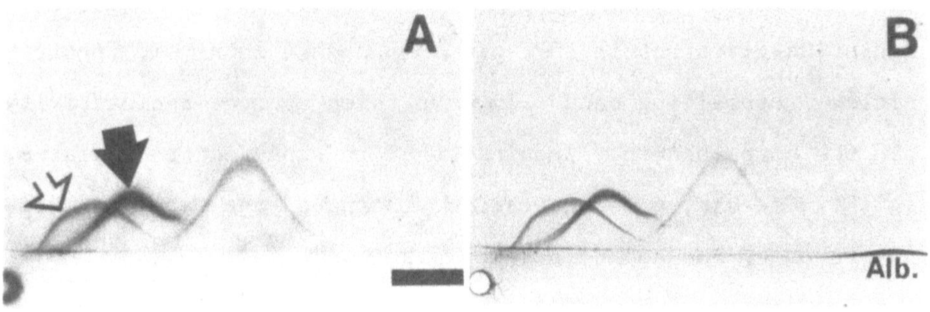

Fig. 2. Crossed radio-immunoelectrophoresis of 50 µg Triton-solubilized human lymphocyte proteins with postelectrophoretic incubation with ^{125}I-labelled lentil lectin.
A. Autoradiograph.
B. The same plate after protein staining.
Three major precipitates selectively reveal radioactivity. The precipitates denoted by open and closed arrows have been identified as HLA-A,B,C and HLA-DR histocompatibility antigens respectively. Antibody concentration: 16 µl/cm^2. Exposure time: 5 days at 22°C. The bar represents 1 cm.

precipitate corresponds to an as yet unidentified glycoprotein which is known from other experiments to be exposed at the outer surface of human lymphocytes

WGA-complexes of erythrocyte membrane proteins detected with labelled anti-WGA antibodies.

Triton-solubilized human erythrocyte membrane proteins were mixed with WGA (ratio 500:1 w/w) and allowed to react for 10 min at room temperature before crossed immunoelectrophoresis was performed in presence of 0.5% (v/v) Triton X-100 (1). After electrophoresis the plates were pressed. Incubation was performed as described for lymphocyte proteins in a solution containing 50 µg/l of ^{125}I-labelled anti-WGA. Subsequent processing was as described for lymphocyte proteins.

After incubation of the immunoplates with ^{125}I-labelled anti-WGA antibodies, two precipitates corresponding to glycophorin (MN-glycoprotein, GP) and the major intrinsic protein complex (containing band III) have taken up more radioactivity than the corresponding precipitates in the control plates, Fig. 3. However, some unspecific binding of the labelled antibody to other immunoprecipitates was found.

Addition of WGA at the present concentration to the solubilized membrane proteins did not change the shape of the glycophorin precipitate. However, after addition of 100 times more lectin, this precipitate appeared with reduced height, indicating precipitation of the lectin-glycophorin complex (2). By affinity

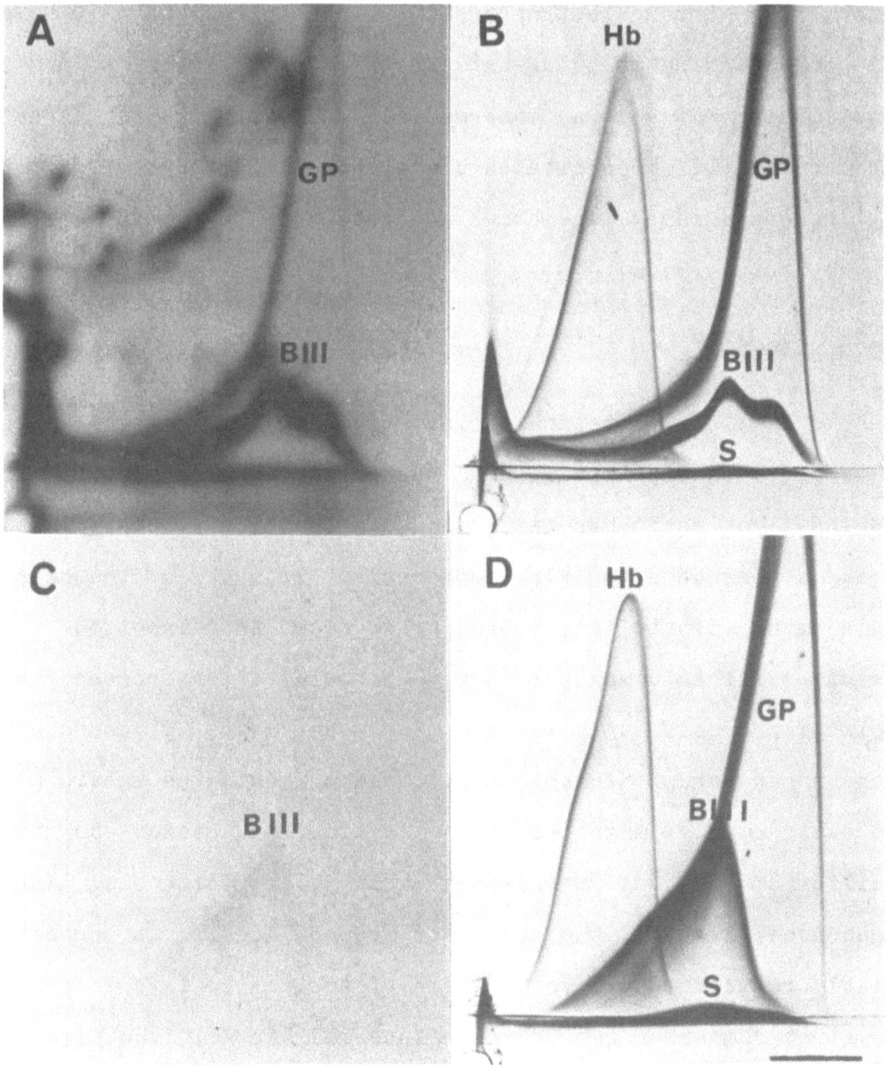

Fig. 3. Crossed radio-immunoelectrophoresis with postelectrophoretic incubation with ^{125}I-labelled anti-WGA antibodies.
A and B. 20 µg Triton-solubilized human erythrocyte membrane proteins mixed with 0.04 µg WGA (B protein stained).
C and D. A control experiment (D protein stained).
GP = glycophorin (MN glycoprotein) and B III = major intrinsic protein complex containing Band III protein Note some unspecific binding of labelled antibody to other precipitates, S = spectrin and Hb = hemoglobin.
Anti-erythrocyte membrane membrane antibodies: 5 µl/cm^2.

interaction immunoelectrophoresis with either WGA (330 ug/ml gel corresponding to 50 ug/cm^2) in the first dimension or as immobilized WGA in an intermediate gel, both the glycophorin and the band III precipitates are affected (2). Also by affinity chromatography, these two antigens have been found to bind to the lectin (2,15).

Discussion

This work is not a systematic investigation but an exploration of some potentialities for identification of glycoproteins. The labelled lectin has been reacted with the glycoproteins before, during of after electrophoresis. In spite of the reversible nature of the lectin binding we found that labelled lectin was still in complex with glycoproteins after crossed immunoelectrophoresis as it was seen for the major glycoproteins from human serum. In this type of experiments the lectin glycoprotein complex migrates in a lectin-free medium so that equilibrium is not maintained during electrophoresis. After immunoprecipitation, the amount of bound lectin is probably greatly reduced. By incorporation of the radioactive lectin in the first dimension gel or in an intermediate gel, the time interval in which conA may dissociate from the glycoproteins is reduced. Presumably, this explains why the concentration of radioactive conA in the gels could be lowered to less than 0.1 mg/ml in the latter two experiments.

In the experiments with postelectrophoretic incubation it was necessary to remove unbound radioactive lectin from the gel by subsequent washes. Non-specific labelling of major immunoprecipitates may be a serious problem, cf. (7), but was not predominant here probably due to the very low lectin concentration (75 μg/l).

In cases where radio-labelling destroys the lectin function, it seems possible to use labelled anti-lectin antibodies. When the labelled antibody was used in rocket immunoelectrophoresis of WGA, as little as 0.5-1.0 ng of WGA could be quantified. Detection of WGA-glycoprotein complexes after immunoelectrophoresis works in principle, see Fig. 3, but it was necessary to optimize the WGA-protein ratio. Furthermore, the anti-lectin antibody showed a relatively high unspecific binding to all immunoprecipitates. Mixing of anti-erythrocyte membrane protein antibodies with anti-WGA antibodies in order to obtain coprecipitation without dissociation of lectin-glycoprotein complexes during the second dimension electrophoresis and subsequent incubation, was unsuccessfull.

The present methodology may represent an advantage when the difference in electrophoretic mobility of lectin and glycoprotein is negligible. Because the extra work connected with the radiolabelling procedure, the methods are recommended only when the lectin is scarce, very expensive or when large series of analyses have to be performed.

Acknowledgments. Ms. Pia Jensen, Kirsten Norup Jensen and Gitte Plesner are thanked for performing immunoelectrophoreses. Financial support has been provided by the Danish Medical Research Council, Bloddonorernes Forskningsfond, Brandt Brandved Foundation, and Kong Christian X's Fond.

REFERENCES

1. Bjerrum, O.J. and Bøg-Hansen, T.C. (1976) Biochim.Biophys.Acta 455, 66-89
2. Bjerrum, O.J., Ramlau, J., Bock, E. and Bøg-Hansen, T.C. (1981) in Jacobs, C. and Cuatrecasas, P. (eds.) Membrane Receptors: Methods for Purification and Characterization. Chapman and Hall, Oxford
3. Bøg-Hansen, T.C. (1981) Scand.J.Immunol.Suppl.11, in press
4. Bøg-Hansen, T.C., Bjerrum, O.J. and Ramlau, J. (1975) Scand.J.Immunol. 4, Suppl. 2, 141-147
5. Bøg-Hansen, T.C., Jensen, P., Hinnerfeldt, F. and Takeo, K. (1981) these proceedings
6. Bøyum, A. (1968) Scand.J.Clin.Lab.Invest.Suppl. 97, 77
7. Brogren, C.H., Peltre, G., Bøg-Hansen, T.C. and Hansen, A. (1976) Prot. Biol. Fluids 24, 781-786
8. Morrison, M. (1974) Methods of Enzymology 31, 103
9. Harboe, M.N.G. and Ingild, A. (1973) Scand.J.Immunol. 2, Suppl. 1, 161-164
10. Hunter, W.M. (1971) in Kirkham, K.E. and Hunter, W.K. (eds.) Radioimmunoassay Methods. Churchill Livingstone, Edinburgh, pp 2
11. Lasky, R.A. and Mills, A.D. (1977) FEBS Lett. 82, 314-316
12. Plesner, T. (1976) Scand.J.Immunol. 5, 1097-1102
13. Plesner, T. (1978) Scand.J.Immunol. 8, 363-367
14. Plesner, T. and Bjerrum, O.J. (1980) Scand.J.Immunol. 11, 341-351
15. Tannner, M.J.A. and Anstee, D.J. (1976) Biochem.J. 153, 267-270
16. Weeke, B. (1973) Scand.J.Immunol. 2, 15-35
17. Wilken, M., Plesner, T., Bjerrum, O.J. and Mørk-Hansen, M. (1981) in preparation

Part III

Glycoprotein studied by Reactions with Lectins

STUDY OF THE MOLECULAR HETEROGENEITY OF ALPHA-FETOPROTEIN BY LECTIN-AFFINITY IMMUNOELECTROPHORESIS

J.P. Kerckaert and B. Bayard
Institut de Recherches sur le Cancer de Lille, U 124 INSERM, (Pr. G. Biserte), BP 311, 59020 Lille Cédex, France

Introduction

The biochemical heterogeneity of alpha-fetoprotein (AFP) has been demonstrated by many workers using a variety of techniques (for review see 1). Since the first demonstration by Smith and Kelleher (2) of the existence of Concanavalin A (Con A)-affinity molecular variants in rat and human AFPs, lectins have been often used to reveal a number of distinct molecular populations in several AFP species. Then, the lectin-binding heterogeneity of AFP is now considered under three different but closely related topics as schematized in figure 1.

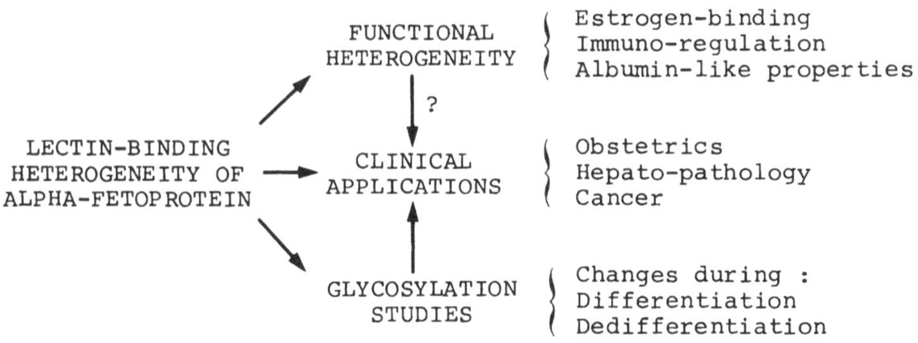

Fig. 1.- The different aspects of the study of the lectin-binding heterogeneity of AFP.

Lectins - Biology, Biochemistry, Clinical Biochemistry, Vol. I

Firstly, does the molecular heterogeneity coïncides with a functional heterogeneity ? To date this question remains open because of the ignorance of the biological role of AFP and because of conflicting results about every of the proposed biological properties of AFP.

The second topic is more fundamental since it concerns the problem of the glycoprotein glycosylation phenomenon and its cellular control. In this view, AFP represents a model because it is an onco-developmental gene product which can be easily purified and characterized at different stages of hepatic cell differentiation and dedifferentiation. In fact little is known of the control of AFP molecular variant synthesis, secretion and degradation, and more generally of the control of peptide glycosylation under malignant transformation.

The third aspect is relevant to the clinical applications of the qualitative analysis of AFP by means of its lectin-affinity profiles in obstetrics (3-5) and cancer differential diagnosis (6). In fact these studies could be connected with the previous one on the glycosylation changes of AFP during fetal development or in pre-malignant and malignant conditions.

From our earlier studies using gel electrophoresis and affinity chromatography with different lectins, it appeared that rat AFP could be fractionated into up to nine molecular variants (7). So with regard to such a range of heterogeneity more sensitive techniques were called for. The lectin-affinity immunoelectrophoresis technique initially presented by Bøg-Hansen (8) seemed to meet our requirements. Here we present results on the lectin-binding heterogeneity of AFP as applications of this new technique in the different AFP research aspects described above.

Material and Methods

Source of AFP. AFP was isolated from amniotic fluids and fetal sera of pregnant (14-18 days gestation) Wistar rats and Swiss mice. Human AFP was obtained from fluids collected in procured abortions (10 weeks pregnancies). Hepatoma-bearing rat sera were collected from rats which have been fed with 3'-MDAB for three weeks.

AFP purification. AFP from rat and mouse were purified by immuno-adsorption on anti rat AFP-Sepharose 4B as described previously (9). Human AFP was prepared in a similar way using anti human AFP-Sepharose 4B.

Lectins. Con A was purchased from Pharmacia Fine Chemicals. The *Ricinus communis* lectin (RCA_I) was obtained as described previously (9). The *Lens culinaris* lectin (LCA) was prepared from crude extracts of dry seeds by affinity chromatography on G-100 Sephadex column (10). After washing the column with a pH 7.2 phosphate buffered saline until the A_{280nm} value returned to the baseline, the lectin was eluted with 0.1 M glucose. Wheat germ, *Ulex europeus* and *Lotus tetragonolobus* lectins were a gift of Dr. H. Debray (Lille, France).

Lectin-affinity immunoelectrophoresis. The method developed by Bøg-Hansen (8) was performed using free lectins in the first dimension gel. Lectins and antibodies were included at 50°C in 1 % Agarose A (Pharmacia) melted in 37 mM Tris, 24.5 mM Veronal, 0.36 mM calcium lactate and 0.2 mM sodium azide solution (pH 8.6). In the first dimension electrophoresis, 1 to 5 µl samples containing 10 to 100 µg/ml AFP were run at 10 V/cm for 2 h in a gel containing the lectin (100 µg/cm²). The second dimension was performed at 2 V/cm for 18 h in an anti-AFP impregnated agarose gel.

Results

Lectin-Binding Heterogeneity of Fetal Rat, Mouse and Human AFP

The conventional crossed immunoelectrophoresis of fetal sera with a second dimension gel impregnated with monospecific anti AFP antiserum gives a single precipitation peak (Fig. 2, control). Including free lectins in the first dimension gel leads to the appearance of several peaks which reflect different degrees of interaction and/or different number of binding sites between the lectin and the glycoprotein (8, 11).

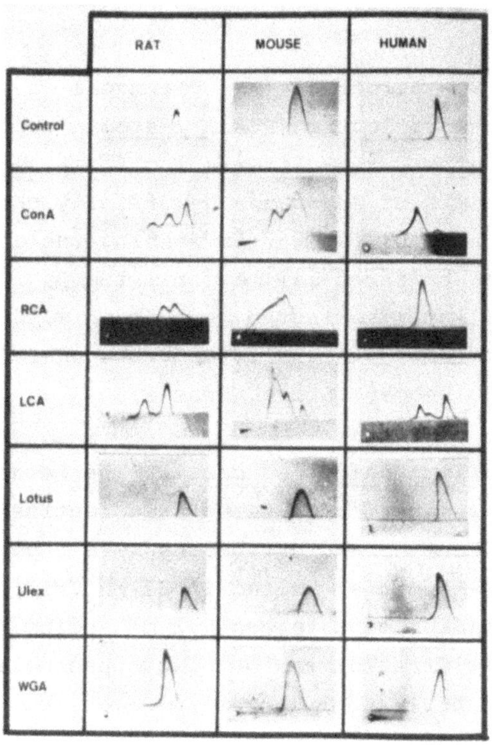

Fig. 2.- Lectin-affinity patterns of rat, mouse and human AFPs obtained with several lectins included in the first dimension gel (100 μg/cm²).

The patterns obtained with several lectins (Fig. 2) reveal a marked microheterogeneity in the rat, mouse and human AFPs from fetal sera. Except with *Lotus*, *Ulex* and Wheat germ lectins which failed to react, three or more peaks can be numbered in each kind of AFP with Con A, RCA and LCA.

Developmental Changes of the Lectin-Binding Heterogeneity of AFP

A typical pattern of the Con A-affinity variants of fetal rat AFP (14-18 days gestation) consists of three fractions : Con A-non reactive (50-60 %), Con A-weakly reactive (30-40 %) and Con A-reactive (10-20 %). However as shown in figure 3, developmental changes can be observed in the distribution of these variants.

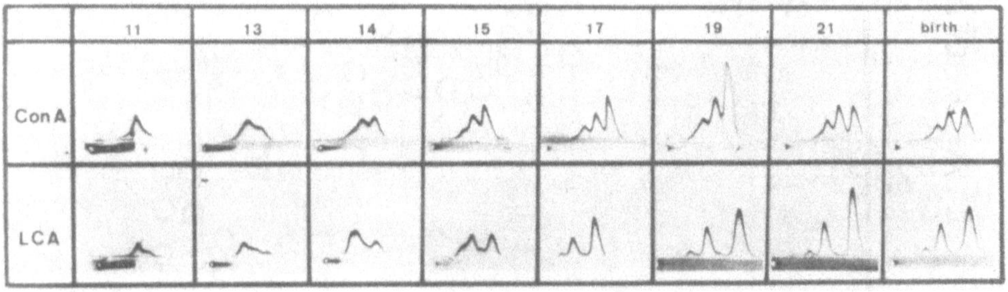

Fig. 3.- Con A and LCA affinity patterns of fetal rat AFP at different stage of development (from day 11 ± 1 of gestation to birth.

In early development the lectin-affinity patterns reveal that AFP is almost entirely Con A or LCA-reactive. Then, from day 13 to day 19 of gestation there is a progressive conversion of the lectin reactive and weakly reactive AFP variants to the lectin-non reactive one. At birth (day 20 ± 1 gestation), there is a significant drop of the Con A-non reactive variant. After birth the lectin-affinity patterns do not vary with age from 1 to 16 days (data not shown).

Variations were also observed in human AFP present in maternal serum. Because of the low AFP concentration range (10 to 300 ng/ml) in maternal serum during gestation, we have adapted (12) the lectin-affinity immunoelectrophoresis technique to the radio-immunologic scale using the procedure described by Axelsen and Nørgaard-Pedersen (13). In these conditions we have been able to compare the LCA-reactivity of maternal AFP in different pregnancies.

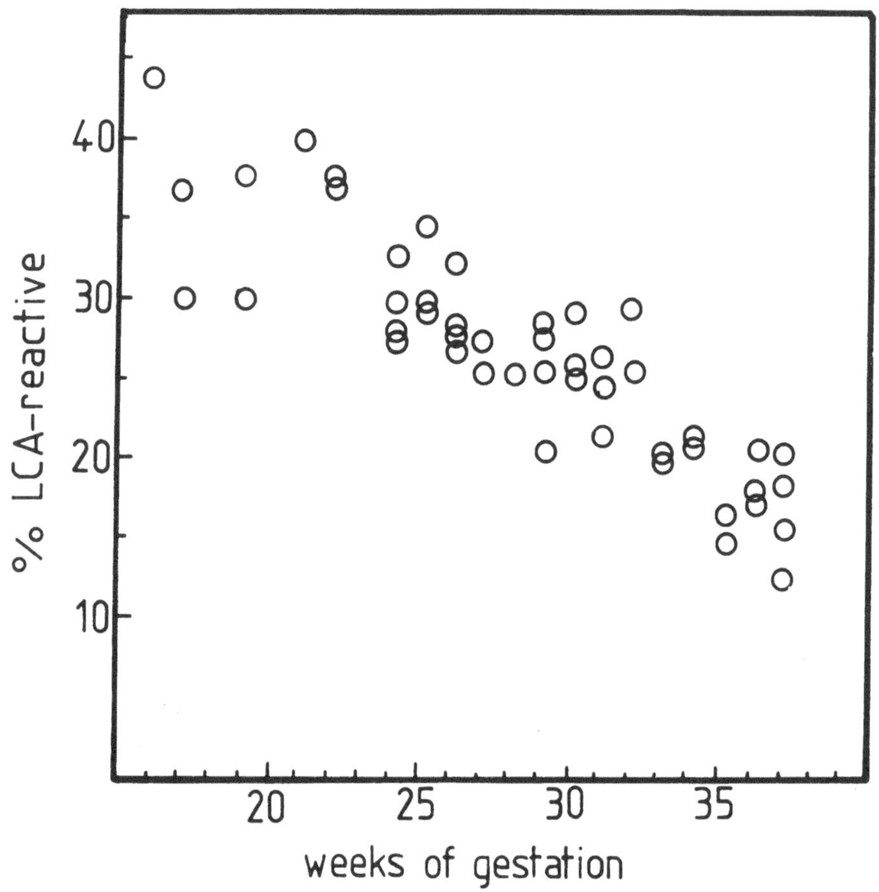

Fig. 4.- Variation of the LCA-reactivity of maternal AFP with gestational age.

Like in fetal serum (Fig. 2), AFP present in maternal serum consists of two LCA-affinity molecular variants. In all the pregnancies studied AFP was found to be mainly LCA-non reactive. However variations can be observed in the proportion of the two variants. These variations have been quantified in figure 4. The percentage of LCA-reactive AFP decreases almost linearly from approximately 40 % at week 15 to 15 % at the end of gestation.

Lectin-Binding Heterogeneity of Tumoral AFP

Although fetal and hepatoma AFPs have been reported immunologically and chemically indistinguishable, some significant differences can be observed in their reactivity with lectins. In figure 5 are compared the patterns of rat and human fetal AFP with rat and human hepatoma derived AFP.

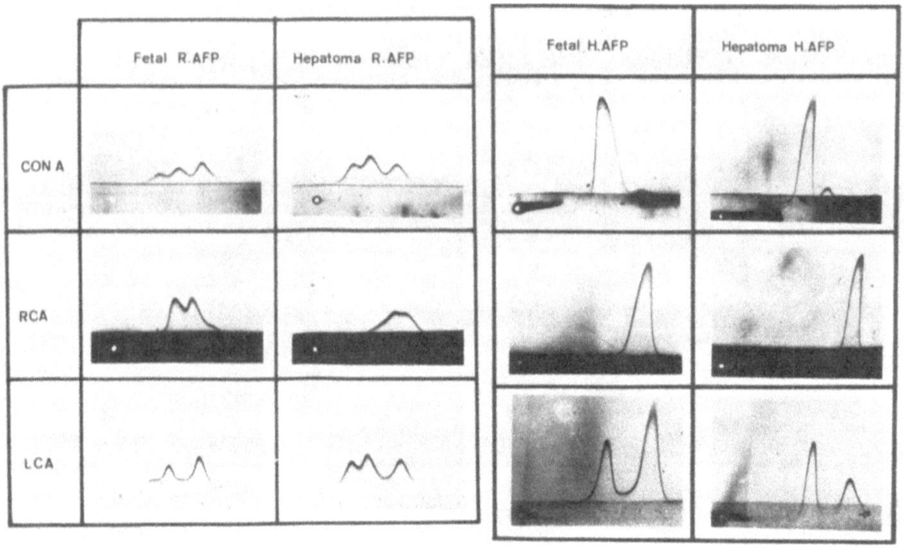

Fig. 5.- Comparison of the lectin-affinity patterns of fetal and hepatoma AFPs from rat and human.

In hepatoma rat AFP, the Con A and LCA-reactive fractions are more important than in the fetal patterns. In human patterns this change appears only with LCA. Indeed, the Con A-affinity patterns of several different human hepatoma and fetal sera look very similar. In turn, the LCA-affinity patterns vary somewhat with individuals independently of AFP concentration (Fig. 6). Moreover, the variations of the AFP lectin reactivity which are supposed to correspond to glycosylation changes, cannot be related with the variations of another glycoprotein like transferrin.

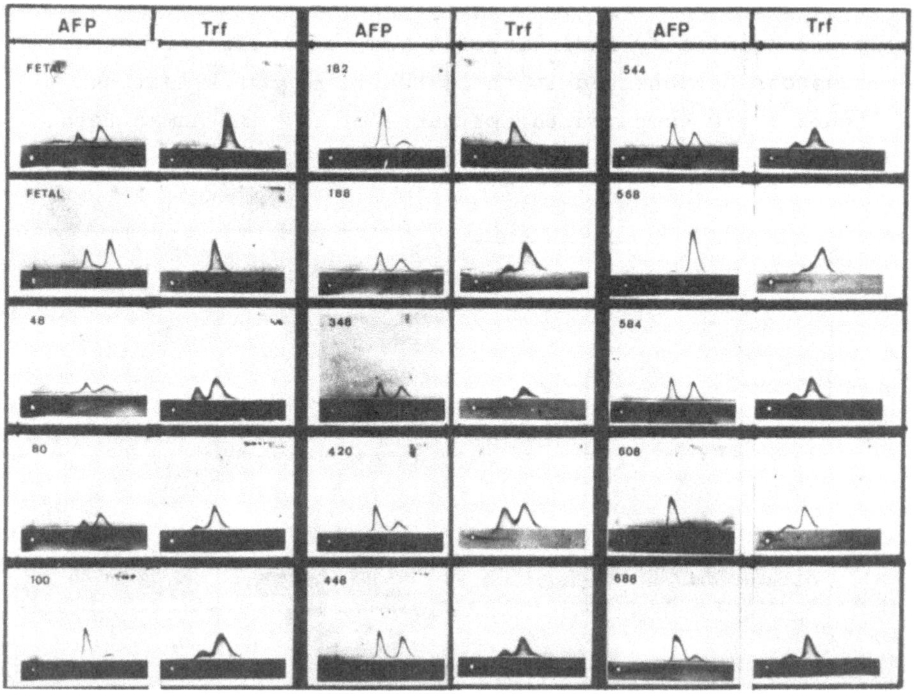

Fig. 6.- LCA-affinity patterns of AFP and transferrin from different hepatoma-bearing individuals. Numbers are AFP concentration (μg/ml) in the sera.

Discussion

Isolation and characterization of the different lectin-binding molecular variants of AFP has not permit until now to reveal specific functional differences between these variants. For example, no significant differences in estrogen binding could be found between the lectin-affinity variants of rat and mouse AFPs (14). Likewise studies on the immunoregulatory potency of the LCA-binding variants did not evidenced a functional heterogeneity of human AFP (15). Thus, the biological reason of such molecular variations is far from being elucidated.

In turn, more encouraging are the studies concerning the understanding of the structural basis of lectin-binding variations of AFP. Recent results on the Con A-binding heterogeneity of rat AFP (to be published) indicate that i) the Con A-reactive AFP variant contains two identical, two-branched glycan units ; ii) the Con A-non reactive AFP possesses two other identical, two-branched glycan units with an additional N-acetylglucosamine residue on the β-mannose. According to this, one could attribute the AFP developmental changes in rat fetuses to the progressive appearance of a glucosaminyl-transferase activity which converts the Con A-reactive AFP to the Con A-non reactive form. The existence of this enzyme, called GlcNAc-Transferase III, which catalyzes the introduction of a GlcNAc residue to the β-mannose has been recently postulated (16).

The changes in Con A-affinity of AFP from hepatoma sources could be also explained by alteration of the same glycosyltransferase activity. Concerning the changes in the LCA-affinity patterns of rat and human AFPs, a similar explanation is not so immediate. Although the presence of some fucose residue could be necessary for LCA-glycopeptide binding (17), more or less specific interactions with the peptidic moiety of AFP could account also in LCA-AFP binding. This observation is confirmed by the fact that the isolated carbohydrate chains of rat AFP are not LCA-reactive (Smith, C.J., Bayard, B. and Kerckaert, J.P., to be published). Therefore, the LCA-affinity changes could be due mainly to polypeptidic conformational changes of AFP molecules and/or to extraneous LCA-reactive material bound on AFP.

In conclusion, lectin-affinity immunoelectrophoresis proves to be a simple and powerful tool in obtaining direct and comparative information on the molecular heterogeneity of AFP. Since the report of abnormally low values of Con A-non reactive AFP in the amniotic fluid in the presence of fetal abnormalities (3), the usefulness of the lectin-binding analysis by this procedure can be consider for the clinical relevance of AFP. This is being illustrated in the present proceedings in obstetric (18) and cancer diagnosis (6).

Acknowledgments

This work was supported by grants from the "Délégation Générale à la Recherche Scientifique et Technique (DGRST n° 78.72643)" and the "Institut National de la Santé et de la Recherche Médicale (CRL n° 78.5.172.2)".

References

1. Smith, C., Kelleher, P. : Biochim. Biophys. Acta 605, 1-32 (1980).
2. Smith, C., Kelleher, P.C. : Biochim. Biophys. Acta 317, 231-235 (1973).
3. Smith, C., Kelleher, P.C., Bélanger, L., Dallaire, L. : Brit. Med. J. 1, 920-921 (1979).
4. Ruoslahti, E., Pekkala, A., Comings, D.E., Seppala, M. : Brit. Med. J. 2, 768-769 (1979).
5. Hindersson, P., Toftager-Larsen, K., Nørgaard-Pedersen, B.: Lancet II, 906 (1979).
6. Breborowicz, J., Mackiewicz, A. : These proceedings.

7. Bayard, B., Kerckaert, J.P. : Biochem. Biophys. Res. Commun. 77, 489-495 (1977).

8. Bøg-Hansen, T.C., Bjerrum, O.J., Ramlau, J. : Scand. J. Immunol. 4, suppl. 2, 141-147 (1975).

9. Kerckaert, J.P., Bayard, B., Debray, H., Sautière, P., Biserte, G. : Biochim. Biophys. Acta 493, 293-303 (1977).

10. Toyoshima, S., Osawa, T., Tonomura, A. : Biochim. Biophys. Acta 221, 514-521 (1970).

11. Bøg-Hansen, T.C. : In "Proceedings of the Third International Symposium on Affinity Chromatography and Molecular Interaction", INSERM Symposia Series, Paris, 86, 399-416 (1979).

12. Kerckaert, J.P., Bayard, B., Puech, F., Codaccioni, X., Biserte, G. : Clin. Chim. Acta 104, 245-249 (1980).

13. Axelsen, N.H., Nørgaard-Pedersen, B. : In "Protides of the Biological Fluids" (Peeters, H., ed.), Pergamon Press, Oxford, 24, 601-604 (1976).

14. Aussel, C., Kerckaert, J.P., Bayard, B. : In "Carcinoembryonic Proteins" (Lehmann, F., ed.), Elsevier, North-Holland Biomedical Press, Vol. II, pp. 297-300 (1979).

15. Soubiran, P., Mucchielli, A., Kerckaert, J.P., Bayard, B., Masseyeff, R. : Scand. J. Immunol. 10, 179-186 (1979).

16. Harpaz, N., Schachter, H. : J. Biol. Chem. 255, 4894-4902 (1980).

17. Debray, H., Montreuil, J. : These proceedings.

18. Toftager-Larsen, K., Nørgaard-Pedersen, B. : These proceedings.

CARBOHYDRATE MICROHETEROGENEITY OF HUMAN ALPHA-FETOPROTEIN. ONCODEVELOPMENTAL ASPECTS.

K. Toftager-Larsen [1], P. Lund Petersen [2] and B. Nørgaard-Pedersen [1]

Department of Clinical Chemistry [1] and Department of Obstetrics and Gynecology [2], Sønderborg Sygehus
DK-64oo Sønderborg, Denmark.

Introduction

Alpha-fetoprotein (AFP) is an oncodevelopmental gene product produced by different fetal organs and by different malignant tumours, especially primary liver cancer and germ cell tumours (7,1o).

During the past decade a large number of studies have described the clinical significance of AFP determinations, both in detection of severe fetal abnormalities, especially neural tube defects (2o), and in monitoring patients with AFP producing tumours (8,12,19). The development of highly sensitive and specific radioimmunoassays for AFP is partly responsible for this rapid development. These methods are based upon the availability of highly purified human AFP and specific antisera to AFP (6,11). One of the few analytical limitations of the AFP radioimmunoassays is that none of the methods described in the literature can determine subfractions or microheterogeneity patterns of AFP from different sources. This carbohydrate heterogeneity was first demonstrated by lectin affinity chromatography studies by Smith and Kelleher (13) for rat AFP and by Ruoslahti et al. (9) for human AFP. The lectin affinity immunoelectrophoresis described by Bøg Hansen (2,3) for analysis of reactivity patterns of different glycoproteins has also been used in several studies (1,5,16,17). This technique has several advantages as reported elsewhere (16,17).

The purpose of this study is to describe the affinity patterns of AFP in serum samples from different AFP producing tumours and in various fetal specimens by affinity immunoelectrophoretical methods with the lectins concanavalin A (con A) and lens culinaris agglutinin (LCA) and to discuss the oncodevelopmental aspects of these findings.

Material and methods.
The samples examined in this study include 6o serum samples from patients with different AFP producing tumours and 16 fetal specimens. All material has been stored at -2o° C.

The fetal specimens include:
1. 4 samples of fetal yolk sac fluid (8-12 weeks of gestation).
2. 7 fetal serum samples (1o-2o weeks) and
3. 5 fetal liver homogenates (1o-12 weeks).

The serum samples from patients with AFP producing tumours include:
1. 16 samples from 7 patients with testicular germ cell tumours
2. 9 samples from 4 patients with ovarian germ cell tumours
3. 6 samples from 2 females and 1 male with germ cell tumours of other localizations (sacrococcygeal, vaginal and mediastinal respectively).
4. 29 samples from 29 patients with primary liver cancer.

The AFP concentration is determined by rocket immunoelectrophoresis (7).

The lectin affinity immunoelectrophoresis is in principle carried out as described by Bøg Hansen (2,3). The first dimension is run in a gel containing 3oo $\mu g/cm^2$ of con A (Pharmacia Fine Chemicals, Uppsala, Sweden) or 5o $\mu g/cm^2$ of purified LCA at a field strength of 5 V/cm until a bromphenol blue stained albumin marker has migrated 3,8 cm (con A) or 7,5 cm (LCA). LCA is purified in our laboratory in principle as described by Toyoshima et al. (18).

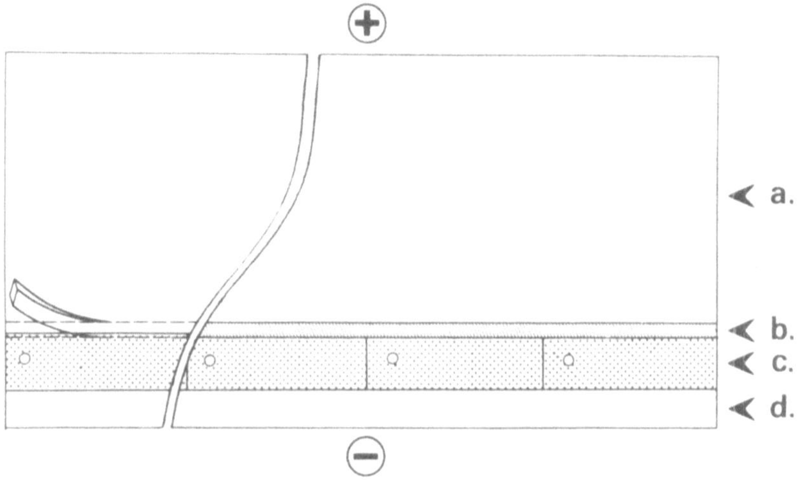

Figure 1. Schematic drawing of Lectin crossed line affinity immunoelectrophoresis (second dimension).
a. antibody containing gel, b. line gel, c. lectin gel slabs from first dimension, d. blank gel.

4 con A or 3 LCA first dimension gel slabs are then transferred to the lower end of a lo x 2o cm glass plate containing 3,75 $\mu l/cm^2$ of anti AFP antibody (Fig. 1) (DAKO Immunoglobulins Ltd., Copenhagen, Denmark). Between the antibody gel and the row of first dimensional gel slabs a 3 mm wide 'line gel' is incorporated in order to obtain a baseline precipitate. This line precipitate makes it much easier to measure the different fractions. After the second dimensional electrophoresis at 2,5 V/cm for 18 hours the precipitates are stained with Comassie Brilliant Blue (Fig. 2 and 3).

The day-to-day reproducibility in terms of coefficient of variation for both LCA and con A fractionations is about 5 per cent.

Results.

The final results of crossed line affinity immunoelectrophoreses are seen in Fig. 2 and 3. With LCA three fractions of AFP can be distinguished, namely a strongly reactive (S-L), a weakly reactive (W-L) and a non-reactive (N-L) fraction. With con A two fractions are seen, a strongly reactive and a non-reactive (N-C) fraction.

The mean values and ranges for the reactivity of AFP with LCA and con A in fetal samples are seen in table 1. In fetal yolk sac fluid the W-L fraction is dominating, comprising close to 5o per cent in most cases, the S-L fraction is about 37 per

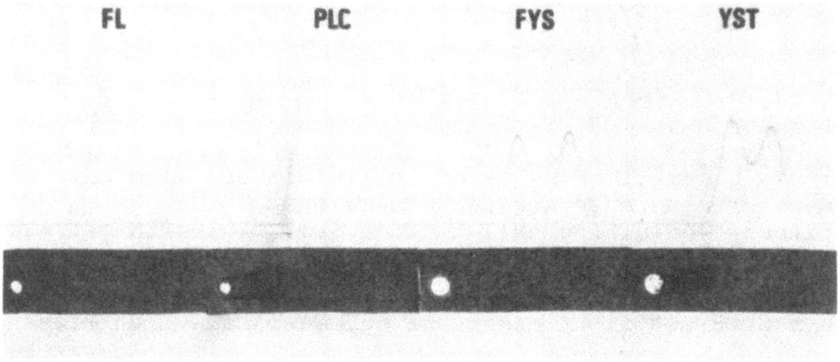

Figure 2. Con A crossed line affinity immunoelectrophoresis. FL: fetal liver homogenate, PLC: serum from a patient with primary liver cancer, FYS: fetal yolk sac fluid, YST: serum from a patient with a yolk sac tumour.

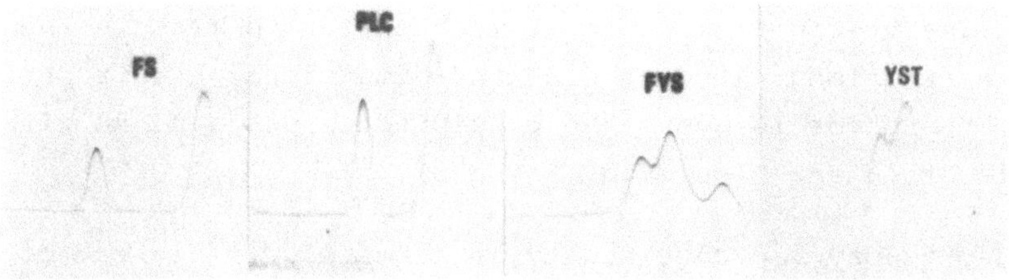

Figure 3. LCA crossed line affinity immunoelectrophoresis. FS: fetal serum, PLC: serum from a patient with primary liver cancer, FYS: fetal yolk sac fluid, YST: serum from a patient with a yolk sac tumour.

cent, and N-L usually below 2o per cent of the total AFP. About 5o per cent of the total AFP consists of the N-C fraction.

In fetal liver the N-L fraction is dominating (about 67 per cent), S-L is about 3o per cent, and W-L only a few per cent. N-C is similarly very low, only a few per cent.

The AFP in fetal serum is most probably produced by fetal liver, and the patterns found in fetal serum are very similar to the reactivity patterns in fetal liver specimens.

The results from the patients with germ cell tumours and patients with primary liver cancers are seen in table 2. AFP in serum from patients with germ cell tumours is characterized by a high W-L fraction (usually above 5o per cent), a S-L fraction usually a little lower, and a very low or undetectable N-L fraction. N-C is about 5o per cent. No difference between the various groups of germ cell tumour patients has been found.

TABLE 1

LCA AND CON A REACTIVITY PATTERNS (PERCENTAGES) OF FETAL AFP

	LCA REACTIVITY			CON A REACTIVITY
SAMPLE *	STRONG	WEAK	NONE	NONE
FETAL YOLK SAC	36.4	45.9	17.8	49.3
N = 4	31.7-42.3	35.8-52.8	10.6-21.8	41.9-57.7
FETAL SERUM	30.3	2.6	67.2	2.3
N = 7	21.0-41.3	0 - 4.5	54.2-76.9	0 - 4.4
FETAL LIVER	30.3	1.9	67.9	3.1
N = 5	23.5-40.9	0 - 3.6	59.1-76.6	1.8-4.8

* The total AFP concentration in all fetal samples was above 4oo mg/l.

In all cases but two a very strong positive correlation (r = o,94) between the W-L and N-C fractions is found. In the two remaining cases - one testicular tumour and one ovarian tumour - high W-L fractions (75,o and 71,4 per cent respectively) is found, but N-C fractions are of the 'usual' percentage (46,7 and 44,7 per cent). In 9 of the patients samples have been ob-

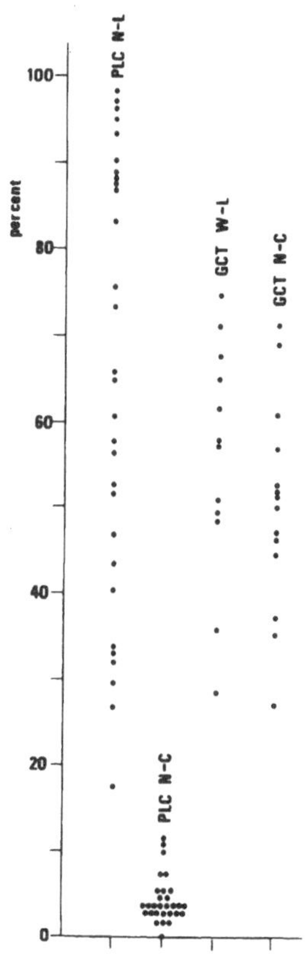

Figure 4

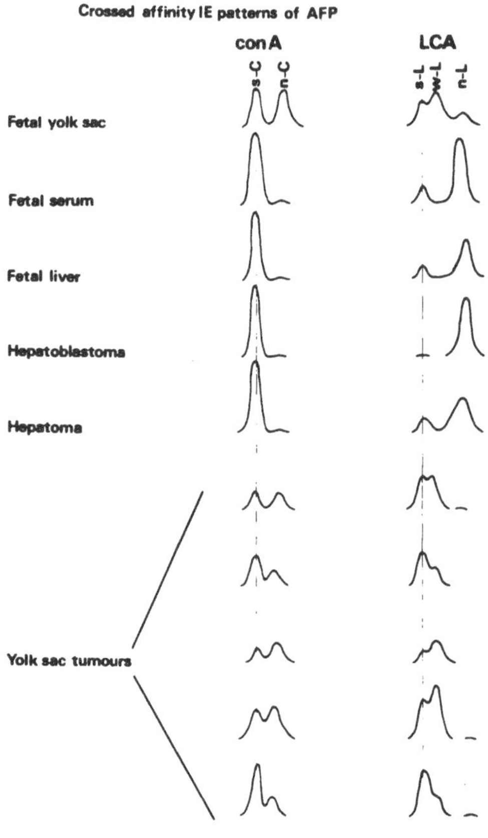

Figure 5

Figure 4. Distribution of characteristic fractions of AFP in patients with primary liver cancer (PLC) and in patients with germ cell (yolk sac) tumours (GCT). For nomenclature of fractions see text.

Figure 5. Typical patterns of con A and LCA affinity immunoelectrophoretical patterns of AFP in fetal specimens and in sera from patients with primary liver cancer and from patients with yolk sac tumours.

tained at different times during the course of the disease, but only small fluctuations in the different fractions are seen. The changes in W-L fractions are followed by equal changes in the same direction in N-C in 8 of the 9 cases.

The patterns seem to be independent of total AFP concentration in the same patient. In one case the total AFP decreased 25-fold, but the lectin reactivity pattern was unchanged. The W-L and N-C fractions both ranged from approximately 25 to 75 per-cent. Except for the two patients mentioned above no definite groups of patients can be identified based on a combination of the two analyses. The distribution of values of the characte-ristic fractions is seen in Fig. 4, and typical patterns are shown in Fig. 5.

As seen in table 2 AFP in serum from patients with primary liver cancer consists of S-L and N-L, usually with most N-L (in two thirds of the patients above 5o per cent, and in most cases no W-L fraction. N-C is low, in most cases 2 - 5 per

TABLE 2
LCA AND CON A REACTIVITY PATTERNS (PERCENTAGES) OF AFP FROM DIFFERENT MALIGNANT TUMOURS

	LCA REACTIVITY			CON A REACTIVITY	CONCENTRATION OF TOTAL AFP
TUMOUR SITE	STRONG	WEAK	NONE	NONE	MG/L
TESTICULAR (GCT) N = 7	42,5 25,0-50,8	57,1 48,6-75,0	0,4 0 - 1,8	50,8 37,6-71,4	12 4 - 39
OVARIAN (GCT) N = 4	46,2 28,6-71,1	53,7 28,2-71,4	0,2 0 - 0,7	45,2 27,2-57,0	65 7 - 175
OTHER (GCT) N = 3	44,7 32,1-64,0	56,3 36,0-67,9	0 -	55,3 35,6-69,3	13 8 - 22
ALL (GCT) N = 14	44,0 25,0-71,1	55,7 28,2-75,0	0,2 0 - 1,8	50,2 27,2-71,4	27 4 - 175
HEPATIC (PCL) N = 29	35,0 1,8-71,1	0,8 0 - 4,1	64,2 17,5-98,2	5,2 0 - 11,4	232 3 - 1440

GCT : GERMINAL CELL TUMOUR PLC : PRIMARY LIVER CANCER

ONLY THE FIRST SAMPLE OBTAINED FROM EACH PATIENT IS USED FOR CALCULATIONS OF MEAN AND RANGE cent (Fig. 4). The LCA and con A patterns of AFP in serum from patients with primary liver cancer seem to be independent of the total AFP concentration.

Discussion.

The occurrence of AFP during early embryogenesis and the recurrence of AFP in various malignant diseases such as in primary liver cancer and in germ cell tumours favour the concept of malignant transformation of cells as a special phase of development. Cancer cells and normal cells have the same genomic content, but express different particular parts of their genomic information. During several phases of development and of cancer the same parts of the gene can be re-expressed as carcinoembryonic proteins or oncodevelopmental gene products. The AFP production by cancer cells of primary liver cancer and germ cell tumours is similar to the AFP production by fetal liver cells and fetal yolk sac cells respectively (4). For germ cell tumours the AFP synthesis is dependent on the presence of yolk sac elements in the tumour tissue as described by Teilum (15). These endodermal sinus tumour (yolk sac) elements have been shown to synthesize AFP (7,8,14,15).

Our results support this oncodevelopmental concept. The pattern of reactivity of fetal yolk sac AFP with LCA and with con A closely resembles the pattern of AFP produced by germ cell tumours, whereas the findings for fetal liver AFP are similar to the findings for AFP produced by primary liver cancers (tables 1 and 2 , and fig. 5). AFP weakly reactive with LCA and AFP not reactive with con A is almost exclusively produced by the fetal yolk sac and by germ cell tumours containing yolk sac elements. We therefore consider these fractions to be relatively 'yolk sac specific'. Similarly, AFP not reacting with LCA is mainly produced by fetal liver and primary liver cancers, and can to some extent be considered as 'Liver specific'. During the course of disease we have found no change in the lectin affinity patterns of AFP, and the patterns were also independent of the total AFP concentration.

Based upon LCA and con A fractionations Breborowicz and Mackiewicz have been able to identify several different profiles (1). We have not been able to make a similar classification, but have found a continuum of varying patterns for germ cell tumours and for primary liver cancers (Figs. 4 and 5). We have found a very strong correlation between the LCA weakly reactive and the con A non-reactive fraction of AFP in all patients with germ cell tumours except two. The significance of the discrepancy in these two cases is not yet known. Only three patients with extragonadal germ cell tumours have been examined. However, the reactivity pattern of AFP was identical to the findings in gonadal tumours.

AFP from patients with primary liver cancer showed a much larger variation than expected in the LCA affinity pattern, but not in the con A affinity pattern. Whether this has any clinical significance remains to be seen.

In summary, we have found that the carbohydrate microheterogeneity of AFP is similar in fetal yolk sac fluid and in serum from germ cell tumours containing yolk sac elements, but distinctly different from the heterogeneity found in samples from fetal liver and primary liver cancer patients. These findings further support the oncodevelopmental concept that the cellular basis for the AFP production in malignant tumours is biochemically analogous to the AFP production by the yolk sac endoderm and fetal liver cells during early embryogenesis.

Acknowledgements.

This study has been supported by the Danish Medical Research Council and 'Fonden for lægevidenskabelig forskning m.v. ved sygehusene i Ringkøbing, Ribe og Sønderjyllands amter'. The expert technical assistance of Mrs. Birgit Andreassen and Mrs.

Jette Rasmussen and secretarial assistance of Mrs. Karin Hansen is gratefully acknowledged.

References.

1.Breborowicz,J., Mackiewicz,A.: These proceedings.

2.BØg-Hansen,T.C.: Analyt. Biochem. 56, 48o-488 (1973).

3.BØg-Hansen,T.C., Bjerrum,O.J., Ramlau,J.: Scand.J.Immunol. 4, suppl. 2, 141-147 (1975).

4.Gitlin,D., Perricelli,A., Gitlin,G.M.: Cancer Res. 32, 979 -982 (1972).

5.Kerckaert,J.P., Bayard,B., Fournier,C., Quief,S., Biserte, G.: Carcinoembryonic Protein, vol. II (ed. F.G. Lehmann), Elsevier/North-Holland Biomedical Press 1979.

6.Nishi,S., Hirai,H.: Gann Monogr. Cancer Res. 14, 79 (1973).

7.NØrgaard-Pedersen,B.: Scand.J.Immunol. 5, suppl. 4, 1-45 (1976).

8.NØrgaard-Pedersen,B., Albrechtsen,R., Teilum,G.: Acta Path. Microbiol. Scand. sect.A, 83, 573-589 (1975).

9.Ruoslahti,E., Engvall,E., Pekkala,A., Seppälä,M.: Int.J.Cancer 22, 515-52o (1978).

1o.Ruoslahti,E., Hirai,H.: Scand.J.Immunol. 8, suppl. 8, 3-26 (1978).

11.Ruoslahti,E., Seppälä,M.: Int.J.Cancer, 8, 374-383 (1971).

12.Schultz,H., Arends,J., Brincker,H., Skov Jensen,T., NØrgaard-Pedersen,B., Sell,A., Steenholdt,S.: Scand.J.Immunol. 8, suppl. 8, 1o9-114 (1978).

13.Smith,C.J., Kelleher,P.C.: Biochim.Biophys.Acta 317, 231-235 (1973).

14.Talerman,A.: Cancer 46, 38o-385 (198o).

15.Teilum,G.: Scand.J.Immunol. 8, suppl. 8, 75-89 (1978).

16.Toftager-Larsen,K., Kjærsgaard,E., Jacobsen,J.C., NØrgaard-Pedersen,B.: Clin.Chem. (198o) (in press).

17.Toftager-Larsen,K., NØrgaard-Pedersen,B. These proceedings.

18.Toyoshima,S., Osawa,T., Tonomura,A.: Biochim.Biophys.Acta 221, 514-521 (197o).

19.Trichopoulos,D., Sizaret,P., Tabor,E., Gerety,R.J., Martel, N., Munoz,N., Theodoropoulos,G.: Cancer 46, 736-74o (198o).

2o.U.K.Collaboratoive Study on Alpha-fetoprotein in relation to Neural-tube Defects: Lancet i, 1323-1332 (1977).

THE DIAGNOSTIC SIGNIFICANCE OF CONCANAVALIN A REACTIVITY PATTERN OF HUMAN ALPHA-FETOPROTEIN IN AMNIOTIC FLUID.

K. Toftager-Larsen and B. Nørgaard-Pedersen
Department of Clinical Chemistry, Sønderborg Hospital
DK-6400 Sønderborg, Denmark

Introduction

Several recent studies have shown that concanavalin A (con A) affinity chromatography (1,9,11) or affinity immunoelectrophoresis (4,7,12) of amniotic fluid alpha-fetoprotein (AFP) is an important confirmatory test in the early diagnosis of neural tube defects (NTD) or other severe fetal malformations. Significantly elevated AFP levels in second trimester amniotic fluid samples are nearly always associated with severe fetal malformations, but borderline elevated AFP values represent a diagnostic dilemma with a risk of aborting a normal fetus (false positive) or not terminating a pregnancy with an abnormal fetus (false negative). In these cases auxiliary testing is necessary. The additional tests include especially ultrasonography, but also several biochemical tests such as degree and type of blood contamination, if any, acetylcholinesterase pattern, demonstration of neuronal membrane protein (D_2) and con A reactivity pattern of amniotic fluid AFP.

The purpose of this presentation is to describe the clinical use and limitations of con A crossed affinity immunoelectrophoresis of AFP in second trimester amniotic fluid samples.

Material and Methods

The amniotic fluid samples examined in this study include normal and pathological samples obtained by amniocentesis in ear-

Lectins - Biology, Biochemistry, Clinical Biochemistry, Vol. I

ly pregnancy. The total number of fluids examined per year in our laboratory is about 2000, and our routine analytical procedure is as follows: Quantitation of total AFP in amniotic fluid, detection (degree and type) of blood contamination, and determination of concanavalin A (con A) reactivity pattern in amniotic fluid with elevated levels of total AFP.

AFP quantitation is carried out by rocket immunoelectrophoresis using 5 µl of undiluted amniotic fluid, and the day to day coefficient of variation is about 5%. This technique and the 95% reference interval for total AFP in amniotic fluid has been described in detail elsewhere (6). This reference interval has recently been confirmed on a larger material (Fig. 1).

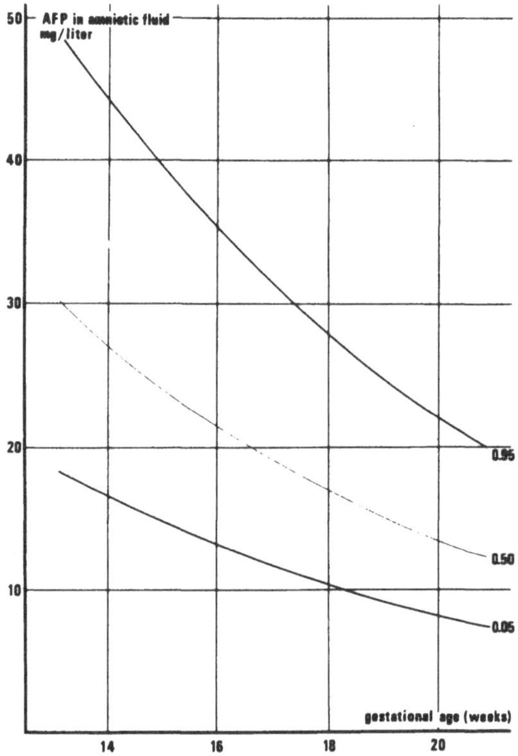

Figure 1. Reference interval for amniotic fluid alpha-fetoprotein. Calculated parametrically from 1281 consequtive samples from normal pregnancies.

Degree of blood contamination, if any, tested on 2 ml of a thoroughly mixed representative sample of amniotic fluid. After centrifugation the size of the red cell precipitate is determined and compared to standards with a known degree of blood contamination. The type of red cells in the precipitate is determined by a slightly modified Kleihauer Test (3). If a fetal blood contamination is detected elevated levels of total AFP in amniotic fluid may be corrected (7).

Con A reactivity pattern of AFP in amniotic fluid is determined by con A crossed line affinity immunoelectrophoresis (7, 12). The affinity immunoelectrophoresis is in principle carried out as described by Bøg-Hansen (2). The first dimension electrophoresis is carried out in a gel containing 3oo $\mu g/cm^2$ of free con A (Pharmacia Fine Chemicals Ltd., Uppsala, Sweden). The electrophoresis is run at 5 V/cm until a bromphenolblue stained albuminmarker has migrated 3,8 cm. Four first dimension gel slabs are then transferred to the lower end of a lo x 2o cm glass plate with an upper gel containing 3,75 $\mu l/cm^2$ of anti-AFP antibody (DAKO Immunoglobulins Ltd., Copenhagen, Denmark). A line gel containing a small amount of AFP is incorporated between the antibody gel and the row of first dimensional gel slabs. The second dimension electrophoresis is run at 2,5 V/cm for 18 hours. After staining of the precipitates with Coomassie Brilliant Blue® the percentage of AFP non-reacting with con A is determined by measuring the height of the two peaks above the AFP line precipitate. The day to day reproducibility in terms of coefficient of variation is about 5%.

The material includes: 1. 519 samples from pregnancies with a normal fetus, irrespective of the total AFP concentration, 2. 22 samples from 11 twin pregnancies, and 3. 34 samples from pregnancies with a severe fetal malformation. Part of this material has been reported in an earlier study (12), namely 275 normal samples, 16 samples from twin pregnancies, and 27 from pregnancies with severe fetal anormalies. All samples have been stored at -2o°C.

Results

The distribution and the mean values of con A non-reactive AFP in 519 amniotic fluid samples from pregnancies with a normal outcome are shown in Fig. 2

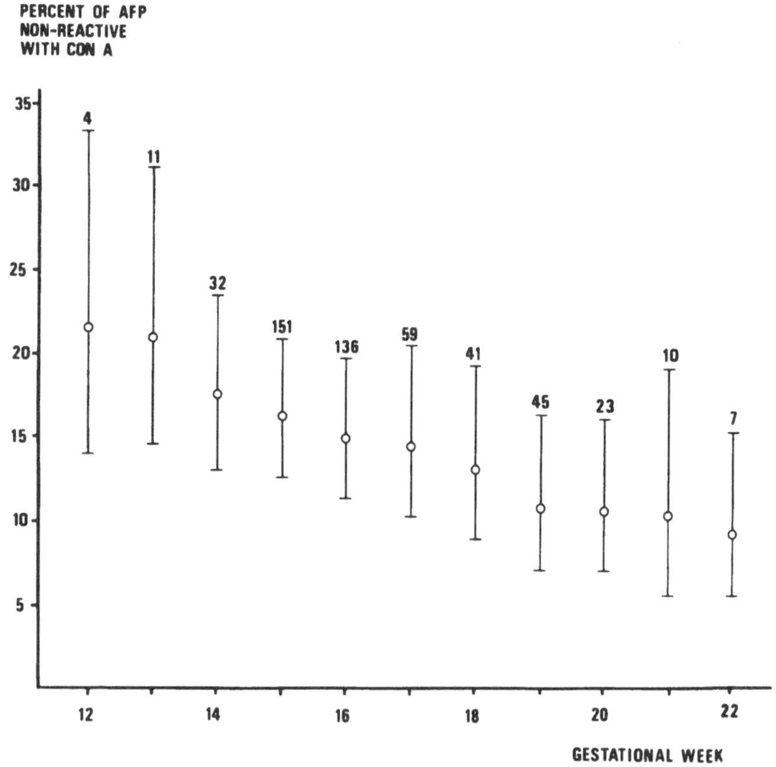

Figure 2. Concanavalin A non-reactive fraction of alpha-feto-protein. Mean ± S.D. for 519 samples from pregnancies with a normal outcome. The values within each week were found to be

log normal distributed. The total number of samples examined in each week is shown in the figure.

A decrease in the con A non-reactive fraction of AFP with advancing gestational age is seen within the observed period. The distribution of values within each of the gestational weeks from 14 to 2o are found to be log normal distributed. The lower 1% and o,1% limit as shown in Fig. 3 have been calculated by regression analyses. Fig. 3 also shows the 34 samples from pregnancies with a severe malformation of the fetus. In all these samples the percentage of con A non-reactive AFP is less than the 1% lower limit, and in 31 of the samples less than o,1%.

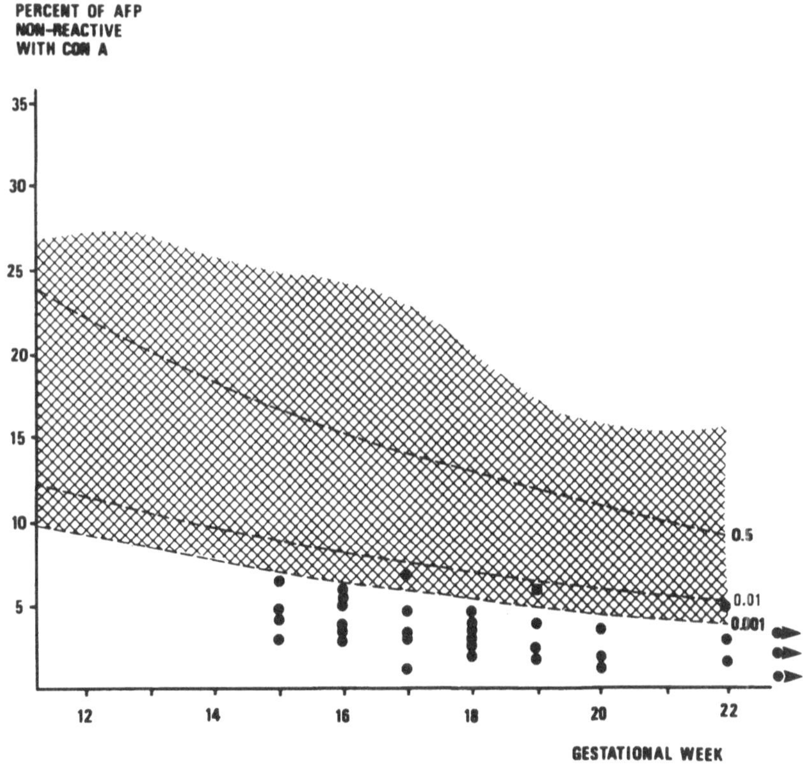

Figure3. Concanavalin A non-reactive fraction of alpha-feto-

protein. Mean and lower 1% and o,1% limits calculated from the material shown in Fig. 2. Dots represent 34 samples from pregnancies with severe fetal malformations.

Fig. 4 shows the total AFP concentrations and con A non-reactive fractions of AFP in corresponding amniotic fluid samples from twin pregnancies. It is seen that the level of con A non-reactive AFP is independent of the total AFP concentration, not only when samples from different twin pregnancies are compared, but also when samples from the same twin pregnancy are compared.

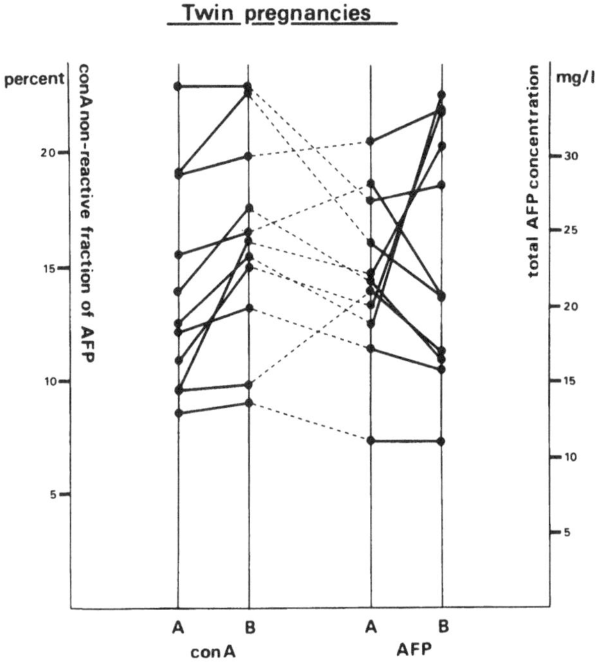

Figure 4. Total AFP concentration and con A non-reactive fraction of AFP in 22 samples from 11 twin pregnancies. Solid lines indicate corresponding values from twin A and twin B. Dotted lines indicate corresponding values for total AFP and con-A non-reactive AFP.

Discussion

The existence of two fractions of AFP differing in carbohydrate content was first shown by Smith et al. in 1973 (1o). They separated rat AFP on a con A sepharose column and found one fraction binding and one fraction not binding to con A. However, it was not before Ruoslahti et al. in 1978 (8) described the differences in the con A reactivity pattern of AFP in human fetal serum and amniotic fluid that it was realized that this heterogeneity might be utilized diagnostically.

It has been shown that the con A non-reactive fraction of AFP produced by the fetal yolk sac is about 5o% (7). The same pattern is found in amniotic fluid samples from very early pregnancy (less than 1o weeks of gestation), indicating that amniotic fluid AFP in early pregnancy mainly originates from the yolk sac. With advancing gestation the con A non-reactive fraction of AFP in amniotic fluid gradually decreases, and after the 2oth week the con A reactivity pattern is more and more similar to the pattern found in fetal liver and fetal serum (7). This gestational age dependency was first described by Ruoslahti et al. (9), and has been confirmed more extensively by Toftager-Larsen et al. (12). As seen in Fig. 2, where the total number of normal samples examined is 519, the decrease per week from 12th to 2oth week varies from about 2% to about 1%. Smith et al. (11) and Brock (1) have not been able to show any gestational age dependency. However, as shown in table 1 their materials are smaller, and the observed period is either 'second trimester' (1) or only from 15th to 18th week of gestation (11). Toftager-Larsen et al. (12) have furthermore demonstrated the same gestational age dependency in serial amniotic fluid samples obtained at different gestational weeks from the same pregnancy.

Table 1. Published results on concanavalin A reactivity pattern of human amniotic fluid AFP.

AUTHOR	SAMPLES FROM PREGNANCIES WITH A NORMAL OUTCOME			SAMPLES FROM PREGNANCIES WITH A PATHOLOGICAL OUTCOME			GESTATIONAL AGE DEPENDENCY	METHOD*
	TOTAL NUMBER	CON A NON-REACTIVE AFP (PERCENT) RANGE	OBSERVED PERIOD (weeks)	TOTAL NUMBER	CON A NON-REACTIVE AFP (PERCENT) RANGE	OBSERVED PERIOD (weeks)		
SMITH ET AL.	20	12.4-37.0	15-18	10	1.2-5.2	14-25	NO	CAC + RIA
RUOSLAHTI ET AL.	74	12 - 39**	11-20	21	6 - 16	14-20	YES	CAC + RIA
BROCK	52	9.1-55.1	?***	6	5.2-8.9	17-21	NO	CAC + RIA
TOFTAGER-LARSEN ET AL.	275	6.0-37.5	12-24	27	0.5-6.3	15-24	YES	CLAIE

* CAC: COLUMN-AFFINITY-CHROMATOGRAPHY
RIA: RADIO-IMMUNO-ASSAY
CLAIE: CROSSED-LINE-AFFINITY-IMMUNO-ELECTROPHORESIS

** APPROXIMATE VALUES TAKEN FROM FIGURE

*** SECOND TRIMESTER

The total AFP concentration is also decreasing with advancing gestational age. However, no correlation between the total AFP concentration and the con A non-reactive fraction of AFP has been found (12). This is further confirmed by the findings in amniotic fluid samples from twin pregnancies (fig. 4). It is also seen that the total AFP concentration and the con A non-reactive fraction of AFP can differ considerably in fluids from the two amniotic fluid cavities in a twin pregnancy.

In cases of certain severe fetal malformations (mainly neural tube defects, but also omphalocele, congenital nephrosis and others) with leakage of fetal serum or cerebrospinal fluid AFP to the amniotic fluid the con A non-reactive fraction of AFP is low as seen in fig. 3. This has been shown by several groups (1,9,11,12), and as seen in table 1 similar results have been obtained by the different laboratories. In this study we have shown that all 34 samples from pregnancies with severe fetal anormalies had levels of con A non-reactive AFP below the 1% lower limit for normal samples, and in 31 of these samples even below the o,1% lower limit (fig. 3).

Among the samples shown in table 1 there were 12 cases of false positives i.e. high total AFP concentrations in normal pregnancies and 9 cases of false negatives i.e. normal total AFP concentrations in pregnancies associated with malformations of the fetus. In 19 of these 21 cases (9o%) the con A non-reactive fraction of AFP was indicating the correct diagnosis i.e. high in normal pregnancies and very low in pregnancies with a malformation of the fetus.

These results have been obtained either by con A column affinity chromatography and AFP radioimmunoassay on the eluted fractions (1,9,11) or by con A crossed line affinity immunoelectrophoresis (12). We have used the immunoelectrophoretical technique which has several advantages (12,13). It is easy to perform, the con A reactivity pattern is directly visualized and furthermore free con A is used in the gel. Kerckaert & Bayard (5) have reported differences in the properties of sepharose-bound con A from different manufacturers in contrast to the corresponding preparation of free con A.

In summary we have found the described con A affinity immunoelectrophoretical technique very reliable and useful as a supplementary test for the analysis of amniotic fluid AFP in detection of severe fetal malformations. Apparently, the final interpretation of the amniotic fluid analyses will be more reliable with less false positive and false negative results.

Acknowledgements

This study has been supported by the Danish Medical Research Council and 'Fonden for lægevidenskabelig forskning m.v. ved sygehusene i Ringkøbing, Ribe og Sønderjyllands amter'. The expert technical assistance of Mrs. Birgit Andreassen and Mrs. Jette Rasmussen and secretarial assistance of Mrs. Karin Hansen is gratefully acknowledged.

References

1. Brock,D.J.H.: Brit.Med.J. 2, 14o2-14o3 (1979).
2. Bøg-Hansen,T.C.: Analyt.Biochem. 56, 48o-488 (1973).
3. Clayton,E.M., Foster,E.B. & Clayton,E.P.: Obstet. Gynecol. 35, 642-645 (197o).
4. Hindersson,P., Toftager-Larsen,K. & Nørgaard-Pedersen,B.: Lancet 2, 9o6, 1979.
5. Kerckaert,J.P. & Bayard,B.: Biochem.Biophys.Res.Commun. 92 95-1o1 (198o).
6. Nørgaard-Pedersen,B.: Scand.J.Immunol. 5, suppl.4, 1-45 (1976).
7. Nørgaard-Pedersen,B., Toftager-Larsen,K., Philip,J. & Hindersson,P.: Clin.Genet. 17, 355-361 (198o).
8. Ruoslahti,E., Engvall,E., Pekkala,A. & Seppälä,M.: Int.J. Cancer 22, 515-52o (1978).
9. Ruoslahti,E., Pekkala,A., Comings,D.E. & Seppälä,M.: Brit. Med.J. 2, 768-769 (1979).
1o. Smith,C.J. & Kelleher,P.C.: Biochim.Biophys.Acta 317, 231 -235 (1973).
11. Smith,C.J., Kelleher,P.C., Belangér,L. & Dallaire,L.: Brit.Med.J. 1, 92o-921 (1979).
12. Toftager-Larsen,K., Kjærsgaard,E., Jacobsen,J.C., Nørgaard-Pedersen,B.: Clin.Chem. (198o) (in press).
13. Toftager-Larsen,K., Lund-Petersen,P. & Nørgaard-Pedersen, B. These proceedings.

APPLICATION OF LECTIN AFFINITY ELECTROPHORESIS FOR STUDIES OF MICROHETEROGENEITY OF HUMAN ALPHAFETOPROTEIN

J. Breborowicz and A. Mackiewicz

Institute of Biostructure, Department of Pathological Anatomy, Academy of Medicine, 60-355 Poznan, Poland

Introduction

Alphafetoprotein (AFP) is a glycoprotein produced by fetal liver and yolk sac, and also to a minor extent by other fetal organs. In various species AFP is either absent in adult tissues or it is present in trace amounts only but it reappears under some neoplastic conditions, especially in hepatoma and in yolk sac tumour. The estimation of AFP is a useful method in diagnosis and monitoring of hepatoma and yolk sac tumour, as well as in prenatal diagnosis of neural tube defects.

Human AFP behaves as a homogenous protein in most biochemical methods but microheterogeneity can be demonstrated in starch gel electrophoresis (9), in extended agarose gel electrophoresis (1), in isoelectrofocusing (1) and in the reaction with lectins (11). Parmelle and coworkers showed that the microheterogeneity of AFP revealed by isoelectrofocusing is due to the binding of fatty acids (8) and the same phenomenon might explain the differences in electrophoretic mobility of AFP revealed by extended agarose gel electrophoresis and by starch gel electrophoresis. The microheterogeneity of AFP detected in the reaction with lectins can be demonstrated by affinity chromatography (11) or affinity electrophoresis (4). Most probably it is due to differences in carbohydrate. Dambuyant and coworkers (3) as well as Ruoslahti and Adamson (10) reported that AFP produced by yolk sac tumours differs in reaction with concanavalin A (con A) from AFP produced by hepa-

Lectins - Biology, Biochemistry, Clinical Biochemistry, Vol. I

tomas. Both groups suggested that these differences might be of importance in clinical diagnosis of cancer. Of practical significance is also the report of Smith who found that estimation of the AFP reactivity with con A is helpful for prenatal diagnosis of neural tube defects (12).

We consider that estimation of the AFP reactivity with lectins might be important in other fields of clinical diagnosis also. Therefore we have undertaken a systematic study of the microheterogeneity of AFP derived from various sources. We used lectin affinity immunoelectrophoresis with con A and lentil lectin (LCA) as it was demonstrated by Kerckaert et al. that LCA is superior to con A in detecting microheterogeneity of AFP (4).

Materials and Methods

By means of radio-rocket immunoelectrophoresis according to Nørgaard-Pedersen (7) and double antibody radioimmunoassay (Swierk, Poland) we studies the AFP concentration in 165 samples. All samples except kidney culture media contained AFP in higher concentration than 20 ng/ml. Culture media were studied only by radiorocket immunoelectrophoresis.

Culture media: 18 samples of media were taken from primary monolayer cultures of fetal liver, intestines, kidneys and yolk sac. The first change of culture medium containing large amounts of AFP, which was present in fetal organs before the culture, was always discarded and replaced by a medium containing either C^{14}-leucine or Se^{75}-methionine. This medium was taken from the culture vessels after 2 - 7 days and used for our studies. The details of the tissue culture technique are described elsewhere (6). The age (week of gestation) of the fetuses whose organs were used for the culture and the number of medium samples from cultures of various fetal organs

Table 1. Number of culture medium samples

Organ used for culture	Weeks of gestation 5	8	9	10	10/11	12	13/14	Total
yolk sac	1							1
liver		1	3	2	1			7
intestine		2		2	1	1		6
kidney				2	1		1	4

are shown in Table 1.

Amniotic fluid: 63 samples of amniotic fluid were studied. 22 samples were obtained by direct puncture of the amniotic cavity at the time of spontaneous delivery (the 34th - 40th week of gestation), 21 samples were obtained by amniocentesis (the 15th - 33rd week of gestation) and 20 samples were taken during artificial abortion (6th - 14th week of gestation).

Fetal serum and neonatal urine: From two artificially aborted fetuses (10th and 14th weeks of gestation) muscle tissue was obtained, chopped with sharp knives and washed with culture medium. Assuming that fetal muscles do not produce AFP, it seems reasonable that AFP present in washings obtained in this way is derived from fetal blood. Three samples of cord serum were obtained at the time of delivery which occurred in the 37th, the 38th and the 40th week of pregnancy. Two other samples were pooled cord serum obtained from deliveries at term. One of these samples was a cord blood serum prepared by IARC as an AFP standard (preparation No. 72/225). One sample of urine was obtained from a male neonatus shortly after delivery in the 35th week of gestation.

Patient sera: 76 AFP positive sera were obtained from patients with various pathological disorders: Benign liver disorders: 17 (acute viral hepatitis: 5, chornic viral hepatitis: 6, liver cirrhosis: 6); hepatoma: 21; yolk sac tumours: 27; and liver metastases of cancer: 11 (stomach cancer: 2, pancreatic cancer: 2, bile duct cancer: 3, breast cancer: 1, cancer originating in the digestive system: 3). The age of patients varied from 2 months to 79 years; 48 patients were older than 40 years. In all patients with cancer and in most patients with benign liver disorders the clinical diagnosis was confirmed by histopathological examination.

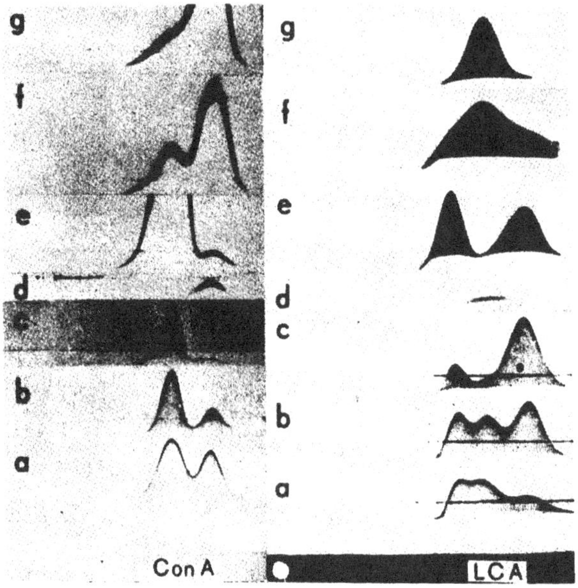

Fig. 1: Lectin crossed immunoelectrophoresis with con A and LCA of culture media and amniotic fluids.
a: Amniotic fluid week 7
b: Amniotic fluid week 14
c: Amniotic fluid week 24
d: Fetal kidney week 10/11
e: Fetal liver week 10/11
f: Yolk sac week 5
g: Fetal intestine week 10/11 of gestation.
Plates a-c stained with Coomassie Brilliant Blue, d-g autoradiograms. Anode at right.

Lectin affinity crossed immunoelectrophoresis with con A and LCA was performed according to Bøg-Hansen et al. (2), for details cf. (5). Assuming the concentration proportional to the area of the precipitate, the relative concentration of AFP variants was determined quantitatively by placing the plates on millimeter paper and counting the number of squares under each peak. The area of the whole AFP precipitate was taken as 100%. The coefficient of the day-to-day variation of this method was 1.6%.

Results

Two AFP variants were observed after crosse- immunoelectrophoresis with con A: AFP reacting with con A (AFP-R.C) and AFP not reacting with con A (AFP-N-C). Both variants can be seen in Fig. 1. The taller cathodal rocket corresponds to AFP-R-C and the smaller anodal rocket corresponds to AFP-N-C (Fig. 1a). With LCA three variants were observed: AFP retarded strongly by LCA (AFP-S-L), AFP retarded weakly by LCA (AFP-W.L) and AFP not retarded by LCA (AFP-N-L). All three LCA variants are seen in Fig. 1b as three peaks separated by two saddles. The cathodal peak corresponds to AFP-S-L, the anodal to AFP-N-L and the intermediate one to AFP-W-L. Each component was quantified and a number of characteristic profiles could be distinguished.

The following three profiles were observed in our samples of amniotic fluid:

Profile AF 1 observed from the 6th to the 9th week of gestation: AFP-R-C ≈ AFP-N-C; AFP-S-L ≈ AFP-W-L > AFP-N-C (Fig. 1a).

Profile AF 2 observed from the 11th to the 18th week of gestation: AFP-R-C > AFP-N-C; AFP-S-L ≈ AFP-W-L < AFP-N-L (Fig. 1b).

<u>Profile L</u> observed after the 20th week of gestation: AFP-R-C ≫ AFP-N-C; AFP-S-L ≥ AFP-W-L ≪ AFP-N-L (Fig. 1C). Profile L was also observed in sera of 10 and 14 week fetuses, in all 5 samples of cord sera and also in the single sample of neonatal urine (35th week of gestation).

The changes of the AFP variants throughout the gestation period are shown graphically in Figs. 2 and 3.

AFP produced in vitro by various fetal organs differed in pro-

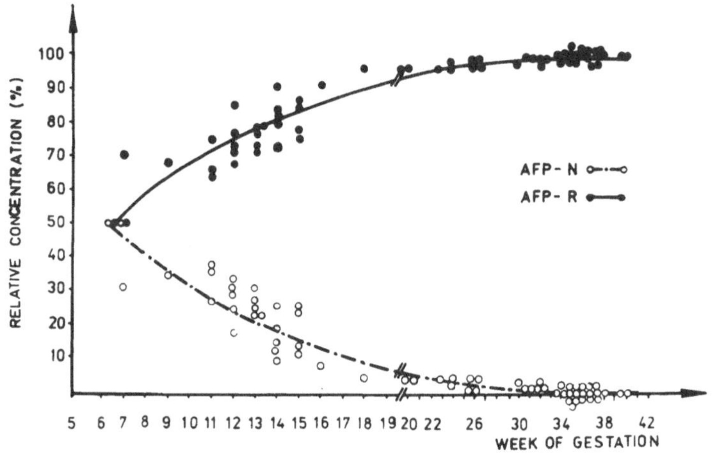

Fig. 2. Relative concentration of con A variants of AFP in amniotic fluid throughout pregnancy.

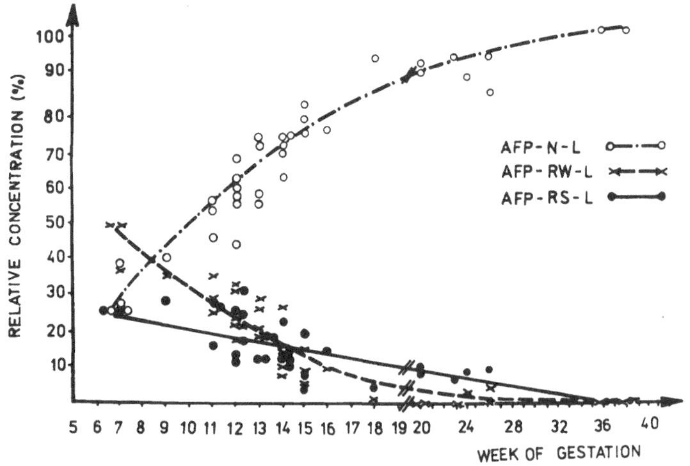

Fig. 3. Relative concentration of LCA variants of AFP in amniotic fluid throughout pregnancy.

files but always an organ produced the same profile. In spite of minor differences, AFP produced by yolk sac, intestines and kidneys was similar:

Profile Yo: AFP-R-C < AFP-N-C; AFP-S-L << AFP-N-L (Fig. 1F).

Profile H1 shown by culture media of liver: AFP-R-C >> AFP-N -C; AFP-S-L >> AFP-W-L << AFP-N-L and AFP-S-L ≈ AFP-N-L.

In patients with hepatoma additional profiles were demonstrated:

Profile H2: AFP-R-C >> AFP-N-C; AFP-S-L >> AFP-W-L < AFP-N-L.

Profile H3: AFP-R-C >> AFP-N-C; AFP-S-L >> AFP-W-L < AFP-N-L and AFP-S-L >> AFP-N-L. Profiles L, H1, H2 and H3 were repeatedly encountered in patients with hepatoma, except in one patient (Fig. 4). In all but one patient with yolk sac tumour the following profiles were observed (see Fig. 5):

Profile Y1: AFP-R-C > AFP-N-C; AFP-S-L >> AFP-W-L ≈ AFP-N-L.

Profile Y2: AFP-R-C > AFP-N-C; AFP-S-L ≈ AFP-W-L >> AFP-N-L.

Profile Y3: AFP-R-C > AFP-N-C; AFP-S-L < AFP-W-L >> AFP-N-L.

Profile Y4: AFP-R.C > AFP-N-C; AFP-S-L << AFP-W-L >> AFP-N-L.

In all patients with benign liver disorders profile L was observed (Fig. 4). In sera of 4 patients with liver metastases of cancer, profile Y3 was detected and in further 4 patients the following profile was found:

Profile M: AFP-R-C >> AFP-N-C; AFP-S-L >> AFP-W-L > AFP-N (Fig. 6). Everyone of the described profiles were demonstrated (each in one patient only) and the profiles coded A 1-5 (Figs. 4, 5 and 6). Profiles A4 and A5 were observed in patients with metastatic cancer of the liver:

Profile A4: AFP-R-C > AFP-N-C; AFP-S-L > AFP-W-L ≈ AFP-N-L.

Profile A5: AFP-R-C >> AFP-N-C; AFP-S-L > AFP-W-L < AFP-N-L.

The frequency of the AFP profiles in our material of culture fluids from fetal organs, of fetal body fluids and of sera of patients with various disorders are summarised in Table 2.

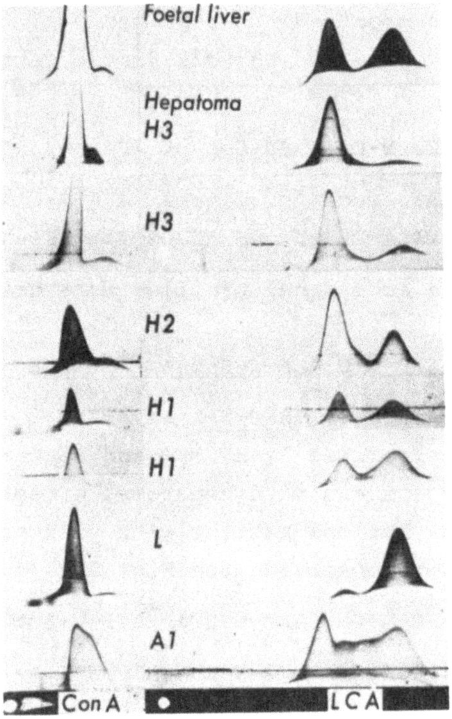

Fig. 4. AFP variants in sera of patients with hepatomas.

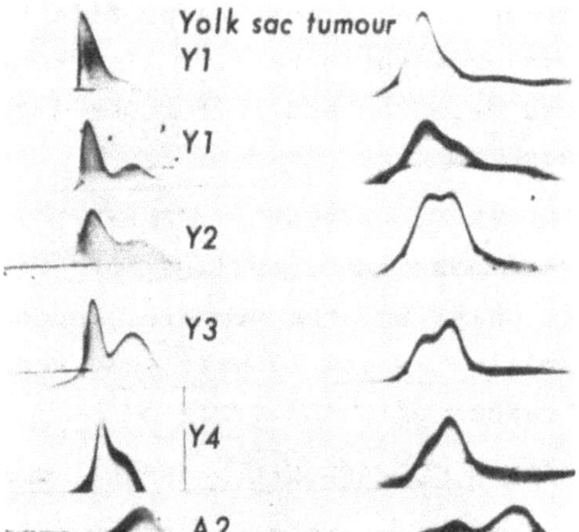

Fig. 5. AFP variants in sera of patients with yolk sac tumours.

Discussion

The results described here confirm previous reports on the value of lectin affinity electrophoresis for studies of the microheterogeneity of AFP in various biological fluids. Application of two lectins (con A and LCA) reveals a higher degree of microheterogeneity than application of only one of these

Table 2. Number of samples containing AFP of various profiles.

Origin of sample	AFP profile												
	AF1	AF2	L	H1	H2	H3	Y0	Y1	Y2	Y3	Y4	M	A1-5
<u>culture media</u> (<u>18</u>)													
yolk sac (1)							1						
liver (7)				7									
intestine (6)							6						
kidney (4)							4						
<u>fetal serum</u> (<u>2</u>)			2										
<u>cord serum</u> (<u>5</u>)			5										
<u>neonatal urine</u> (<u>1</u>)			1										
<u>amniotic fluids</u> (<u>63</u>)													
6th-9th week (4)	4												
11th-18th (24)		24											
18th-40th (37)			37										
<u>patients' sera</u> (<u>76</u>)													
benign liver disorders			17										
hepatoma (21)			6	6	3	5							1
yolk sac tumour (27)								3	7	10	6		1
liver metastases (11)										4		4	3

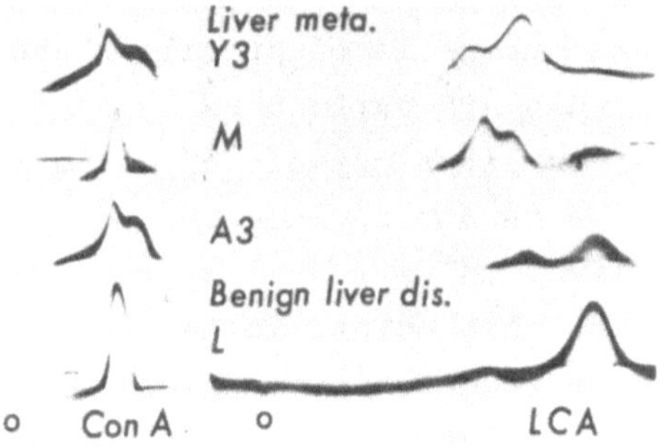

Fig. 6. AFP variants in sera of patients with benign liver disorders and secondary liver deposits of cancer.

lectins. This observation agrees with our findings that AFP variants detected with con A in crossed immunoelectrophoresis do not correspond to the variants demonstrated with LCA (5).

Only 5 out of 17 different AFP profiles were demonstrated in fetal life and in non-neoplastic conditions. The number of AFP profiles observed in neoplastic conditions may be reduced from 15 to 4 by changing the criteria. However, at present it is convenient to distinguish between the larger number of profiles as it facilitates correlation of the microheterogeneity of AFP and the histological appearance of hepatomas.

The demonstrated microheterogeneity of AFP from various fetal organs and body fluids helps to understand the observations of Smith and coworkers who reported the decrease of AFP-N in amniotic fluids of ova with neural tube defects (12), cf. also B. Nørgaard-Pedersen and coworkers as reviewed in (13). Our findings support the suggestion that a change in AFP profile (instead of profile AF2, profile L is observed) is due to leakage of fetal serum into the amniotic fluid. Amniotic fluid in the first trimester of pregnancy contains AFP which is produced mainly by two organs: Yolk sac (profile Yo) and fetal liver (profile H1). Therefore, as expected, the first trimester profile AF1 is an intermediate between profiles Yo and H1. However, it is difficult to explain why AFP in fetal serum at the end of the first trimester and from the later periods of gestation as well as in amniotic fluid in the late pregnancy and in neonatal urine is of profile L and not of H1. It is possible either than AFP variant RS-L is quickly metabolized or that fetal liver in later periods of gestation is producing AFP of profile L. We cannot exclude the possibility that AFP produced by fetal liver in vivo is different from AFP produced in vitro. To clarify this point, it is necessary to examine fetal urine and serum from various periods of gestation as well as AFP produced by fetal liver in the second and in the third trimesters of pregnancy.

It was expected that profiles of AFP produced by tumours originating in various organs will resemble the profiles of AFP produced by the respective fetal organs. Generally this expectation was confirmed. However, we found that while every fetal organ produces AFP of only one profile, the cancers of the respective organs produce AFP differing considerably with respect to variants but having some features in common. The hepatomas produce AFP with L, H1, H2 and H3 profiles. AFP-L is also present in the sera of patients with liver cirrhosis and with other inflammatory conditions of the liver and AFP-H1 is produced by fetal liver. Profiles L, H1, H2 and H3 are similar in low content of AFP-N and of AFP-RW-L. Preliminary results (material not shown) indicate that hepatomas with low cellular atypia are found in patients with AFP profile L and H1, and that hepatoma with more prominent cellular atypia are present in patients with AFP profiles H2 and H3. The last observation must be confirmed in a larger group of patients. It is interesting from the standpoint of clinical diagnosis that at least for two thirds of patients with hepatoma it might be possible to exclude AFP positive benign liver disorders based on studies of the reaction of AFP with lectins. Again, examination of larger groups of patients is necessary.

A variety of AFP profiles was observed also in patients with yolk sac tumours. In these patients profiles Y1, Y2, Y3 and Y4 were observed while fetal yolk sac produces Yo. All profiles coded with the letter Y have as common features a relatively high concentration of AFP-N and a high concentration of AFP-RW-L. Attempts to correlate AFP profiles with the histological appearance or clinical course of yolk sac tumours have not been undertaken yet.

Most patients who suffered from metastatic cancer of the liver and were AFP positive, differed in AFP profiles from those observed in patients with hepatoma and yolk sac. As this group is small and heterogeneous, we cannot draw any conclusions yet as to the clinical value of the estimation of AFP profiles in such patients. It was already suggested by Dambuyant and co-workers (3) and by Ruoslahti and Adamson (10) that studies of the reaction of AFP with con A might be of importance for clinical diagnosis. Our results confirm those of previous workers and in addition they indicate that LCA is more useful for the determination of AFP heterogeneity than con A and that most details are obtained when both lectins are used.

In conclusion we find that studies of reactions of AFP as well as other clinical markers with lectins may be valuable for clinical diagnosis.

References

1. Alpert, E., Drysdale, K., Isselbacher, K.J., Schur, P.J.: J.Biol. Chem. 247, 3792-3798 (1972).
2. Bøg-Hansen, T.C., Bjerrum, O.J., Ramlau, J.: Scand. J. Immunol. 4, Suppl. 2, 141-147 (1975).
3. Dambuyant, C., Sizaret, Ph., Martel, N.: Scand.J. Immunol. 8, 323-330 (1978).
4. Kerckaert, J.P., Bayard, B., Bisert, G.: Biochim. Biophys. Acta 576, 99-108 (1979).
5. Mackiewicz, A., Breborowicz, J.: The same proceedings.
6. Mackiewicz, A., Breborowicz, J.: Oncodevelopmental Biology and Medicine (in print).
7. Nørgaard-Pedersen, B., Lindsten, J., Philip, J.: Clin. Genet. 7, 170 (1975).
8. Parmelee, D.C., Evenson, M.A., Deutsch, H.F.: J. Biol. Chem. 253, 2114-2119 (1978).
9. Purves, L.R., van der Merve, E., Bersohn, I.: Lancet 2, 464-465 (1970).
10. Ruoslahti, E., Adamson, E.: Bioch. Biophys. Res. Comm. 85, 1622-1630 (1978).
11. Smith, C.J., Kelleher, P.C.: Biochim. Biophys. Acta 317, 231 (1973).
12. Smith, C.J., Kelleher, P.C., Belanger, L., Ballaire, R.: Br. Med. J. 1, 920 (1979).
13. Toftager-Larsen, K., Nørgaard-Pedersen, B.: These proceedings.

THREE-DIMENSIONAL AFFINITY ELECTROPHORESIS OF HUMAN ALPHAFETO-PROTEIN.

A. Mackiewicz and J. Breborowicz
Department of Pathological Anatomy, Academy of Medicine,
60-355 Poznan, Poland

INTRODUCTION:

Quantitative immunoelectrophoresis introduced by Laurell (7) was popularized in a manual by Axelsen *et al*. (1) and have been the basis for several new techniques. One of these techniques is lectin affinity immunoelectrophoresis developed by Bøg-Hansen (as reviewed in (3)) by which sensitive, quantitative analytical studies of the reactions of carbohydrate-containing proteins with lectins can be performed easily. Generally, results obtained with con A in affinity electrophoresis are comparable with the results obtained in affinity chromatography (2, 6). However, lectin affinity chromatography seems more laborious and expensive and requires larger amounts of material than lectin affinity immunoelectrophoresis. Therefore we became interested in the application of lectin affinity immunoelectrophoresis for characterization of alphafetoprotein (AFP) produced under various physiological, pathological and experimental conditions (5, 8). Kerckaert *et al*. demonstrated that human AFP is separated into two variants in affinity immunoelectrophoresis with con A and into three variants with lentil lectin (LCA) (6). The different

Lectins - Biology, Biochemistry, Clinical Biochemistry, Vol. I

reactions of con A and LCA with AFP could not be predicted as both lectins have the same specificity: MAN > GLC > DGlcNAc. We have termed the variants in the following way: AFP-N-C (AFP non-reactive with con A); AFP-R-C (AFP reactive with con A); AFP-S-L (AFP retarded strongly by LCA); AFP-W-L (AFP retarded weakly by LCA); and AFP-N-L (AFP not retarded by LCA). In order to find the correspondance between these variants we developed an electrophoretic system in which the sample is run first in agarose containing one lectin (the first dimension), then in agarose containing the second lectin (the second dimension) and, finally, in agarose containing anti-AFP antibodies (the third dimension). The third dimension is simply a quantitative immunoelectrophoresis as an easy way to detect the quantitative location of AFP.

MATERIALS AND METHODS

Samples. Seven samples containing AFP were used for studies. AFP concentration was determined by double antibody radioimmunoassay (Swierk, Poland) and by rocket immunoelectrophoresis. A good agreement of measurements by both methods was obtained.

1. Amniotic fluid, 6th week of gestation; 70 µg AFP/ml.
2. Culture medium of primary monolayer culture of human fetal liver obtained from a fetus aborted in the 10th week of gestation; culture conditions described elsewhere (8); 4 µg AFP/ml.

3. Serum of 2.5 year old girl with yolk sac tumour; 1,4 μg AFP/ml.
4. Serum of 12 years old girl with hepatoma; 8 μg AFP/ml.
5. Serum of 35 years old man with hepatoma; 1,8 μg AFP/ml.
6. Serum of 28 years old woman with liver metastases of breast cancer; 0.1 μg AFP/ml.
7. Serum of 62 years old man with liver metastases of bilary ducts cancer; 0.62 μg AFP/ml.

Depending on the concentration of AFP, the volume for electrophoresis varied from 10 to 50 μl.

Chemicals and Reagents. All electrophoretic experiments were performed in 1% HSA agarose (Mr = -0.13) from Litex, Glostrup, Denmark, in 0.02 M barbital buffer pH 8.6.
The agarose was cast on glass plates. Extracts of <u>lens culinaris</u> seeds were prepared according to Razavi (9) and dialysed with barbital buffer pH 8.6. 200 μl of this extract had the same reactivity in affinity immunoelectrophoresis as one mg of purified LCA. Purified LCA (lot No. EC 12627) and purified con A (lot CL 3366) were from Pharmacia Fine Chemicals, Sweden. Horse anti human AFP antiserum was kindly supplied by dr. Hidematsu Hirai, Sapporo, and was salted out to obtain gamma globulins with approximately the same titer as commercial rabbit anti human AFP gamma globulin from DAKO Immunoglobulins, Denmark (cat.No. 100008).

Conventional two-dimensional lectin affinity immunoelectrophoresis according to Bøg-Hansen et al. (2). The first dimension gel contained either 150 µg of purified con A or 20 µl of extract of Lens Culinaris seeds per cm^2. Conditions of electrophoresis as described below.

Three-dimensional lectin affinity immunoelectrophoresis. Technical details are shown in Fig. 1. The sample was first run in agarose containing con A (150 µg/cm^2) at 10 V/cm, and

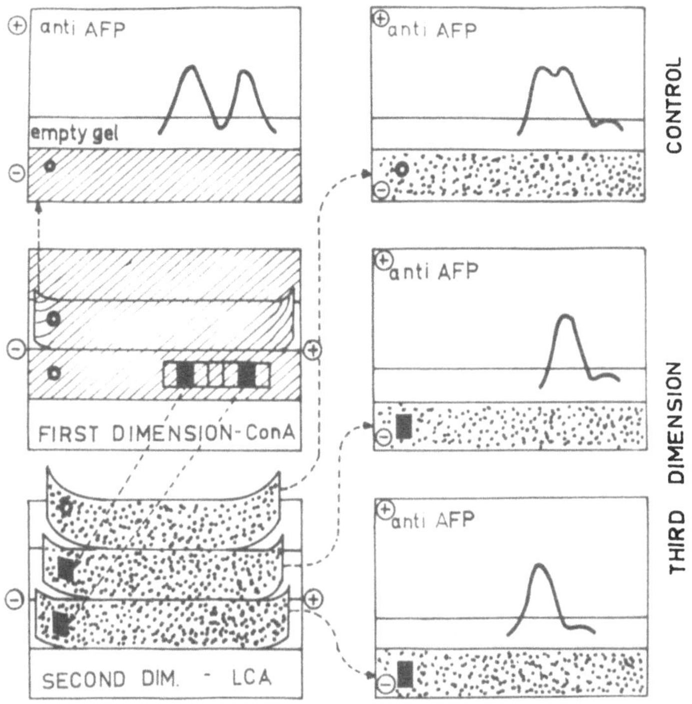

Fig. 1. Technical procedure of three-dimensional affinity crossed immunoelectrophoresis. For simplification only 3 of 8 gel strips are shown on the second dimension plate.

after albumin labeled with bromophenol blue had migrated 8 cm, the gel strip (the first dimension gel) was cut into pieces 2 x 5 x 1.5 mm and single pieces were placed in a gel containing LCA (20 µl extract/cm^2).

The electrophoresis in the first dimension was always run in two parallel gels (Fig. 1, middle panel left). While one gel was cut into pieces after the first dimension electrophoresis and used for second dimension gels (Fig. 1, lower panel left), the second gel was used conventionally (Fig. 1, upper panel left) in order to estimate the variants present. In the second dimension electrophoresis 8 gels were run in parallel: In the first gel (Fig. 1, lower panel left, upper lane) the whole sample was applied as a control. Pieces cut from the first dimension gel were included into the remaining 7 gels (Fig. 1, lower panel left, lower lanes). The electrophoresis with LCA was run for about 120 min at 10 V/cm (albumin migration 8 cm). The second dimension gels were then transferred to the final glass plates and antibody gels (as well as an extra empty gel between LCA gel and the antibody gel) were cast (Fig. 1, left panels). The third dimension electrophoresis was run overnight at 2 V/cm. Depending on the concentration of AFP in the sample the amount of anti AFP gamma globulin preparation included in the antibody gel was varied from 0.1 to 1 µl/cm^2. All samples were run in this electrophoretic system. One sample (amniotic fluid, week 6) was studied also in the reversed system, i.e. LCA included in the first dimension gel and con A in the second dimension gel.

Autoradiography. The concentration of AFP in samples 3, 6 and 7 was so low that the AFP precipitates could not be demonstrated by staining with Coomassie Brilliant Blue. Therefore radioactive AFP was included below the lectin-containing gel, and precipitates were detected by autoradiography. Details of this technique will appear elsewhere (5).

RESULTS

When we tested various samples in affinity crossed immunoelectrophoresis with con A and LCA, we found that most of our samples (1, 3, 5, 6 and 7) contained the 5 known variants of AFP: AFP-R-C (AFP reacting with con A), AFP-N-C (AFP not reacting with con A), AFP-S-L (AFP reacting strongly with LCA), AFP-W-L (AFP reacting weakly with LCA) and AFP-N-L (AFP not reacting with LCA). In lectin crossed immunoelectrophoresis

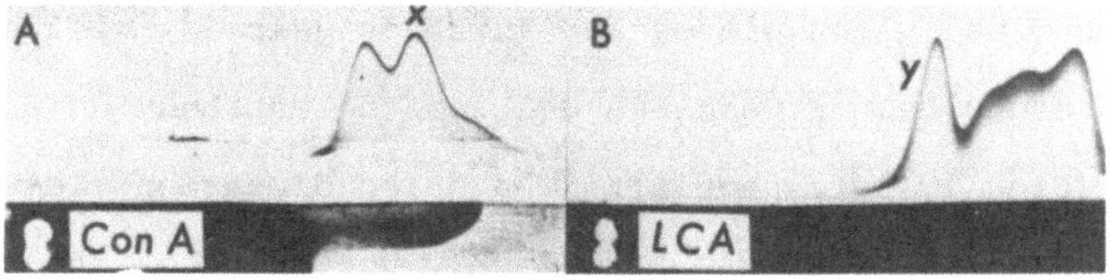

Fig. 2. AFP variants present in serum of hepatoma patient (sample 5):
A. Affinity crossed immunoelectrophoresis with con A. Apart from usual two AFP variants (R and N) additional variant x is present.
B. Affinity crossed immunoelectrophoresis with LCA. Apart from the three usual AFP variants (S-L, W-L, N-L) the additional variant y is present. For identification of all AFP variants see Fig. 4f.

the reacting variants are seen as retarded peaks between the original and the normal position without lectin. Serum of a hepatoma patient (sample 5) contained two additional variants, which we termed X and Y (Fig. 2). Samples 2 and 4 contained both con A variants but only two LCA variants (AFP-S-L and AFP-N-L).

In separate experiments, the first dimension gel with con A was cut into pieces and the pieces were incorporated in gels

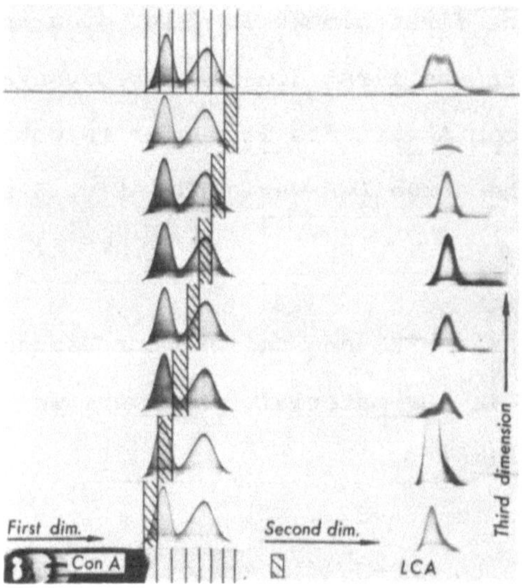

Fig. 3. AFP variants in serum of patient with yolk sac tumour (sample 3) demonstrated by three-dimensional affinity crossed immunoelectrophoresis.
Top left: Affinity crossed immunoelectrophoresis with con A, control experiment performed simultaneously with the first dimension experiment.
Top right: Affinity crossed immunoelectrophoresis with LCA, control experiment performed simultaneously with the second dimension electrophoresis.
All seven left profiles are reproductions of the top one. The hatched area indicates which part of the first dimension, con A containing gel, was used as a sample for the second dimension electrophoresis. All seven LCA profiles below the top one were obtained in the three dimensional electrophoresis. Results presented in this figure are summarised in Fig. 4, d.

with LCA for the second dimension electrophoresis. Fig. 3 shows the results with AFP in serum from a patient with yolk sac tumour. A control plate was run in parallel (Fig. 3, top right). The con A variants present in the individual gel pieces were separated into LCA variants and their correspondence determined.

Results were the same irrespective of the order in which the lectins were used in the first dimension (Fig. 4, a and b). We chose to use con A in the first dimension because a good separation of the two con A variants is easier to obtain than a good separation of the three LCA variants. Fig. 4 summarizes our results.

Generally there is not a one-to-one uniform correspondence of con A and LCA variants in our material. However, we observed the following regularities:

1. AFP-R-C corresponds to AFP-S-L.
2. AFP-R-C corresponds to AFP-N-L in amniotic fluid (sample 1), culture medium of fetal liver (sample 2), and serum from patients with hepatoma (samples 4 and 5).
3. AFP-R-C corresponds to AFP-W-L in other samples.
4. AFP-N-C corresponds to AFP-W-L and AFP-N-L.
5. AFP-N-C corresponding to AFP-S-L was not observed.

As seen in Fig. 4f variant X found with con A in serum of one hepatoma patient was separated into three variants with LCA: Y, AFP-W-L and AFP-N-L. However, the separation of X from

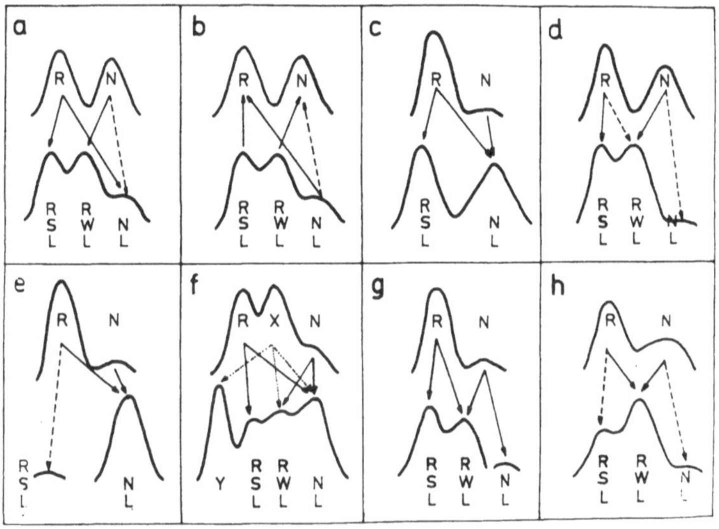

Fig. 4. Correspondence of con A and LCA variants. Profiles obtained with con A are always on top and those obtained with LCA are always at the bottom.

a - amniotic fluid,

b - amniotic fluid in the reversed experiment, i.e. LCA was used in the first dimension and con A in the second dimension;

c - AFP produced in vitro by fetal liver;

d - serum of yolk sac tumour patient (cf. Fig. 3);

e - serum of hepatoma patient (sample 4);

f - serum of hepatoma patient (sample 5);

g - serum of patient with metastatic breast cancer;

h - serum of patient with metastatic cancer of biliary duct;

Continuous arrows: substantial amounts in two variants correspond to each other.

Broken arrows: trace amounts of variants correspond to each other;

Dotted arrows indicate only correspondence of variant X to LCA AFP.

AFP-N-C was not satisfactory. Therefore it is possible that variant X corresponds to variant Y only. This is strengthened by the normal correspondence of AFP-N-C to AFP-W-L and AFP-N-L.

These experiments indicate existence of at least 6 molecular variants of AFP: AFP-R-C/AFP-S-L; AFP-R-C/AFP-N-L; AFP-N-C/AFP-S-L; AFP-N-C/AFP-W-L; AFP-N-C/AFP-N-L; AFP-X possibly identical to AFP-Y.

In spite of the term 'three-dimensional electrophoresis' the results are obtained on two-dimensional plates. A truly three dimensional picture may be obtained by superposition of all plates obtained with the consecutive gel pieces. Fig. 5 is a drawing of the three-dimensional pattern obtained with amniotic fluid, week 6 (sample 1), to show the simplest correspondence of the con A and LCA variants.

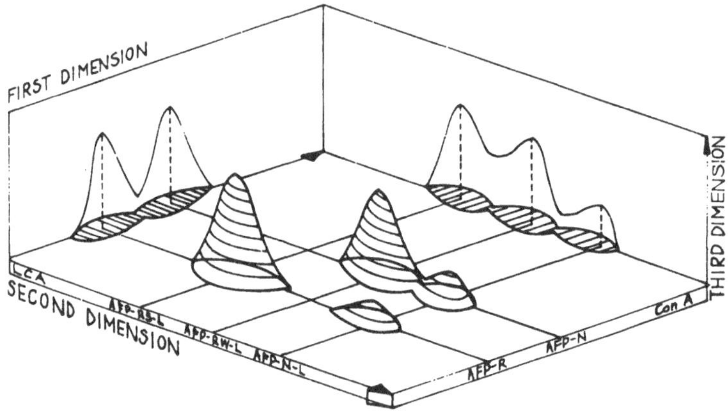

Fig. 5. 'Three dimensional' drawing based on results obtained with 10th week amniotic fluid. Compare with Fig. 4a and b.

DISCUSSION

The three-dimensional electrophoresis presented here permits a comparison of results obtained with different lectins. The third dimension immunoelectrophoresis provides two advantages: 1. It enables studies of protein mixtures instead of purified protein preparations. 2. It provided a quantitative assessment.

The method can be performed with simple equipment and requires relatively small amounts of chemicals (e.g. lectin preparations). We found it very important to run control experiments simultaneously. The method may be generalized for identification of bands obtained in polyacrylamide electrophoresis (preliminary results, data not shown).

The results obtained in our three-dimensional system indicate that there is not a simple correspondence of con A and LCA variants of AFP derived from different sources. This conclusion is in agreement with earlier findings that AFP reacts differently with con A and with LCA (5, 6). Our results indicate that there exist at least six molecular variants of AFP. This heterogeneity is most probably due to differences in carbohydrate content and configuration, and it will be studied further by affinity chromatography on insolubilized lectins. Possibly it reflects differences in glycosylation among different fetal and neoplastic cells, but another explanation might be ligand-binding. However, we demonstrated that binding of arachidonid acid does not influence the reaction with con A or LCA (5).

Therefore lectin affinity immunoelectrophoresis might be an important tool in studies of differentiation of neoplastic cells. In addition, the characteristic patterns of AFP present in body fluids are used in prenatal diagnosis of neural tube defects (10) and in differential diagnosis of cancer (4, 5).

Two variants are of special interest: AFP-R-C/AFP-N-L may be a candidate for AFP specific for liver tissue as it was observed only in samples containing culture medium of fetal liver, in amniotic fluid and in sera of hepatoma patients. However, it may be premature to accept this AFP variant as a specific product of hepatocytes. AFP-X is a new variant which has not been observed earlier.

REFERENCES

1. Axelsen, N.H., Krøll, J., Weeke, B.: Scand.J.Immunol. 2, Suppl. 1, 1973.
2. Bøg-Hansen, T.C.: Anal.Biochem. 56, 480, 1973.
3. Bøg-Hansen, T.C. in Affinity Chromatography. Les Colloques de l'INSERM (ed. J.M. Egly) Vol. 86, pp. 399-416, INSERM, Paris, 1980.
4. Breborowicz, J., Mackiewicz, A.: These proceedings.
5. Breborowicz, J., Mackiewicz, A., Breborowicz, D.: Scand. J.Immunol. (in press).
6. Kerckaert, J.P., Bayard, B., Bisert, G.: Biochem.Biophys. Acta 576, 99-108, 1979.
7. Laurell, C.B.: Anal. Biochem. 10, 358, 1965.
8. Mackiewicz, A., Breborowicz, J.: Oncodevelopmental Biology and Medicine (in press).
9. Razawi, L.: Proc.Soc.Exp.Biol.Med. 118, 717-719, 1965.
10. Smith, C.J., Kelleher, P.C., Belanger, L., Ballaire, R.: Br.Med.J. 1, 920, 1979.

MURINE α-FETOPROTEIN - DEMONSTRATION AND DYNAMICS OF FOUR MOLECULAR FORMS IN CONCANAVALIN A CROSSED AFFINO-IMMUNOELECTROPHORESIS

J. Hau, P. Svendsen

Laboratory Animal Unit, University of Odense
523o Odense M, Denmark

B. Teisner

Institute of Medical Microbiology, University of Odense
523o Odense M, Denmark

G. Thomsen Pedersen
Institute of Obstetrics and Gynaecology, University of Odense
523o Odense M, Denmark

Introduction

Molecular microheterogeneity of human α-fetoprotein (AFP) was demonstrated by Alpert and co-workers (1). Several reports indicate that the level of the concanavalin A (Con A) non-reactive form of AFP is lower in the amniotic fluid of fetuses with certain malformations than in the amniotic fluid of normal fetuses (2, 3, 4, 5).

Molecular heterogeneity of murine AFP (m-AFP) has been demonstrated using several different techniques (6, 7, 8, 9, 1o). During fetal development in mice there seems to be an increase in the degree of glycosylation of m-AFP (11) and it has been suggested that the yolk sac is the main source of the less glycosylated forms of m-AFP, which dominate during early development in utero (1o, 12).

In this study we examined the level of total m-AFP in maternal

Lectins - Biology, Biochemistry, Clinical Biochemistry, Vol. I

serum during pregnancy and compared it to the fetal and placental mass during pregnancy. The levels of four molecular forms of m-AFP with different affinity to Con A were measured by quantitative crossed affino-immunoelectrophoresis in whole body saline homogenates of fetal and neonatal mice.

Materials and Methods

Experimental animals. Random-bred primigravid SPF mice of the strain NMRI/Bom were used. Conception time was determined by the presence of a vaginal plug and designated Day-1 of pregnancy. The animals were anaesthetized by intraperitoneal injection of o.3 mg pentobarbitone sodium and bled between 9 and 11 a.m. on each day of pregnancy. The fetuses and placentae were weighed. Ten juveniles were killed on days 0, 1, 2, 3, 4, 5, 7, lo, 13 and 16 after birth. The whole bodies of the fetuses and the neonatal mice were washed with isotonic NaCl, homogenized in isotonic NaCl and centrifuged at 4,ooo g. The supernatant was used the same day or stored at -2o^{0} C until use. Homogenates of fetal liver, spleen and brain were treated as whole body homogenates

Production of antisera against m-AFP. Serum pooled from mice 11-18 days pregnant was fractionated by gel filtration chromatography on a Pharmacia K 26/loo column packed with ACA-34 (L'Industrie Biologique Francaise). The protein separation profile consisted of three peaks and the fractions from the intermediate peak were pooled, concentrated, mixed with Freund's incomplete adjuvant and injected subcutaneously into goats, at doses of 5oo µg protein every two weeks.

The goat antiserum was absorbed with a 4o% volume of a pool of serum from female mice for 24 hours and ultracentrifuged at 7o,ooo g for 1 hour. The supernatant was tested for antibody activity by crossed immunoelectrophoresis using sera from female and pregnant female mice as antigens.

Rocket line immunoelectrophoresis. Rocket line immunoelectrophoresis (13) for quantification of m-AFP was performed on 11 x 2o cm glass slides in a 1% agarose gel (Indubiose A 37, L'Industrie Biologique Francaise) with a TRIS barbital buffer, ionic strength o.o2o, pH 8.3 at 2.5 V/cm for 18 hours. The antibody-containing gel contained 1.5 µl/cm^2 of a di-specific antibody preparation against m-AFP and pregnancy-associated murine protein 2.

A neutral gel with wells was placed on the cathodic side of the antibody-containing gel.

Cathodic to this gel was the antigen-containing gel which contained 1.5 µl/cm^2 of a pool of serum from pregnant females.

One hundred arbitrary units (AU) refer to the amount of m-AFP in a serum pool from mice in late pregnancy.

Rocket affino-immunoelectrophoresis. Using the same conditions as for quantitative rocket immunoelectrophoresis a Con A-containing intermediate gel of 1 x 2o cm was cast between the antibody-containing gel and a neutral gel in which the wells were cut. The intermediate gel contained 3oo µg Con A/cm^2 corresponding to 1.82·lo^{-5} M Con A. The samples were examined also in control experiments without Con A in the intermediate gel. The antiserum was directed against m-AFP and four pregnancy-associated murine proteins.

Quantitative crossed affino-immunoelectrophoresis. The crossed immunoelectrophoresis with free Con A added to the first dimension gel was performed as described by Bøg-Hansen et al. (14) on 7 x 7 cm glass slides. The buffer was TRIS barbital buffer with ionic strength o.o2o and pH 8.5. The wells received 5 µl antigen samples. The first dimension gel contained 3oo µg Con A/cm^2 corresponding to 1.82·lo^{-5} M Con A and it was run at lo V/cm until a bromphenol blue labelled albumin marker had migrated 4 cm from the antigen well. An intermediate gel of lo mm was cast between the first dimension gel and the antibody-containing second dimension gel. The second dimension electrophoresis was performed in an antibody-containing gel at 2.5 V/cm for 18 hours. As control the antigen samples were tested in crossed immunoelectrophoresis without Con A. The slides were stained with Coomasie Brilliant Blue. The slides were magnified lo times by projection on paper and the areas of the different peaks measured by planimetry. Five identical electrophoreses were performed using the same antigen samples and the variation of the planimetry results was found to be within $\pm$ lo%.

Results

The maternal serum level of m-AFP was measured by rocket line immunoelectrophoresis in 22o serum samples from 22o primigravid mice. m-AFP was detectable from Day-lo of pregnancy, increased exponentially until Day-19 and decreased before parturition (Fig. 1 a). The half-life of m-AFP was 15 hours, determined by the changes of m-AFP concentration in serum of male and non-pregnant female mice injected with 5oo AU of m-AFP.

The fetal weight was measured in the 9o mice in which the circulating maternal m-AFP concentration could be measured (Fig. 1 b). The correlation coefficient of maternal AFP level and fetal weight during pregnancy was o.99, which is significant at $p < o.ool$ (Student's t-test). Fig. 1 c shows the placental weight of the 9o mice and the correlation coefficient of maternal m-AFP level and placental weight during pregnancy was o.52, significant at $p < o.o5$ (Student's t-test). All values in Fig. 1 represent mean $\pm$ 1 s.e.m., each point represents mean values of lo samples on each day.

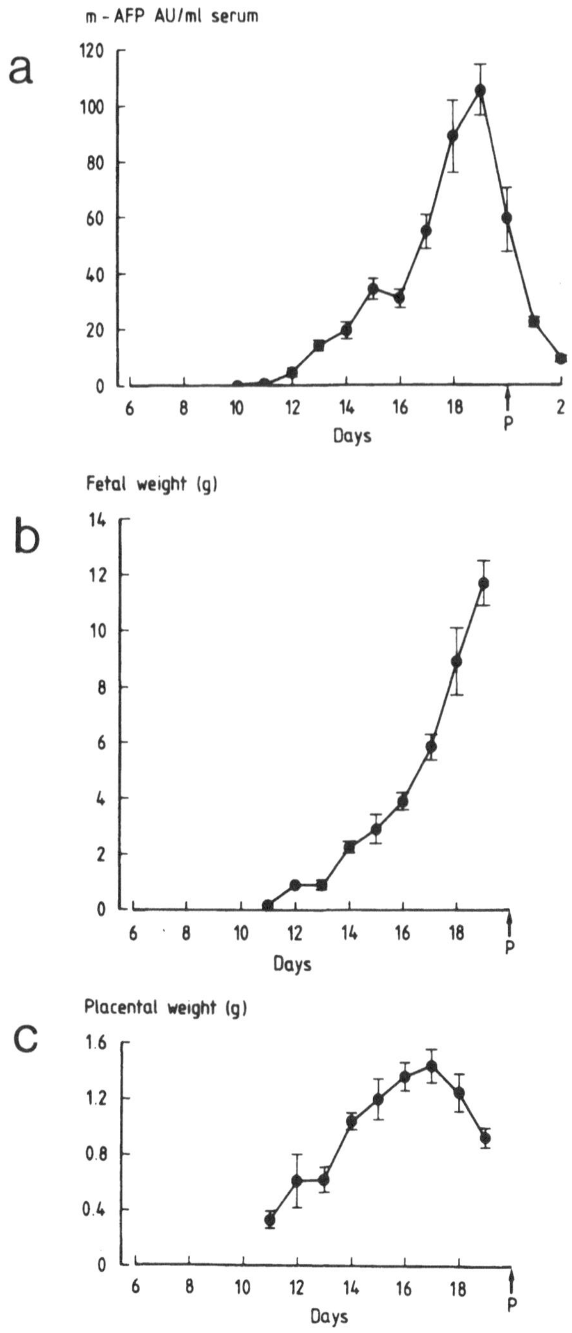

Fig. 1: See text for explanation.

Four molecular forms of m-AFP were detected in crossed affino-immunoelectrophoresis in fetal and juvenile tissue (Fig. 2). They were designated 0, 1, 2 and 3 according to increasing affinity to Con A, form 0 being non-reactive. Fig. 2 a shows m-AFP in crossed immunoelectrophoresis with fetal homogenate as the antigen. Figs. 2 b, c, d show crossed affino-immunoelectrophoreses with 3oo µg Con A/cm^2 in the first dimensional gel and with b) 18-day fetal homogenate as the antigen, c)

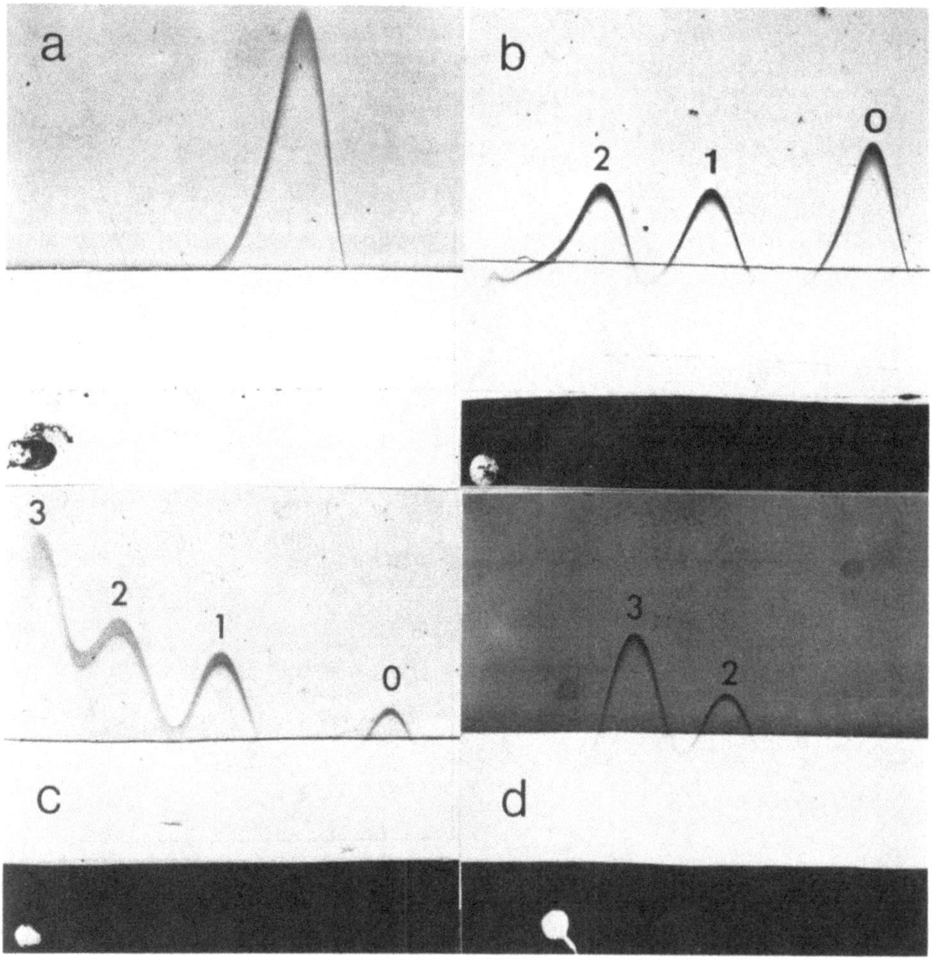

0-day neonatal homogenate as the antigen and d) 4-day neonatal homogenate as the antigen.

The fetal homogenate which was injected into non-pregnant animals to calculate the half-life of m-AFP was examined in crossed affino-immunoelectrophoresis and consisted of molecular forms 0 and 1 only. The consecutive blood samples which were drawn during the following 24 hours contained only m-AFP in the molecular forms 0 and 1.

The existence of a molecular form of fetal m-AFP which could be completely precipitated by free Con A was examined by comparison of rockets in affino-immunoelectrophoresis with rockets in immunoelectrophoresis without Con A. The rocket heights were identical in the two types of electrophoresis, indicating that no completely precipitable form of m-AFP was present.

The different Con A-binding molecular forms of m-AFP in fetal homogenate were also present in fetal serum, fetal brain extract, fetal spleen extract, fetal liver extract and amniotic fluid, whereas only forms 0 and 1 could be detected in serum from females 16 to 19 days pregnant.

The total m-AFP level was measured by rocket line immunoelectrophoresis in fetal and juvenile homogenates.

The levels of the four molecular forms of m-AFP were measured by quantitative crossed affino-immunoelectrophoresis during fetal and neonatal development (Fig. 3). The concentration of each molecular form was measured by its percentual contribution to the total m-AFP level. In 12-day-old fetuses the two forms with weakest affinity to Con A (0 and 1) dominate, whereas in homogenates of juveniles the two forms with strongest affinity to Con A (2 and 3) dominate. At the time of birth all four forms of m-AFP are present and contribute approximately 25% each to the total AFP level.

Discussion

The level of m-AFP in maternal serum during pregnancy increased exponentially throughout pregnancy (Fig. 1 a) and the correlation coefficient between fetal weight (Fig. 1 b) and maternal m-AFP level was o.99. The correlation between maternal m-AFP level and placental weight (Fig. 1 c) was much lower (r = o.52).

The very close correlation between maternal m-AFP level and fetal weight seems to make mice a relevant choice as a model for the study of various treatments affecting fetal growth and development.

Four electrophoretically different molecular forms of m-AFP were demonstrated by using Con A in the first dimension gel in crossed affino-immunoelectrophoresis, in homogenates of fetuses and juveniles. Only molecular forms 0 and 1 could be detected in maternal blood, and when the molecular forms 0 and 1 were injected into non-pregnant animals, only the molecular forms 0 and 1 could be detected until the protein was catabolized.

We were unable to detect any form of m-AFP with more than one Con A-binding carbohydrate residue. If such forms were present in our material, they would be precipitated by the Con A in the first dimension gel in crossed affino-immunoelectrophoresis (15). Accordingly, rocket heights of fetal AFP in rocket affino-immunoelectrophoresis with Con A would have been smaller than in the rocket immunoelectrophoresis without Con A in the intermediate gel.

It has been found that AFP in early murine fetuses contain less sialic acid, than in later stages of development (7, 11). This group was able to detect up to five electrophoretic forms of m-AFP by polyacrylamide gel electrophoresis. These forms were designated Fp1 and Fp5, where Fp1 contained less than 1

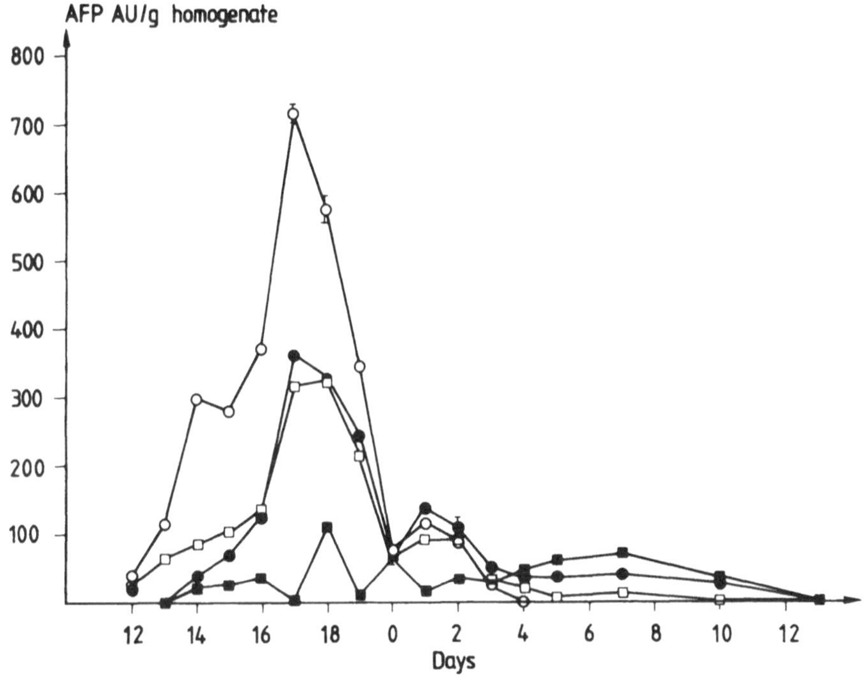

Fig. 3: The level of the four molecular forms of m-AFP from 12-day fetuses to lo-day juveniles. Each point represents mean values of lo samples ± 1 s.e.m. O-O represent molecular form 0, □—□ represent molecular form 1, ●—● represent molecular form 2 and ■—■ represent molecular form 3.

residue of sialic acid and Fp5 contained 2.2-2.7 residues of sialic acid (7). We interpret the changes from Con A non-reactive and weakly reactive forms to strongly binding forms of AFP as an expression of completion of the enzymatic system for glycosylation of AFP.

The concept of a gradual change towards more glycosylated AFP forms during murine fetal development is in accordance with results concerning AFP during human fetal development (16). However, during fetal development in the rat, AFP interaction with Con A reveals a gradual change from mainly Con A reactive forms towards a Con A non-reactive form (17, 18). An explanation of the appearance of increasingly glycosylated m-AFP could be that the yolk sac primarily secretes the less glycosylated forms and the liver secretes the more glycosylated forms of m-AFP (12). Another possible reason might be a developmental step-by-step establishment of the complete membrane-bound multiglycosyltransferase system of the liver cell. This would account for the increasing amount of the more glycosylated forms of m-AFP during development.

Although the different molecular forms of m-AFP found in fetal homogenate were present in amniotic fluid, fetal serum and extracts of fetal brin, spleen and liver, only the molecular forms 0 and 1 could be detected in samples of maternal serum. This indicates that fetal m-AFP is changed or that the molecular forms 0 and 1 are selectively transferred during the transplacental passage. If the different forms of m-AFP are not transferred in the same proportions as they exist in amniotic fluid and fetal serum, it complicates the direct usefulness of the Con A-binding pattern of AFP in maternal serum as a clinical indicator of fetal abnormalities as suggested by Kelleher and Smith (19). However, although the potential biological significance of the heterogeneity of AFP, revealed by Con A and other lectins, is not known yet, the present results demonstrate the potential diagnostic value of the heterogeneous nature of AFP in detection of fetal malformations (2o) and various cancer diseases (21, 22).

Acknowledgements

This work was supported by the Danish Medical Research Council (Project No. 512-2oo84) and F. L. Smidth & Co.'s Jubilæumsfond.

References

1. Alpert, S., Drysdale, J. W., Isselbacher, K. J., Schur, P. H.: Human α-Fetoprotein. Isolation, Characterization and Demonstration of Microheterogeneity. J. Biol. Chem. 247, 3792-3798.(1972)

2. Smith, C. J., Kelleher, P. C., Billanger, L. and Dallaire, L.: Reactivity of amniotic fluid alpha-fetoprotein with concanavalin A in diagnosis of neural tube defects. Br. Med. J. 1, 92o-921 (1979).

3. Hindersson, P., Toftager-Larsen, K. and Nørgaard-Pedersen, B.: Concanavalin A reactivity of amniotic fluid alpha-foetoprotein in diagnosis of neural tube defects. Lancet 11, 9o6 (1979).

4. Ruoslahti, E.,Pekkala, A., Connings, D. E. and Seppälä, M.: Determination of subfractions of amniotic fluid alpha-fetoprotein in diagnosing spina bifida and congenital nephrosis. Br. Med. J. 3, 768-769 (1979).

5. Nørgaard-Pedersen, B., Toftager-Larsen, K., Philip, J. and Hindersson, P.: Concanavalin A reactivity pattern of human amniotic fluid AFP examined by crossed affino-immunoelectrophoresis. A definite test for neural tube defect? Clin. Genet. 17, in press. (198o)

6. Gustine, D. L. and Zimmerman, E. F.: Amniotic fluid proteins. Evidence for the presence of foetal plasma glycoproteins in mouse amniotic fluid. Amer. J. Obstet. Gynecol. 114, 553-56o. (1972)

7. Gustine, D. L. and Zimmerman, E. F.: Developmental changes in microheterogeneity of foetal plasma glycoproteins of mice. Biochem. J. 132, 541-551. (1973)

8. Kerckaert, J.-P., Bayard, B. and Biserte, G.: Microheterogeneity of rat, mouse and human α_1-fetoprotein as revealed by polyacrylamide gel electrophoresis and by crossed immuno-affino-electrophoresis with different lectins. Biochem. Biophys. Acta 576, 99-1o8. (1979)

9. Higgins, P. J.: Heterogeneity, immunological comparison and concentration profiles of alpha-foetoprotein derived from late-gestational and early postnatal mouse tissue. J. Reprod. Immunol. 1, 75-87. (1979)

1o. Ruoslahti, E. and Adamson, E.: Alpha-fetoproteins produced by the yolk sac and the liver are glycosylated differently. Biochem. Biophys. Res. Comm. 85, 1622-163o. (1978)

11. Zimmerman, E. F. and Madappally, M. M.: Sialyltransferase: Regulation of alpha-fetoprotein microheterogeneity during development. Biochem. J. 134, 8o7-81o. (1973)

12. Wilson, J. R. and Zimmerman, E. F.: Yolk sac: Site of developmental microheterogeneity of mouse alpha-fetoprotein. Develop. Biol. 54, 187-2oo. (1976)

13. Krøll, J.: Rocket-line immunoelectrophoresis. Scand. J. Immunol. Vol. 2., Suppl. 1, 83-87. (1973)

14. Bøg-Hansen, T. C., Bjerrum, O. J. and Ramlau, J.: Detection of biospecific interaction during the first dimension electrophoresis in crossed immuno-electrophoresis. Scand. J. Immunol. 4, Suppl. 2, 141-147. (1975)

15. Bøg-Hansen, T. C.: Some characteristic reactions between glycoproteins and lectins in analytical affinity electrophoresis. In: Protides of the Biological Fluids. Proceedings of the 27th Colloq. (H. Peeters, ed.) 659-664.(198o)

16. Ruoslahti, E. R., Engvall, E., Pekkala, A. and Seppälä, M.: Developmental changes in carbohydrate moity of human alpha-fetoprotein. Int. J. Cancer, 22, 515-52o. (1978)

17. Fournier, C., Kerckaert, J. P., Bayard, B., Collyn, M. and Biserte, G.: Crossed lectin-immuno-electrophoresis of in vivo and in vitro produced rat alpha-foetoprotein. In: Protides of Biological Fluids, 27th Colloq. (H. Peeters, ed.) 623-626. (1979)

18. Kerckaert, J. P. and Bayard, B.: These proceedings.

19. Kelleher, P. C. and Smith, C. J. P.: Origin of maternal serum alpha-fetoprotein. Lancet, Dec. 15., 13o2-13o3. (1979)

2o. Toftager-Larsen, K. and Nørgaard-Pedersen, B.: These proceedings.

21. Breborowicz, J. and Mackiewicz, A.: These proceedings.

22. Mackiewicz, A. and Breborowicz, J.: These proceedings.

THE DETECTION OF HORMONE-ASSOCIATED VARIATIONS IN α_1-ACID GLYCOPROTEIN USING CONCANAVALIN A CROSSED IMMUNO-AFFINOELECTROPHORESIS

C. Wells, E.H. Cooper, M.R. Glass
The Unit for Cancer Research and Department of Obstetrics and Gynaecology, University of Leeds, Leeds, England

T.C. Bøg-Hansen
The Protein Laboratory, University of Copenhagen, Copenhagen, Denmark

Introduction

Crossed immunoelectrophoresis is a semi-quantitative technique comprising of a normal electrophoretic step in the first dimension, followed by electrophoresis into an antibody containing gel in the second dimension(1). Crossed immuno-affinoelectrophoresis (CIAE) is a technique in which lectin is incorporated into the first dimension gel, thus retarding, either partially or completely, any glycoproteins that interact with the lectin (2). When human serum was electrophoresed on this system, using the lectin concanavalin A (con A), several of the constituent proteins displayed microheterogeneity, including α_1-acid glycoprotein (α_1-AGP) (2). α_1-AGP is a globulin of molecular weight 40,000. It has a very high carbohydrate content, comprising 45% by weight of the glycoprotein. Yamauchi and Yamashina (3), using DEAE-cellulose chromatography and starch gel electrophoresis, isolated various polymorphic forms of α_1-AGP which had very similar amino acid compositions but differed in their relative contents of neutral sugars. More recently, using 360-MHz 'H NMR spectroscopy and permethylation analysis, α_1-AGP has been shown to possess five different types of heteroglycans (4).

In the present study, the heterogeneity of the carbohydrate part of α_1-AGP, as revealed by con A CIAE (2), was studied under various conditions associated with alterations in the serum concentration of female sex hormones.

Lectins - Biology, Biochemistry, Clinical Biochemistry, Vol. I

Materials

Normal human sera and pregnancy sera were supplied by The Regional Transfusion Centre, Leeds, post partum sera by the Department of Obstetrics & Gynaecology, The University of Leeds and sera from patients with prostatic cancer were drawn from the serum bank at the Unit for Cancer Research, The University of Leeds.

Sheep antiserum for human α_1-acid glycoprotein (code No. BA20) was obtained from Seward Laboratory, U.A.C. House, Blackfriars Road, London, U.K. Concanavalin A and agarose (Type II: Medium EEO) were obtained from the Sigma Chemical Co. Ltd., Poole, Dorset, U.K.

Methods

Crossed immunoelectrophoresis was carried out essentially as described by Weeke (5) and modified by Bøg-Hansen for the incorporation of con A into the first dimension gel (2). The concentration of con A incorporated into the first dimension gel was 70 $\mu g/cm^2$ which corresponds to 4.0 $\mu mol/l$. Antiserum specific to α_1-AGP was incorporated into the second dimension gel to give a concentration of 1.0 $\mu l/cm^2$. The gel was poured to a depth of 1.5 mm. A sample volume of 2 μl of human serum was used on each occasion.

17 β-Oestradiol, progesterone and prolactin were measured by radioimmunoassay in the Department of Steroid Biochemistry, The University of Leeds and in the Department of Nuclear Medicine, Leeds General Infirmary (6).

Results

The patterns obtained by crossed immunoelectrophoresis of normal human serum in the absence and presence of con A in the first dimension gel are shown in Fig. 1. The single peak observed in the absence of lectin

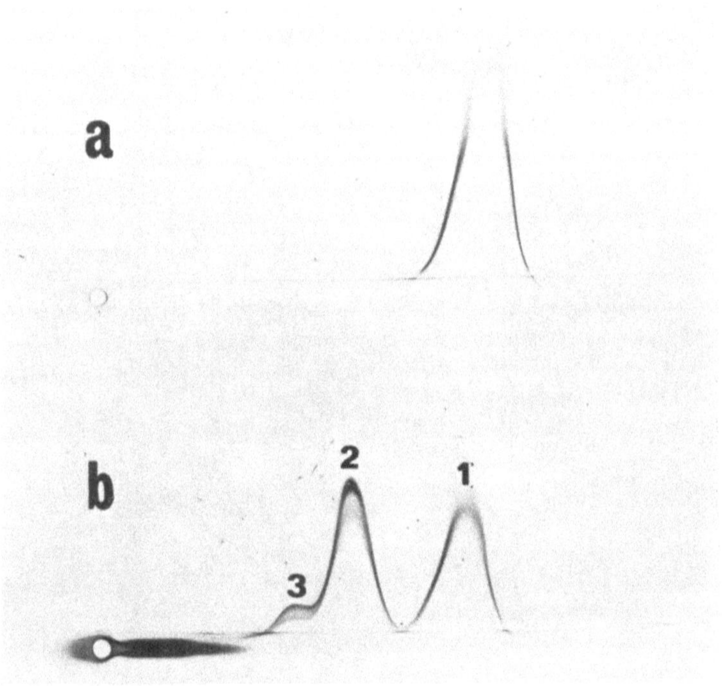

Fig. 1 - Crossed immunoelectrophoresis of normal human serum (2 μl) against antiserum to α_1-AGP in the second dimension. The first dimension gel contained a) no con A and b) 70 $\mu g/cm^2$ con A.

separates into three species in the presence of con A. This is consistent with previous findings (2,7). The relative proportions of components 1, 2 and 3 (Fig. 1) are 49% ($\pm$ 4.2%), 43% ($\pm$ 3.1%) and 8% ($\pm$ 2.8%) respectively, the proportions being calculated by measuring the areas under the peaks. The extent of retardation in the first dimension is indicative of the affinity of the species for con A. Component 1 does not react weakly and component 3 reacts relatively strongly with the lectin.

Pregnancy sera, however, display a different pattern on con A CIAE. Fig. 2 shows the electrophoresis on this system of sera taken during pregnancy at a) 8 weeks and b) 34 weeks gestation. In early pregnancy, the three components are present in the same proportions as in normal sera. In late pregnancy, peak 3 is no longer present and the relative amounts of peaks 1 and 2 are approximately 77% and 23% respectively, thus indicating a smaller

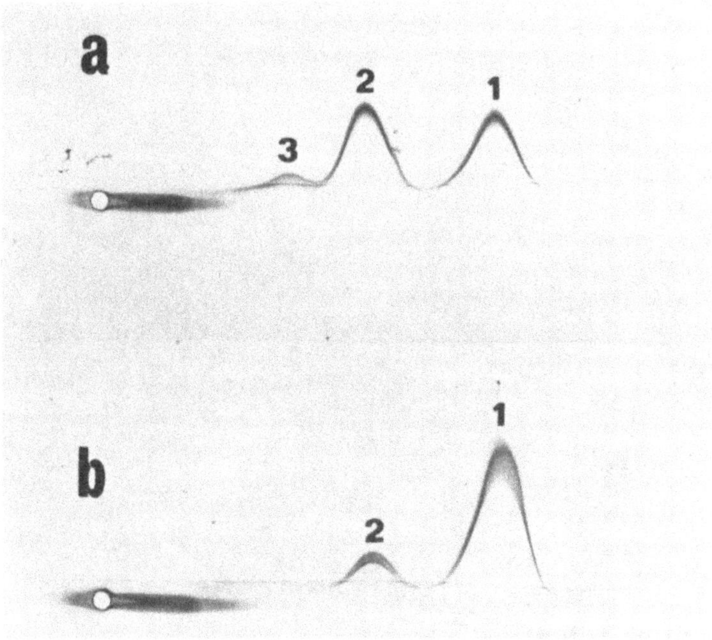

Fig. 2 - Con A crossed immuno-affinoelectrophoresis, against antiserum to α_1-AGP in the second dimension, of human pregnancy sera (2 μl) at a) 8 weeks and b) 34 weeks gestation.

amount of con A binding material than in normal sera. This change in pattern occurs between 6 and 16 weeks, during which time intermediate patterns are observed. A total of 65 pregnancy sera were studied and all shown to conform with this change of pattern leading to the concept of a "pregnancy-type" α_1-AGP pattern of con A binding. Peaks 1 and 2 of the normal pattern and the pregnancy-type pattern were shown to coincide by mixing normal sera and late pregnancy sera and electrophoresing the mixed sample.

A pregnancy-type con A binding pattern for α_1-AGP was also seen when patients with prostatic cancer were treated with oestrogen therapy. Fig. 3 shows the patterns obtained before and after treatment with estramustine phosphate .

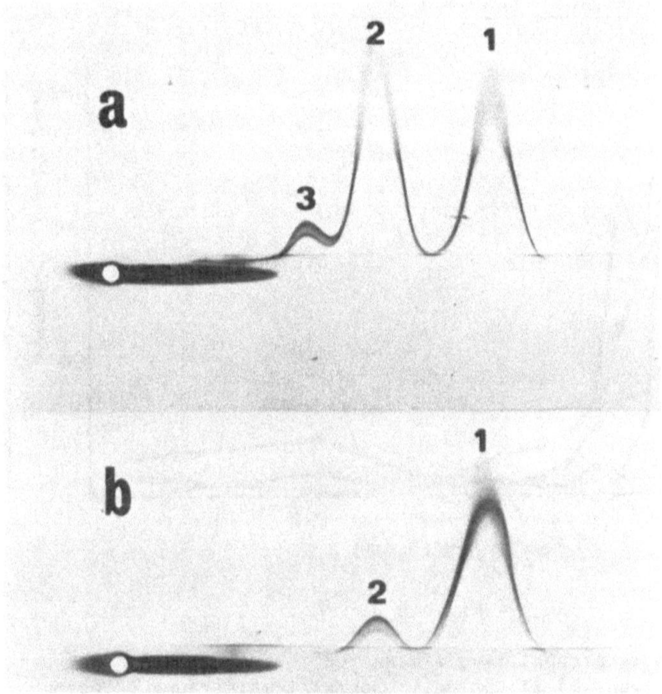

Fig. 3 - Con A crossed immuno-affinoelectrophoresis, against antiserum to α_1-AGP in the second dimension, of sera (2 1) from a patient with prostatic cancer a) before treatment and b) after treatment with estramustine phosphate.

Post partum studies were carried out in an attempt to find a cause for this change in the con A binding pattern of α_1-AGP. Following delivery, 17 β-oestradiol, progesterone and prolactin levels fall to within the normal range in a few days in the non-lactating woman, as shown in Fig. 4a. Prolactin levels remain raised for a longer period in the lactating woman (Fig. 4b). Con A CIAE carried out on a longitudinal series of post partum sera from two lactating and two non-lactating women, however, showed that the pregnancy-type pattern of α_1-AGP binding to con A persisted for at least 42 days post partum in all four cases.

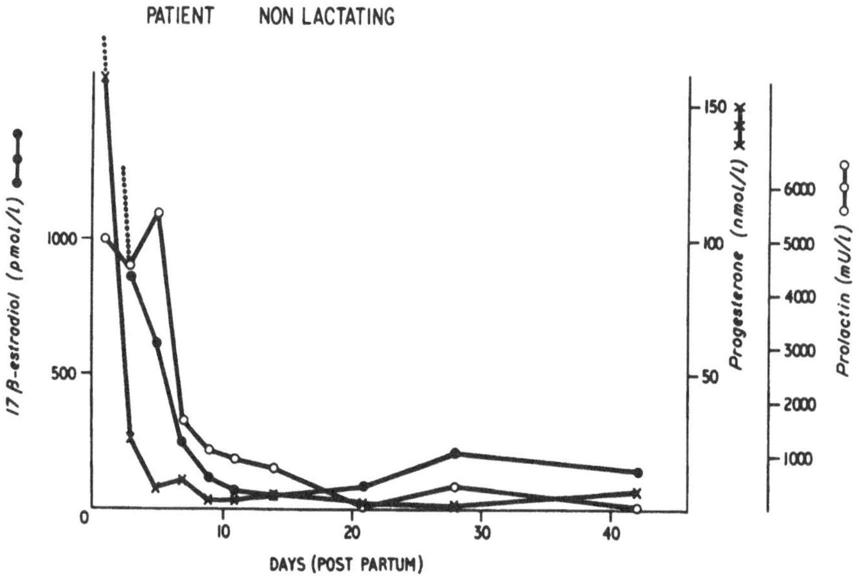

Fig. 4a

Fig. 4 a & b - Hormonal levels measured in two patients post partum a) non-lactating b) lactating. Normal female range of hormones during menstrual cycle: 17 β-oestradiol 200 - 2000 pmol/l; progesterone <2 - 80 nmol/l and prolactin <500 mU/l.
The pregnancy-type con A binding pattern of α_1-AGP persists throughout the time courses shown.

Discussion

α_1-Acid glycoprotein was shown to be homogeneous on a normal crossed immunoelectrophoretic system, but displayed a degree of heterogeneity when con A was incorporated into the first dimension. The lectin con A binds to mannose and glucose residues and therefore reacts with glycoproteins containing these residues in an exposed position. Theoretically differences in carbohydrate structure may give rise to multiple peaks, the position of which is determined by their affinity to the lectin (8). Thus, the behaviour of α_1-AGP on con A CIAE is in agreement with the heterogeneity in the carbohydrate part of this protein shown by other workers (3, 4). The three components are found in normal sera and

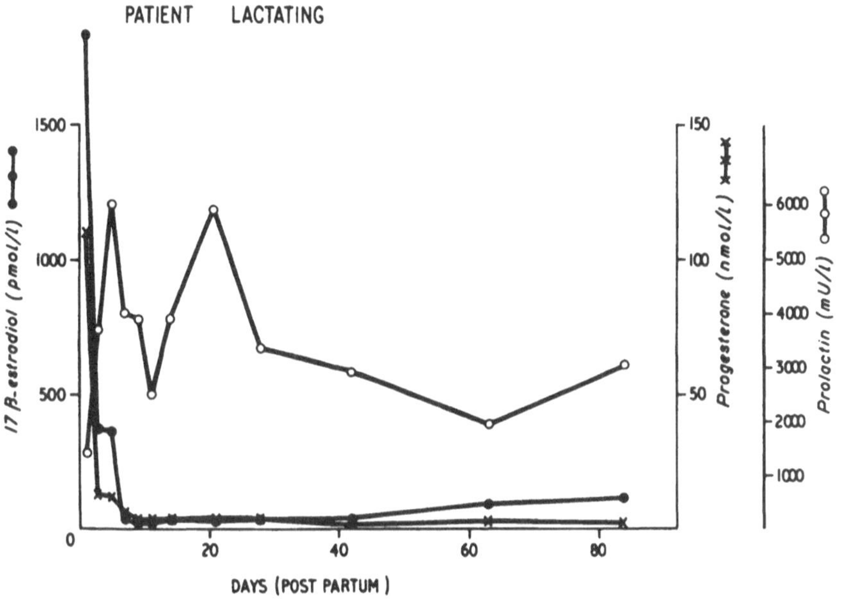

Fig. 4b

therefore may be related to the normal function and metabolism of the protein.

The change in pattern of α_1-AGP on con A CIAE, observed during pregnancy and following oestrogen therapy, strongly suggests that female hormones affect the carbohydrate composition of α_1-AGP and therefore possibly its metabolism or function. Progesterone (9) and oestradiol (10) have been shown to bind to α_1-AGP and it may be that the affinity of this glycoprotein for these hormones depends on the composition of the carbohydrate moiety. Post partum studies, however, show that the pregnancy-type pattern is retained for many weeks after hormonal levels return to normal, despite the half-life of the protein being 5.5 days (11).

Thus, a change in the carbohydrate part of α_1-AGP has been observed, which may affect the metabolism or function of the protein. This change may be induced by an increase in female hormones, as in pregnancy or oestrogen therapy. The results presented here, however, do not support a direct hormonal effect, as post partum studies show that the pattern is not readily returned to normal once the hormone balance is restored. The carbohydrate composition of the different components has yet to be elucidated.

References

1. Clarke, H.G.M., Freeman, T.: Clin. Sci. 35, 403-413 (1968).
2. Bøg-Hansen, T.C., Bjerrum, O.J., Ranlau, J.: Scand. J. Immunol. 4, Suppl. 2, 141-147 (1975).
3. Yamauchi, T., Yamashina, I.: J. Biochem. (Tokyo) 66, 213-223 (1969).
4. Fournet, B., Montreuil, J., Strecker, G., Dorland, L., Haverkamp, J., Vliegenthart, J.F.G., Binette, J.P., Schmidt, K.: Biochemistry 17, 5206-5214 (1977).
5. Weeke, B.: Scand. J. Immunol. 2, Suppl. 1, 47-56 (1973).
6. Martin, R.H., Glass, M.R., Chapman, C., Wilson, G.D., Woods, K.L. : Clin. Endocrinol. (in press).
7. Bøg-Hansen, T.C., Prahl, P., Løwenstein, H.: J. Immunol. Meth. 22, 293-307 (1978).
8. Bøg-Hansen, T.C., Takeo, K.: J. Electrophoresis 1, 67-71 (1980).
9. Westphal, U., Ashley, B.D., Sheldon, G.L.: Arch. Biochem. Biophys. 92, 441-448 (1961).
10. Kerkay, J., Westphal, U.: Biochim. Biophys. Acta 170, 334-337 (1968).
11. Winzler, R.J. : Clin. Chem. 11, 329-347 (1965).

IDENTIFICATION AND CHARACTERIZATION OF PLATELET GLYCOPROTEINS BY CROSSED AFFINITY IMMUNOELECTROPHORESIS

I. Hagen and G. Gogstad
Research Institute for Internal Medicine, Section on Hemostasis and Thrombosis, University of Oslo, Oslo, Norway.

Introduction

Blood platelets are small, anucleate cells which circulate in the blood vessels. They play an essential role in the arrest of bleeding (hemostasis) by adhering to the exposed subendothelial tissue at the site of vessel wall injury, followed by aggregate formation and secretion of substances stored in intracellular granules. Platelets also take actively part in the development of pathological conditions such as thrombosis, probably through the same reactions as those involved in hemostasis. Profound morphological and biochemical alterations take place during these processes, which are still far from clarified (for review see 1,2).

There is increasing evidence that platelet glycoproteins are important for a normal platelet adhesion and aggregation (3-6). This evidence is mainly based on studies of platelets from patients with certain specific platelet disorders, i.e. the Bernard-Soulier syndrome and Glanzmann's thrombasthenia (7). In the first-mentioned disorder the platelets show an impaired ability to adhere to the damaged vessel wall. SDS PAGE of such platelets have revealed a strongly reduced intensity or absence of a 150,000 molecular weight glycoprotein band (GP I) compared to normal platelets. Two membrane glycoproteins (GP II and GP III) are absent or appear in strongly reduced amounts in Glanzmann's thrombasthenia. The main characteristic

Lectins - Biology; Biochemistry, Clinical Biochemistry, Vol. I of these platelets is their inability to form aggregates upon ADP-stimulation. Recently, it has been shown that an α-granule located glycoprotein (GGP or TSP), which is released upon activation of the platelets, is present in strongly reduced amounts in platelets from patients with the Gray platelet syndrome (8). In these platelets the number of α-granules is less than normal, whereas the other intracellular organelles are as in normal platelets.

There are few reports on the use of lectins to characterize platelet glycoproteins. In most studies the identification of platelet glycoproteins has been based on the positive PAS staining after SDS PAGE. However, McGregor and coworkers (9) and Judson and Anstee (10) have been able to identify several platelet glycoproteins by incubation of the SDS polyacrylamide gels with radioactively labeled lectins followed by autoradiography. Lectins have also been used for the isolation of certain platelet glycoproteins (11,12). We have introduced the method of crossed immunoelectrophoresis as described by Bjerrum and Bøg-Hansen (13) to characterize platelet proteins (14,15). This technique permits the rapid identification of glycoproteins by insertion of insolubilized lectins between the first dimension gel and the gel containing the antibodies or by including free lectins in the first dimension gel (13, 16). In the present investigation we have examined platelet glycoproteins by these techniques.

Materials and Methods

Washed human platelets (14) and isolated α-granules (17) were solubilized in Triton X-100 (1% v/v) to a final concentration of 2-10 mg/ml. Anti-platelet antibodies were obtained by immunizing rabbits with a suspension of whole platelets (14). Glycocalicin is an easily soluble platelet glycoprotein which is solubilized from the membrane probably by a Ca^{2+}-activated protease (12,18). It shows immunochemical identity with GP I (18). Monospecific antiserum to glycocalicin was kindly provided by Dr. N. O. Solum, Oslo. Lentil lectin-Sepharose 4B and wheat germ lectin-Sepharose 6MB were from Pharmacia Fine Chemicals, Uppsala, Sweden, and lectin from Ricinus communis was obtained from Dr. E. Saltvedt, Tønsberg, Norway, or from Sigma Chemical Co., St. Louis, USA. Agarose, type HSA, was from Litex, Glostrup, Denmark. Crossed affinity immunoelectrophoresis using either insolubilized lectin in the intermediate gel or free lectin in the first dimension gel was essentially performed as described previously (13,14,16).

Results and Discussion

The reference pattern obtained by crossed immunoelectrophoresis of Triton-solubilized platelet proteins against anti-platelet antibodies is shown in Fig. 1. The arcs are numbered according to their position relative to the anode as described previously (14). Several of the immunoprecipitates have been identified by use of either monospecific antiserum, purified antigens or SDS PAGE of immunoprecipitates which had been cut out from unstained immunplates (Table I).

Table I. Identification of some of the immunoprecipitates.

Immunoprec. No.	Identified as	Method of identification
1	Platelet f.4	Monospec. antiserum
6	Albumin	Monospec. antiserum
13	GP I	Monospec. antiserum (15)
14	Granule glycopr.	Purif. antigen (19)
16	GP II/GP III	SDS PAGE of immunoprec.
19	Factor XIII	Monospec. antiserum (19)
24	Fibrinogen	Monospec. antiserum (14)

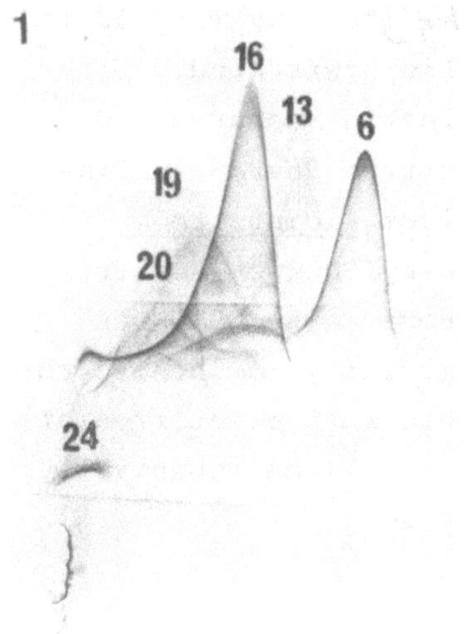

Fig. 1

Crossed immunoelectrophoresis of Triton-solubilized platelet proteins (150 µg) using whole platelet antibodies (600 µg/cm^2). The first dimension electrophoresis was run at 10 V/cm for 45 min., and the second dimension overnight at 1-2 V/cm. The electrophoresis buffer consisted of Tris (0.038 M), glycine (0.1 M), pH 8.6. An intermediate gel with no antibodies was inserted beteen the first dimension gel and the gel containing antibodies to improve the resolution.

Immunoprecipitate 13 represents the membrane glycoprotein GP I, and is not detected by crossed immunoelectrophoresis of solubilized Bernard-Soulier platelets. Immunoprecipitate 16 has been shown to consist of a complex between the membrane glycoproteins GP II and GP III. This immunoprecipitate is not seen by crossed immunoelectrophoresis of solubilized platelets from thrombasthenic (type I) patients (15). All these glycoproteins contain sialic acid as evidenced by their reduced electrophoretic mobility after neuraminidase treatment (12,14). Crossed immunoelectrophoresis of solubilized platelet proteins against anti-glycocalicin antiserum revealed a double peak immunoprecipitate (Fig. 2A). The major, slow-migrating component corresponds to GP I (immunoprecipitate 13), whereas the minor, fast-migrating peak represents its split product glycocalicin (12,15). Both these components were completely retained in an intermediate gel containing wheat germ lectin-Sepharose 6MB (Fig. 2B). This affinity of GP I and glycocalicin for wheat germ lectin has been utilized for preparative purposes (12). Immunoprecipitate 16 interacted with Concanavalin A (14), lentil lectin and the lectin from <u>Ricinus</u>

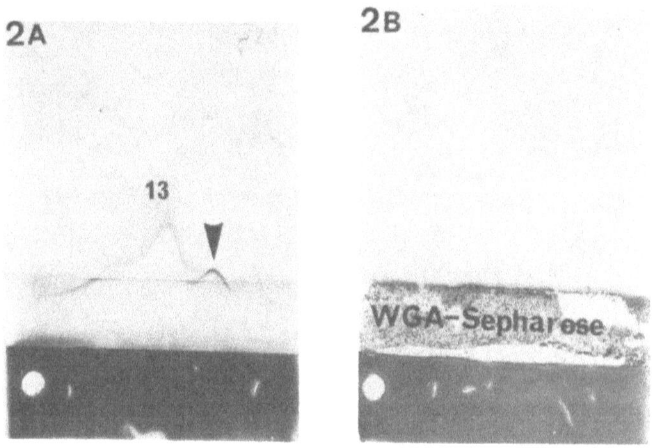

Fig. 2

Crossed affinity immunoelectrophoresis of 150 µg solubilized platelet proteins against anti-glycocalicin antiserum (28 µl/cm^2). (A) Control, no lectins included; (B) Wheat germ lectin-Sepharose 6MB was included in the intermediate gel (a 50% slurry of wheat germ lectin-Sepharose 6MB was mixed with an equal volume of 1% w/v agarose and casted). The arrow indicate the position of glycocalicin.

communis (Fig. 3A-C). However, the presence of 14 µg/cm^2 of wheat germ lectin in the first dimension gel did not affect the position of this immunoprecipitate (14).

The reference pattern obtained by crossed immunoelectrophoresis of proteins solubilized from isolated α-granules has been extensively characterized (19) (Fig. 4A). The high molecular weight granule glycoprotein (GGP or TSP) has been identified by use of purified antigen to be contained in immunoprecipitate 14 (19). This immunoprecipitate was influenced by lentil lectin, but not by Concanavalin A or wheat germ lectin (19). This is in accordance with the findings of McGregor and coworkers (9). Immunoprecipitate 16 was also identified in isolated α-granules, and constituted a component of the granule membrane. This antigen showed immunochemical identity with the plasma membrane component, and exhibited a similar

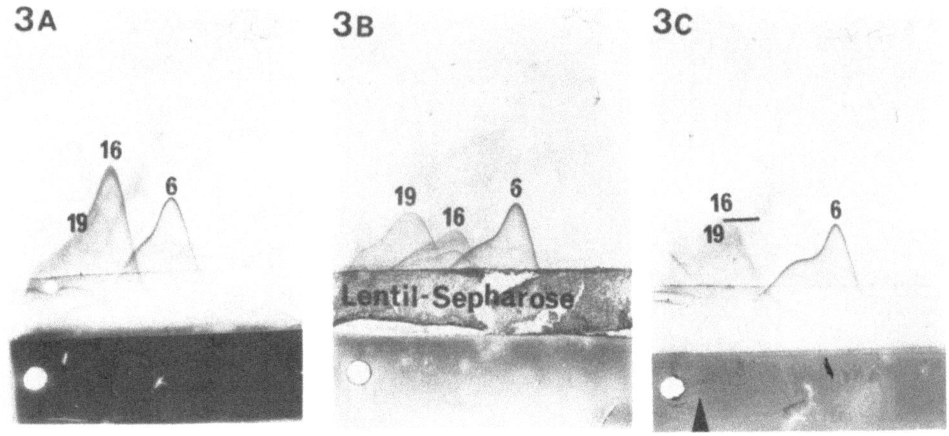

Fig. 3

Crossed affinity immunoelectrophoresis of 150 µg of solubilized platelet proteins against anti-platelet antibodies (600 µg/cm^2). (A) Control, no lectins included; (B) Lentil lectin-Sepharose 4B was included in the intermediate gel (a 50% slurry of lentil lectin-Sepharose 4B in Tris/glycine buffer was mixed with an equal volume of 1%w/v agarose and casted); (C) The first dimension gel contained 51 µg/cm^2 of the lectin from Ricinus communis. The area of immunoprecipitate 16 is reduced when the antigens have to pass a gel of lentil-Sepharose 4B. The bar indicates the shift in position of immunoprecipitate 16 when Ricinus lectin is included in the first dimension gel. Note the affino-precipitate (arrow).

change in electrophoretic migration after neuraminidase treatment (19). When solubilized α-granules were examined, immunoprecipitate 16 showed a shift in migration when wheat germ lectin was included in the first dimension gel (Fig. 4B). Preliminary experiments indicated that this was not the case with the corresponding plasma membrane component. In order to study this in more detail, crossed affinity immunoelectrophoresis with wheat germ lectin was performed of solubilized plasma membranes as well as α-granules. The samples were electrophoresed in parallel, and three different concentrations of the lectin were used (33, 66 and 99 µg/cm^2, respectively). The ratio between the migration distance of No. 16 and an internal standard (No. 6 which represents albumin) was

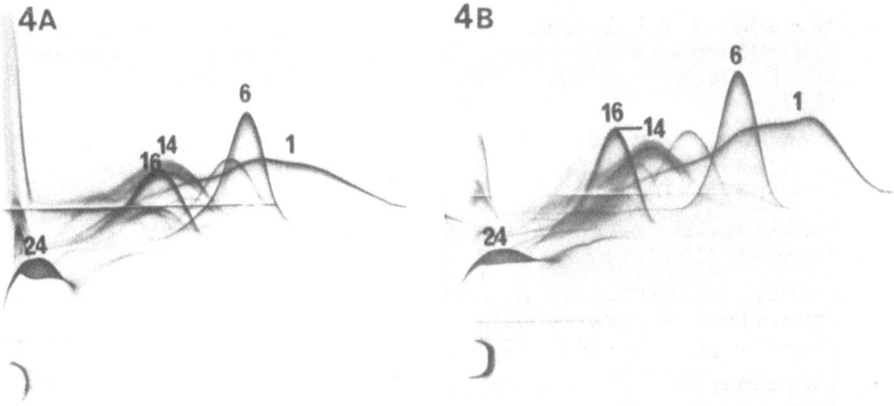

Fig. 4

Crossed affinity immunoelectrophoresis of Triton-solubilized proteins from isolated α-granules (100 μg) using anti-platelet antibodies (1 mg/cm^2). (A) Reference pattern, no lectins included; (B) wheat germ lectin (33 μg/cm^2) was included in the first dimension gel. The bar indicates the shift in the position of immunoprecipitate 16.

calculated for each lectin concentration and compared to the control values. In this study, no pronounced differences between No. 16 in plasma membranes and α-granules could be observed. However, there was a slight tendency of the α-granule component to be more influenced at the lowest concentration of wheat germ lectin tested.

In the present study several platelet glycoproteins have been identified by use of crossed affinity immunoelectrophoresis. The results obtained are in good agreement with those reported by McGregor et al. (9) and Judson and Anstee(10). Furthermore, the results have been used for preparative purposes.

References

1. Mustard, J. F. and Packham, M. A. Pharmacol. Rev. 22, 97-187 (1970).

97-187 (1970).

2. Holmsen, H., Salganicoff, L. and Fukami, M. In Haemostasis, Biochemistry, Physiology and Pathology (D. Ogston and B. Bennett, eds.) John Wiley & Sons, London 1977.

3. Caen, J. P., Nurden, A. T., Jeanneau, C., Michel, H., Tobelem, G. and Levy-Toledano, S. J. Lab. Clin. Med. 87, 586-596 (1976).

4. Phillips, D. R. and Agin, P. P. J. Clin. Invest. 60, 535-545 (1977).

5. Degos, L., Dautigny, A., Drouet, J. C., Colombani, M., Ardaillou, N., Caen, J. P. and Colombani, J. J. Clin. Invest. 56, 235-240 (1975).

6. Jamieson, G. A., Okumura, T., Fishback, B., Johnson, M., Egan, J. J. and Weiss, H. J. J. Lab. Clin. Med.

7. Lusher, J. M. and Bernhart, M. I. In Seminars in Thrombosis and Hemostasis (E. F. Mammon, ed.). Stratton Intercont. Book Corp., New York 1977.

8. Gerrard, J. M., Phillips, D. R., Rao, G., Flow, E. P., Walz, D. A., Ross, R., Harker, L. A. and White, J. G. J. Clin. Invest. 66, 102-109 (1980).

9. McGregor, J. L., Clemetson, K. J., James, E., Greenland, T. and Dechavanne, M. Thromb. Res. 16, 825-831 (1979).

10. Judson, P. A. and Anstee, D. J. In Protides of the Biological Fluids (H. Peeters, ed.) 27, 871-874 (1979).

11. Clemetson, K. J., Pfueller, S. L., Luscher, E. F. and Jenkins, C. Biochim. Biophys. Acta 464, 493-508 (1977).

12. Solum, N. O., Hagen, I., Filion-Myklebust, C. and Stabæk, T. Biochim. Biophys. Acta 597 235-246 (1980).

13. Bjerrum, O. J. and Bøg-Hansen, T. C. Biochim. Biophys. Acta 455, 66-89 (1976).

14. Hagen, I., Bjerrum, O. J. and Solum, N. O. Eur. J. Biochem. 99, 9-22 (1979).

15. Hagen, I., Nurden, A. T., Bjerrum, O. J., Solum, N. O. and Caen, J. J. Clin. Invest. 65, 722-731 (1980).

16. Bøg-Hansen, T. C. In Proceedings of the Third Int. Symp. on Affinity Chromatography and Molecular Interaction (J. M. Egly, ed.) Inserm Symp. Ser. 86, 399-416 (1979).

17. Gogstad, G. Thromb. Res. In press.

18. Solum, N. O., Hagen, I. and Sletbakk, T. Thromb. Res. 18, 773-785 (1980).

19. Gogstad, G., Hagen, I., Korsmo, R. and Solum, N. In prep.

LECTINS AND THE RECEPTOR FOR EPIDERMAL GROWTH FACTOR

E. Nexø
Department of Clinical Chemistry ML, Rigshospitalet (University Hospital), DK-2loo Copenhagen, Denmark

M.D. Hollenberg
Department of Pharmacology and Therapeutics, University of Calgary, Calgary, Canada.

Epidermal growth factor (EGF) is a 6,ooo molecular weight polypeptide that is found in many mammals including man. Though its biological role remains to be elucidated it has a number of known effects such as inhibition of gastric acid secretion, inhibition of palate fusion in vitro and stimulation of cellular growth and differentiation (For survey see (1)).

The first step in the action of EGF is binding to specific receptors present on cellular membranes. The receptor is a 15o,ooo molecular weight glycoprotein that will bind to lectins such as PHA, ConA, WGA and RCA (2), and this ability has proved a most important tool for its further purification. Affinity chromatography of solubilized placental membranes on WGA-Sepharose results in a 3o-fold purification (3), but even more importantly adsorption of the solubilized receptor to insolubillized lectin results in a restoration of its ability to bind EGF, thus making it possible to detect the receptor through various steps of purification (3,4).

The interaction of lectins with the intact receptor has only been studied to a limited extent (5), and the purpose of the present paper is to characterize the effect of ConA upon the binding of EGF to human placenta membranes.

Lectins - Biology, Biochemistry, Clinical Biochemistry, Vol. I

Materials and methods

Human placenta membranes were prepared by differential centrifugation (6).

EGF was purified from male mouse submaxillary glands (7) and iodinated by the Iodogen (Pierce) method (8,9) to a specific activity of approximately 1 µCi per pmol. Concanavalin A (ConA) was from Sigma.

Experimental procedure. loo µl placenta membranes diluted to a protein concentration of approximately o.1 mg membrane protein per ml with buffer A was mixed with loo µl labelled EGF (approximately o.1 pmol) diluted in buffer A. The mixture was incubated for 1 hour at room temperature prior to separation of free and bound peptide by filtration on o.1 µ EGWP millipore filters under reduced presure. All assays were done in duplicate and corrected for a blank, i.e. the same sample incubated with loo pmol unlabelled EGF. 5o µl buffer B or 5o µl ConA (o.ol-lo mg/ml) dissolved in buffer B was added as indicated in the individual experiments. Estimation of binding isotherms were studied by adding unlabelled EGF in a concentration from o.25 to 2o pmol/l. The data were treated according to Scatchard (6). Dissociation curves were obtained by adding loo pmol EGF to each tube at time zero.

Buffer A: o.1 mol/l phosphate, o.15o mol/l sodium chloride, o.1 g/loo ml bovine albumin, pH 7.4.

Buffer B: o.o5 mol/l Tris, 1 mmol/l calcium chloride, 1 mmol/l manganese chloride, pH 7.4.

Results and discussion

The binding of EGF to the placenta receptor has previously been characterized (5). The receptor, found in a density of about 2o pmol per mg protein, binds EGF with an apparent equilibrium constant for association of approximately 2 x lo^9 l/mol. Preincubation of the membranes with lectins such as PHA, WGA and ConA resulted in a decreased binding of EGF (5).

At concentrations of ConA of 1.5 mg/ml the binding of EGF is decreased about 75 %; that is, ConA is not able to inhibit the binding of EGF completely (Fig. 1A). Addition of ConA after labelled EGF has equilibrated with the receptor results in an increased binding of EGF (Fig. 1B).

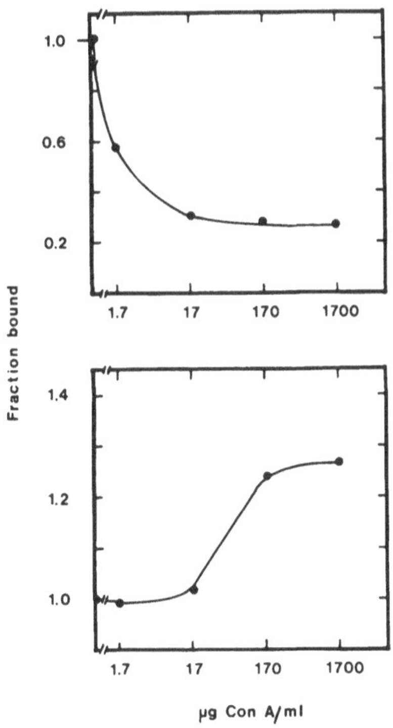

Figure 1. Influence of ConA on the binding of EGF to placenta membranes. A. 5o µl ConA was added to the membranes 3o min prior to labelled EGF. B. 5o µl ConA was added 6o min after addition of labelled EGF. The samples were incubated for an additional 3o min. Abscissa: concentration of ConA in mg/ml on logarithmic scale. Ordinate: fraction bound relative to control sample, i.e. a sample incubated as above with buffer alone.

These apparently contradictory results are explained by the data obtained for the binding isotherms. Fig. 2 presents the results calculated according to Scatchard. Addition of ConA prior to EGF results in a 3-fold decrease in the apparent association constant and an increase in the number of binding sites of about 2o %. Addition of ConA after incubation with EGF results in a similar increase in the number of binding sites, whereas the apparent association constant is alter-

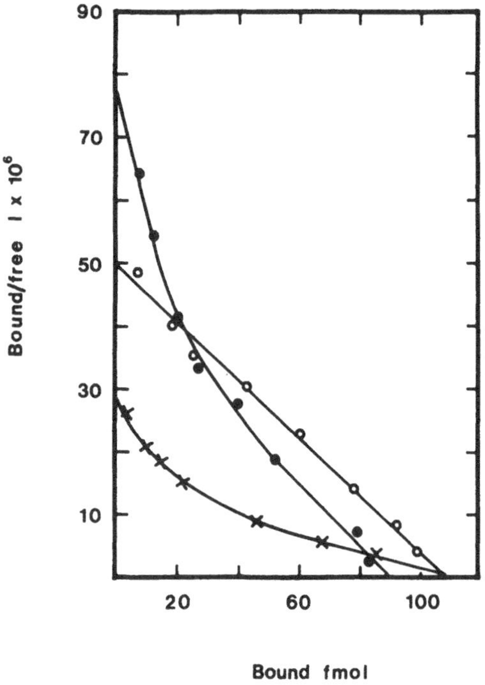

Figure 2. Scatchard plots for the binding of EGF to human placental membranes in the absence or presence of ConA. o——o: Con A not added. x——x: ConA (1 mg/ml) added 1/2 hour prior to addition of labelled EGF. o——o: ConA (1 mg/ml) added 1/2 hour after 1 hours incubation of the membranes with labelled EGF. Abscissa: EGF bound (fmol). The intercept indicates the number of binding sites present on the membranes. Ordinate: EGF bound divided by the molar concentration of free EGF (unit: liter x lo^6). The apparent association constant is the slope of the lines.

ed only to a minor extent. In a further experiment it was shown that the dissociation rate was slowed considerably, when ConA was added after binding of EGF to the receptor (Fig. 3).

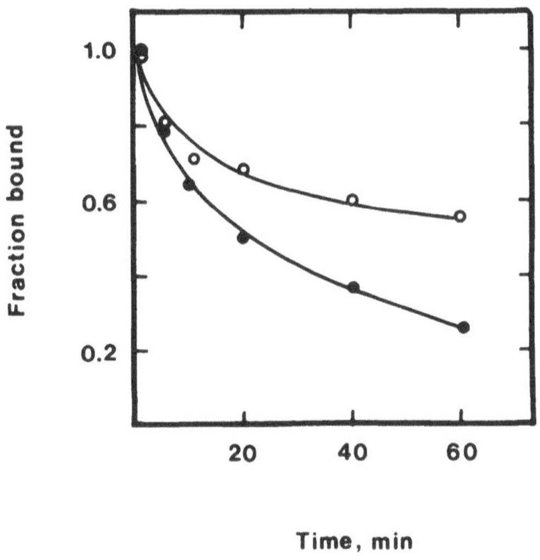

Figure 3. Dissociation of EGF from the receptor in the absence or presence of ConA. o: samples incubated 1 1/2 hour with labelled EGF prior to addition of unlabelled EGF. o: samples incubated 1 hour with labelled EGF and then 1/2 hour with ConA (1 mg/ml) prior to addition of unlabelled EGF. Abscissa: time in minutes. Ordinate: fraction bound relative to amount bound at time zero.

Based on the above results the model schematized in Fig. 4 can be proposed. Two populations of receptors are present. One that is hidden under normal experimental conditions but is unfolded upon exposure to lectin. It is, in this connection, of interest that DNA-virus-transformed cells have been demonstrated to possess a decreased number of receptors for EGF; receptors that are no longer recognized by ConA (11). The lectin added to the membranes prior to EGF binds to the receptor in a manner that decreases the ease of binding of EGF (Fig. 4B). ConA added after binding of EGF stabilizes binding (Fig. 4C).

The physiological role of lectin binding to the EGF receptor is at most speculative. However, the observation has a direct impact on the interpretation of results obtained by receptor assays for EGF (12,13). Results from samples containing lectin like substances should be interpreted with caution.

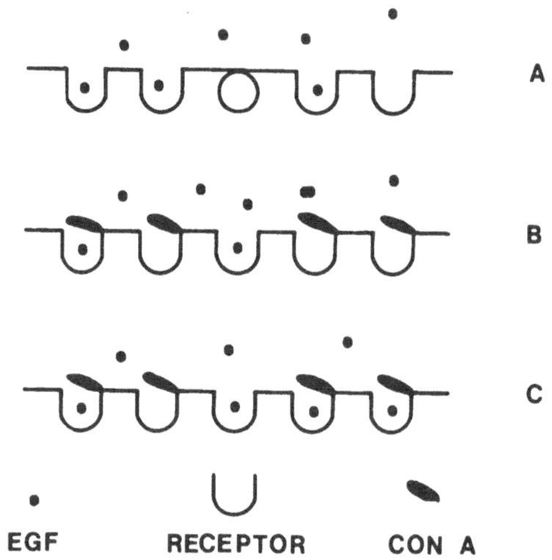

Figure 4. Model for the interaction of ConA with the receptor for EGF. 4A: no ConA added. Some receptors are hidden. 4B: ConA added prior to EGF. The hidden receptors are exposed. The binding of EGF is hindered by the presence of ConA. 4C: ConA added after EGF. The hidden receptors are exposed and binding of EGF is stabilized.

Acknowledgment

The technical assistance of Birgit Walther Jørgensen is warmly appreciated.

References

1. Hollenberg, M.D.: Vitam. Horm. 37, 69-11o (1979).

2. Hock, R.A., Nexø, E., Hollenberg, M.D. Nature 277, 4o3-4o5 (1979).

3. Hock, R.A., Nexø, E., Hollenberg, M.D.: J. Biol. Chem. (198o), in press.

4. Nexø, E., Hock, R.A., Hollenberg, M.D.: J. Biol. Chem. 254, 874o-8743 (1979).

5. Hock, R.A., Hollenberg, M.D.: J. Biol. Chem. (198o), in press.

6. Cuatrecasas, P.: Proc. Natl. Acad. Sci. USA 69, 1277-1281 (1972).

7. Savage, C.R., Jr., Cohen, S.: J. Biol. Chem. 247, 76o9-7611 (1972).

8. Spivak, J.L., Small, D., Hollenberg, M.D.: Proc. Natl. Acad. Sci. USA 74, 4633-4635 (1977).

9. Rees, A.R., Adamson, E.D., Graham, C.F.: Nature 281, 3o9 (1979).

1o. Scatchard, G.: Ann. NY Acad. Sci. 51, 66o-672 (1949).

11. Bernhanu, P., Hollenberg, M.D.: Arch. Biochem. Biophys. 2o3, 134 (198o).

12. Ladda, R.L., Bullock, L.P., Gianopoulis, T., McCormick, L.: Anal. Biochem. 93, 286-294 (1979).

13. Nexø, E., Lamberg, S.I., Hollenberg, M.D.: Comparison of a receptor binding assay with a radioimmunoassay for human epidermal growth factor in urine. In preparation.

IDENTIFICATION OF HUMAN B-LYMPHOCYTE HISTOCOMPATIBILITY ANTIGEN HLA-DR IN CROSSED IMMUNOELECTROPHORESIS OF TRITON X-100 SOLUBILIZED LYMPHOCYTE MATERIAL

T. Plesner, M. Wilken, M. Mørk Hansen
Department of Clinical Chemistry and Department of Medicine, The Finsen Institute, Strandboulevarden 49, DK-2100 Copenhagen, Denmark

O.J. Bjerrum
The Protein Laboratory, University of Copenhagen, Sigurdsgade 34, DK-2200 Copenhagen, Denmark

Abstract

A human B-lymphocyte alloantigen of interest in classification of leukemias and malignant lymphomas has been has identified in crossed immunoelectrophoresis of detergent solubilized chronic lymphocytic leukemia lymphocytes. The following properties were demonstrated: (1) glycoprotein with affinity for Lentil lectin, (2) exposed on the surface of intact B-lymphocytes, (3) consists of two peptides (MW 29,000 and 32,000) and (4) amphiphilic membrane protein. Antisera raised against the isolated immunoprecipitate reacted with B-lymphocytes.

Introduction

Human B-lymphocyte associated alloantigen (HLA-DR) is an example of a cell membrane glycoprotein which is only present on specific, functional subgroups of lymphocytes (2, 11, 24). The demonstration of such antigens has been of great interest in basic immunology and is gaining importance in clinical medicine where leukemias and malignant lymphomas may be subdivided accor-

Lectins - Biology, Biochemistry, Clinical Biochemistry, Vol. I

ding to the cell membrane antigen phenotype (23, 28). Besides contributing to diagnosis, immunological classification carries important prognostic implications and may therefore influence the choice of therapy (6, 15). This paper **gives some preliminary results of identification and isolation of human HLA-DR antigen and the production of heteroantisera.**

HLA-DR is a well characterized glycoprotein with affinity for Lentil lectin (24). It is exposed on the surface of B-lymphocytes including the monoclonal leukemic cells from patients with chronic lymphocytic leukemia (CLL). It consists of two peptide chains, both with a molecular weight close to 30,000 (24). The peptides penetrate the lipid bilayer of the plasma membrane and thus the protein is an integral (amphiphilic) membrane protein (9).

Materials and Methods

CLL lymphocytes: Leukemic B-lymphocytes were isolated from blood of patients with untreated CLL by gradient centrifugation on Lymphoprep ^{R} (Pharmacia Fine Chemicals). After 3 times repeated washing in Hanks Balanced Salt Solution (pharmacy) the cells were solubilized in Triton X-100 (Rohm and Haas, scintillation grade) containing (20 ml/l), Tris buffer (pH 8.2) at a cell: buffer ratio of 1 : 2 (21). The cell lysate was sonicated for 2 x 2 seconds and the supernatant obtained after centrifugation at 3000 x g was saved for further studies. Samples were stored at -20° C.

Affinity chromatography: Glycoproteins were isolated from the cell extract by affinity chromatography on Lentil lectin Sepharose ^{R} (Pharmacia Fine Chemicals) after removal of low molecular weight carbohydrate by gel filtration on Sephadex G 25 (19). Proteins corresponding to 2 ml of cell extract was applied on a 5 ml Lentil lectin **Sepharose** collum. Isotonic phosphate buffered saline (pH 7.4) with Triton X-100 (2 ml/l) was used as chromatography buffer. Glycoproteins were eluted with 0.1 mol/l α-methyl-D-mannoside (Sigma) in the same buffer (19). The glycoprotein containing fractions were identified by fused rocket immunoelectrophoresis against a rabbit anti human lymphocyte antiserum (14,20), pooled and concentrated by ultrafiltration on a PM 10 membrane (Amicon) to the original volume.

Production of anti-glycoprotein antisera: Rabbits were immunized subcutaneously with glycoproteins corresponding to 200 μl of the original cell extract (approximately 2×10^8 lymphocytes) together with the same volume of Freunds incompleate adjuvans. Immunization was repeated 5 times at 4 week intervals. Then the rabbits were bled and the IgG fraction was isolated from serum (12).

Crossed immunoelectrophoresis: CLL lymphocyte extract was analyzed in crossed immunoelectrophoresis as previously described (21) applying the rabbit anti CLL lymphocyte glycoprotein antiserum. The electrophoresis buffer was composed as follows: Tris 73 mmol/l, veronal 24 mmol/l, sodium azide 2 mmol/l, Triton X-100, 5 ml/l, pH 8.6. After electrophoresis the plates were pressed and washed 3 times, dried and stained with Coomassie Brilliant Blue (26).

Lentil lectin affinity electrophoresis: Lentil lectin Sepharose R was equilibrated with electrophoresis buffer, mixed with an equal volume of 20 g/l agarose (type HSA, Litex, Glostrup, Denmark) at 56° C and cast on a glass plate for subsequent affinity electrophoresis (8). In the second dimension immunoelectrophoresis 1 cm antibody-free intermediate gel was interposed between the lectin containing first dimension gel and the antibody containing gel to avoid precipitation in the first dimension gel. The first dimension gel was removed after electrophoresis since it will fragmentize during pressing and washing.

α-methyl-D-mannoside 0.1 mol/l was applied in the second dimension gels to release glycoproteins which have been bound to the Sepharose-immobilized lectin during first dimension electrophoresis (21). In control experiments 0.1 mol/l mannoside was also present during first dimension electrophoresis, to ensure that the observed retardation in electrophoretic mobility was due to a specific lectin-glycoprotein interaction (21).

The affinity of lymphocyte glycoproteins for Lentil lectin was also studied by autoradiography after binding of ^{125}I labelled lectin to the precipitates obtained in crossed immunoelectrophoresis (5).

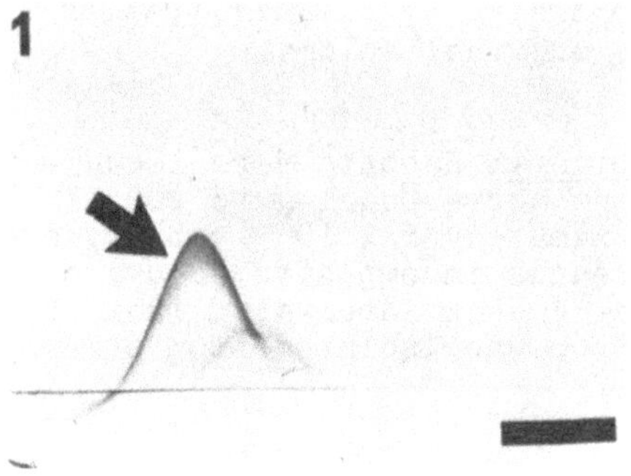

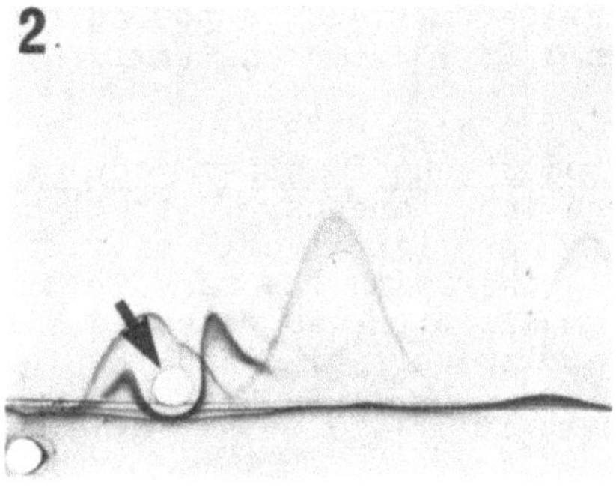

Fig. 1: Crossed immunoelectrophoresis of detergent solubilized CLL- lymphocytes (5 µl, corresponding to approximately $5x10^6$ cells) against a rabbit anti CLL-glycoprotein antiserum (13 µl/ cm^2)(see text). First dimension electrophoresis at 10V/cm for 75 minutes (anode to the right), 2nd dimension electrophoresis at 1.5V/cm overnight (anode at the top). Stained with Coomassie Brilliant Blue. The arrow denotes the putative HLA-DR antigen. The bar indicates 1 cm.

Fig. 2: Crossed immunoelectrophoresis of detergent solubilized CLL-lymphocytes (5 µl, corresponding to approximately $5x10^6$ cells) against a rabbit anti human lymphocyte antiserum (10 µl/ cm^2)(20). Experimental conditions as in Fig. 1. The arrow denotes an application well where 5 µl rabbit anti human HLA-DR antiserum has been applied. Compared to Fig. 1 the precipitation line of HLA-DR is deflected due to the anti HLA-DR antiserum.

Results

4 precipitates were seen after crossed immunoelectrophoresis of CLL lymphocyte extracts against a rabbit anti CLL lymphocyte glycoprotein antiserum (Fig. 1). One of these precipitates was also visible in the unstained gel (Fig. 1, arrow). The corresponding protein was assumed to be HLA-DR since it reacted with a monospecific rabbit anti human HLA-DR antiserum (Fig. 2).

This protein was present in all extracts from more than 50 patients with CLL except in 1 case which was shown by the E-rosette test (13) to be a T-cell leukemia (to be published).

The proteins visualized in Fig. 1 should all be glycoproteins with affinity for Lentil lectin, since the antiserum was raised against such proteins. The affinity of the putative HLA-DR molecule for this lectin was confirmed by Lentil lectin affinity electrophoresis (Fig. 3). The protein is retarded during first dimension electrophoresis due to interactions between glycoprotein and immobilized lectin (Fig. 3A) since retardation can be specifically inhibited by application of α-methyl-D-mannoside (Fig. 3B).

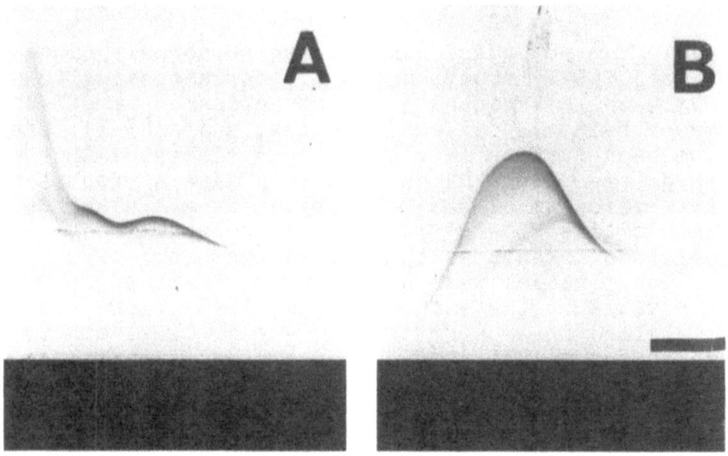

Fig. 3: Lectin-affinity crossed immunoelectrophoresis of detergent solubilized CLL-lymphocytes (5 μl, corresponding to approximately $5x10^6$ cells) against a rabbit antiserum to CLL-glycoproteins (18 $\mu l/cm^2$)(see text). A: The first dimension gel contains Lentil lectin Sepharose (0.4 ml/ml gel. B: Corresponding control with α-methyl-D-mannoside. Stained with Coomassie Briliant Blue.

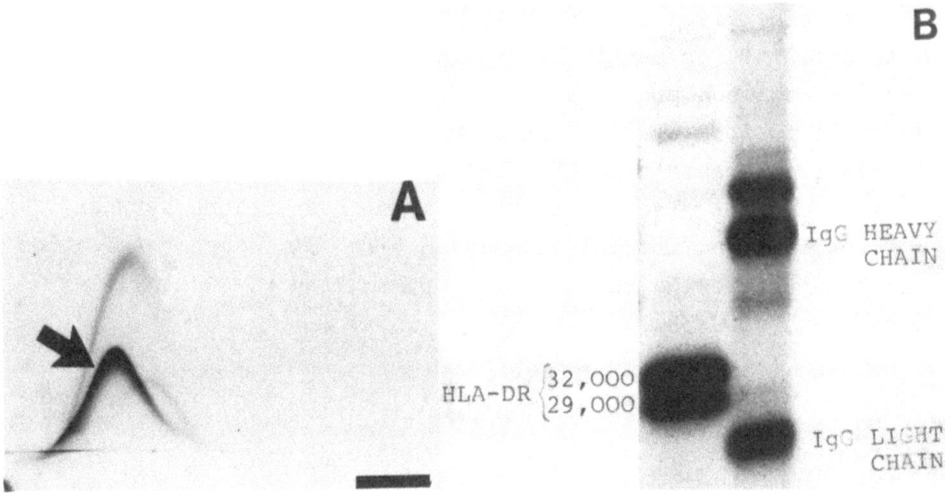

Fig. 4A: Autoradiogram of crossed immunoelectrophoresis of detergent solubilized CLL-lymphocytes (20 µl, corresponding to approximately $2x10^7$ cells) against a rabbit antiserum to CLL-lymphocyte glycoproteins (13 $\mu l/cm^2$)(see text and Fig. 1). The lymphocytes have been surface labelled with ^{125}I by the lactoperoxidase method (see text). Exposure time 5 days at room temperature. B: Autoradiogram of SDS-polyacrylamide gel electrophoresis of the putative HLA-DR precipitate isolated after crossed immunoelectrophoresis as shown in A (exposure time 5 days at -80°C). Two peptides with a molecular weight of 29,000 and 32,000 are visible on the autoradiogram (left track). These peptides were labelled in situ on the surface of CLL-lymphocytes prior to detergent solubilization and crossed immunoelectrophoresis (see Fig. 4A and text). Molecular weight markers: ^{125}I labelled IgG heavy and light chains (and albumin) are seen in the right **lane.**

Autoradiography after crossed immunoelectrophoresis of ^{125}I surface labelled CLL lymphocyte membrane proteins (18) demonstrated that the putative HLA-DR antigen was exposed on the surface of these cells during labelling (Fig. 4A). SDS polyacrylamide gel electrophoresis of the peptides from the precipitate (5, 17, 22) demonstrated two radiolabelled peptide chains with a molecular weight of 29,000 and 32,000 respectively in accordance with published data for HLA-DR (Fig. 4B)(2, 24).

Corresponding agarose gel precipitates obtained with unlabelled cells were used for immunization of rabbits (25, 27): The putative HLA-DR precipitate was cut out of the gel and injected subcutaneously into rabbits together with 100 µl of Freunds incompleate adjuvant per precipitate and residual agarose. Immunization with protein corresponding to one precipitate ($2x10^7$ lymphocytes) was initiated with 5 immunizations at one week intervals followed by boosters at four week intervals and successive bleedings one week later. The antisera thereby obtained were applied in crossed immunoelectrophoresis, in a microcytotoxic assay (16) and in immunofluorescens after conjugation of the isolated IgG fraction with fluorescein isothiocyanate (10). They were specifically cytotoxic for B-lymphocytes in dilutions higher than (1:32) (personal communication from Dr. A. Svejgaard, The Tissue Typing Laboratory, Rigshospitalet, Copenhagen) and stained 10-20 percent of unfractionated blood lymphocytes from healthy individuals in direct immunofluorescense (personal communication from Dr. A. Wiik, Department of Medicine TA, Rigshospitalet, Copenhagen). This fraction of stained cells corresponds well with the expected fraction of B-cells. The antiserum gave rize to a very faint precipitate in crossed immunoelectrophoresis (Fig. 5). The corresponding protein had an electrophoretic mobility which was similar to the putative HLA-DR molecule.

The amphiphilicity of the putative HLA-DR molecule was studied by hydrophobic interaction immunoelectrophoresis and charge shift immunoelectrophoresis (1, 4, 21, 29). It was best demonstrated in hydrophobic interaction electrophoresis (results not shown) where the protein was retarded in a manner very similar to that shown in Fig. 3 for lectin affinity electrophoresis. Likewise the control experiment in Fig. 3B the retardation could be inhibited by application of Triton X-100 in the first dimension gel (4).

Fig. 5: Crossed immunoelectrophoresis of detergent solubilized CLL-lymphocytes (5 µl, corresponding to approximately 5×10^6 cells) against a rabbit anti human HLA-DR antiserum (13 µl/cm^2). A faint Coomassie stained precipitate is seen corresponding to the electrophoretic position of HLA-DR. Experimental conditions and designations as in Fig. 1.

Discussion

Crossed immunoelectrophoresis is a very useful method when studying complex mixtures of proteins since individual proteins may be identified by their electrophoretic migration, the morphology of the precipitate etc. The recent developements in application of probes (e.g. lectins) showing biospecific interaction with proteins during electrophoresis has further increased the quantity of information that may be obtained from these technically very simple studies (7). Introduction of detergent in the agarose gel allows these studies to be extended to hydrophobic (amphiphilic) proteins such as integral membrane proteins (3).

The experiments presented here were aimed at identifying a tissue typing antigen (HLA-DR) present on human B-lymphocytes in order to produce B-cell specific antisera for clinical studies of leukemias and lymphomas. We took advantage of the knowledge, that HLA-DR is a B-lymphocyte glycoprotein with affinity for

Lentil lectin and raised a rabbit antiserum against such proteins. When this antiserum was applied in crossed immunoelectrophoresis 4 precipitates appeared, one of which reacted with a rabbit anti human HLA-DR from another laboratory. The corresponding protein was present in 50 typical cases of B-cell CLL but absent in one case of T-cell CLL. We have not yet obtained sufficient amounts of normal human B- and T-lymphocytes to demonstrate the same distribution of this antigen on such cells. The affinity of the protein for Lentil lectin was confirmed by affinity electrophoresis. Similar results have been obtained previously with another lymphocyte membrane glycoprotein, the HLA-A, B, C alloantigens (21). Besides contributing to the identification of membrane proteins of specific interest such studies may add to our understanding of why some subgroups of lymphocytes respond by blast transformation and proliferation in vitro after "stimulation" with lectins while others do not. The affinity for Lentil lectin of HLA-DR (this paper) and HLA-A, B, C antigens (21) from CLL lymphocytes suggests that the lack of response seen after lectin stimulation of CLL lymphocytes in vitro (Dr. M. Thomsen, personal communication) is not due to decreased binding of lectin but rather to a block in transmembraneous or intrinsic signal transmission.

The limited number of precipitates obtained in crossed immunoelectrophoresis with the anti glycoprotein antiserum was an advantage in these studies, since the molecular weight of the (HLA-DR) peptides could be determined (after prelabelling of the cells) without significant contamination from neighbour precipitates and the isolated precipitate could be used for production of specific anti B-cell antisera.

Acknowledgements

The technical assistance of Mrs. Gitte Plesner is gratefully acknowledged. The work was supported by The Brandt Brandtved Foundation and The Danish Cancer Society. The specificity of the HLA-DR antiserum was tested by Dr. A. Svejgaard at The Tissue Typing Laboratory, Rigshospitalet, Copenhagen and Dr. A. Wiik, Department of Medicine TA, Rigshospitalet, Copenhagen. Monospecific rabbit anti human HLA-DR antiserum was a gift from Dr. P.A. Peterson, University of Uppsala.

References

1. Bhakdi, S., Bhakdi-Lehnen, B., Bjerrum, O.J.: Biochem. Biophys. Acta 470, 35-44 (1977).
2. Billing, R.J., Safani, M., Peterson, P.: J. Immunol. 117, 1589-1593 (1976).
3. Bjerrum, O.J.: Biochem. Biophys. Acta 472, 135-195 (1977).
4. Bjerrum, O.J.: Analyt. Biochem. 90, 331-348 (1978).
5. Bjerrum, O.J., Bøg-Hansen, T.C., Plesner, T., Wilken, M.: pp. - in Bøg-Hansen, T.C. (ed.): Proceedings of The Third Lectin Meeting in Copenhagen, Walter de Gruyter, Berlin. New York (1980).
6. Bloomfield, C.D., Gajl-Peczalska, K.J., Frizzera, G., Kersey, J.H., Goldman, A.I.: N. Eng. J. Med., 301, 512-518 (1979).
7. Bøg-Hansen, T.C.: pp. 399-416 in Egly, J.M. (ed.): Proceedings of The Third International Symposium on Affinity Chromatography and Molecular Interaction, Inserm Symposia Series, Paris 1979.
8. Bøg-Hansen, T.C., Brogren, C.-H.: Scand. J. Immunol. 4, Suppl. 2, 135-139 (1975).
9. Crumpton, M.J., Snary, D., Walsh, F.S., Barnstable, C.J., Goodfellow, P.N., Jones, E.A., Bodmer, W.F.: Roy.Soc. B. 202, 159-175 (1978).
10. Forni, L.: pp. 151-167 in Lefkowits, I & Pernier, B. (eds.): Immunological Methods, Academic Press, New York 1979.
11. Habeshaw, J.A.: Cancer Immunol. Immunother. 7, 37-43 (1979)

12. Harboe, N. & Ingild, A.: pp. 161-164 in Axelsen, N.H., Krøll, J. & Weeke, B. (eds.): A Manual of Quantitative Immunoelectrophoresis. Methods and Applications, Universitetsforlaget, Oslo 1973.

13. Jondahl, M.: Scand. J. Immunol. 3, 739-747 (1974).

14. Just Svendsen, P.: pp. 69-70 in Axelsen, N.H., Krøll, J. & Weeke, B. (eds.): A Manual of Quantitative Immunoelectrophoresis. Methods and Applications, Universitetsforlaget, Oslo 1973.

15. Kadin, M.E., Billing, R.J.: Blood, 51, 813-823 (1978).

16. Kissmeyer-Nielsen, F., Kjerbye, K.R.: pp. 381-383 in Curtoni (ed.): Histocompatibility Testing, Munksgaard, Copenhagen 1967.

17. Norrild, B., Bjerrum, O.J., Vestergaard, B.F.: Anal. Biochem. 81, 432-441 (1977).

18. Philips, D.R., Morrison, M.: Biochemistry 10, 1766-1771 (1971).

19. Plesner, T.: Scand. J. Immunol., 5, 1097-1102 (1976).

20. Plesner, T.: Scand. J. Immunol., 8, 363-367 (1978).

21. Plesner, T., Bjerrum, O.J.: Scand. J. Immunol., 11, 341-351 (1980).

22. Shapiro, A.L., Vinuela, E., Maizel, J.V.: Biochem. Biophys. Res. Commun. 28, 815-820 (1967).

23. Siegal, F.P., Filippa, D.A., Koziher, B.: Am. J. Path. 90, 451-459 (1978).

24. Springer, T.A., Kaufman, J.F., Terhost, C., Strominger, J.L.: Nature 268, 213-218 (1977).

25. Vestergaard, B.F.: Scand. J. Immunol. 4, Suppl. 2, 203-206 (1975).

26. Weeke, B.: pp. 15-35 in Axelsen, N.H., Krøll, J, Weeke, B. (eds.): A Manual of Quantitative Immunoelectrophoresis. Methods and Applications, Universitetsforlaget, Oslo 1973.

27. Welsh, K.E., Turner, M.J.: Tissue Antigens 8, 197-205 (1976).

28. Wernet, P. Schunter, F., Wilms, K., Walter, H.D.: pp. 333-338 in Thierfelder, S., Rodt, H., Thiel, E. (eds.): Immunological Diagnosis of Leukemias and Lymphomas, Springer Verlag, Berlin 1976.

29. Wilken, M., Plesner, T., Bjerrum, O.J., Mørk Hansen, M.: Manuscript in preparation.

LECTIN BINDING PROPERTIES OF CHICKEN MAJOR HISTOCOMPATIBILITY COMPLEX CODED LYMPHOCYTE PLASMA MEMBRANE GLYCOPROTEINS.

C.-H. Brogren and S. Bisati
Institute for Experimental Immunology, University of Copenhagen, DK-2100 Copenhagen Ø, Denmark.

Introduction

This paper is part of our interest to characterize and isolate class I (B-F) and class II (B-L) major histocompatibility complex (MHC) coded alloantigens from chicken lymphocyte plasma membranes (1). The aim of the present work is to study the binding properties of a series of lectins to the individual MHC-alloantigens, which have all been described to be the major lymphocyte cell surface glycoproteins (2,3).

It has been shown in biochemical studies of human histocompatibility antigens (4), that identical core sugar chains containing glucosamine, mannose, and glucose residues are added to asparagine amino-groups immediately after the polypeptide formation during the process of biosynthesis in the endoplasmic reticulum. Thereafter, in the Golgi apparatus, some sugar residues are removed first leaving behind some core sugar residues (mannose) to which new residues of e.g. galactose, galactosamine, glucosamine, fucose, and sialic acids are added forming a complex heterogeneous terminal sugar chain on the glycoprotein. The primary and terminal glycosylation together determine the lectin binding specificities.

It seems that both class I and II chicken lymphocyte MHC-proteins contain mannose structures in a steric configuration, which can bind to lectins like concanavalin A (Con A) and Lens culinaris hemagglutinin (LcH) (5, ibid.), both widely used to purify class I

Lectins - Biology, Biochemistry, Clinical Biochemistry, Vol. I

and II alloantigens from other animal species. A lectin with specificity exclusively for terminal sugars can separate class I and II MHC-alloantigens. In the present work affinity of various lectins to MHC-antigens were studied to select suitable lectins for fractionation of class specific MHC-antigens.

Two analytical methods were used in these studies. Crossed immunoaffinoelectrophoresis, originally described by the editor of this book (6) and line-absorption immunoelectrophoresis were employed to study the lectin binding specificities of chicken MHC-alloantigens (7). Both methods involved the electroimmunoprecipitation of detergent solubilized plasma membrane proteins in agarose gel.

Materials and Methods

1) Lectins. Various lectins coupled to Sepharose 4B - Concanavalin A (Con A-S), Lens culinaris hemagglutinin (LcH-S), and Sepharose 6MB immobilized lectins - Wheat Germ agglutinin (WGA-S), Helix Pomatia agglutinin (HPA-S), were obtained from Pharmacia Fine Chemicals, Uppsala, Sweden. Sepharose 4B coupled Ricin (RCA-II-S), Ricinus communis agglutinin I (RCA-I-S), Pisum Sativum (PEA-S), and Lotus Tetragonolobus agglutinin (Lotus A-S) were purchased from E-Y Laboratories Inc., San Mateo, U.S.A. Several lectins were coupled to Magnogel AcA 44 (Pharmindustrie, Clichy, France) and used in this study. The immobilization of lectins was done according to the procedure of Guesdon, Courcon, and Avrameas (8). Approximately 2 mg protein was bound per ml Magnogel. The lectins used for coupling to Magnogel beads were obtained from a variety of sources. Lentil lectin (LcH) was obtained from Pharmacia Fine Chemicals, Uppsala, Sweden. Following lectins - Limulin, Solanum Tuberusum agglutinin (STA), and Concanavalin A (Dr. M. Monsigny, Orleans, France), Peanut agglutinin (PNA) (Dr. R. Lotan, Rehovot, Israel), Ricin (Dr. S. Olsnes, Oslo, Norway), and Vicia Villosa agglutinin (VVA) (Dr. A. Kimura, Uppsala, Sweden) - were received as gift samples.

2) Antigens. Lymphocyte plasma membrane was isolated from pooled spleen and peripheral blood lymphocytes according to a modified version of the procedure of Blanchet for isolation of erythrocyte plasma membranes (9). Further details will be published elsewhere (10). Four to six months old White Leghorn chickens, maintained in closed flocks at our institute, were used as the source. These chickens are homozygous for MHC B^{13} or B^{6} haplotype according to Briles nomenclature (11).

3) Antibodies. Rabbit antiserum against NP-40 solubilized whole chicken lymphocyte plasma membrane was prepared according to the procedure of Bjerrum (12). A dose of 1.5 mg protein per injection was used. Antiserum against the Lentil lectin binding fraction of the lymphocyte plasma membrane glycoproteins was raised as described in a separate paper in this book (5). Isolation of the immunoglobulin fraction was done according to procedure of Harboe and Ingild (13). Goat anti-chicken immunoglobulin (light chain specific), code 64-397, were obtained from Miles-Yeda, Rehovot, Israel, and coupled to Magnogel (8). Alloantisera with selective specificities for chicken MHC-products ($B\text{-}F^{13}$ and $B\text{-}L^{13}$) were a gift from Dr. K. Hála, Prague, Czechoslovakia.

4) Crossed immuno-affinoelectrophoresis. The procedure as described by Bøg-Hansen (6) was used. All steps were performed in presence of detergent. Agarose gel (Type HSB, Litex, Glostrup, Denmark) and buffer (Tris-veronal, I=0.02) containing 0.2% v/v NP-40 were used (2). First dimension electrophoresis was carried out at 10 Volts/cm for 90 minutes and 2nd dimension for 18-24 hours at 2-3 Volts/cm. Electrophoretic equipment was obtained from Holm-Nielsen ApS, Herlev, Denmark, and LKB-Produkter AB, Bromma, Sweden (Multiphor).

5) Line-absorption immunoelectrophoresis. The method as described by Krøll and Andersen (14) and modified for our studies (7) was used. The version shown here was performed on a 12 x 25.5 cm glass plate with a longitudinal line gel containing the

NP-40 solubilized lymphocyte plasma membrane sample (CLM-B^{13}). The immobilized lectins (lectin-Sepharose and lectin-Magnogel), anti-chicken immunoglobulin light chain-Magnogel, and test alloantisera samples were applied in wells, punched in an antibody free gel between the line gel and the antibody containing gel. The electrophoresis was carried out overnight for about 20-24 hours at 2-3 Volts/cm.

Results and Discussion

The crossed immuno-affinoelectrophoretic experiments were carried out with 7 different Sepharose bound lectins as shown in Fig. 1. Immunoelectrophoretic pattern of the lymphocyte plasma membrane glycoproteins was developed by rabbit antibodies against the LcH binding proteins of the membrane. Four peaks appeared in the pattern. The identification of the classical transplantation antigen (B-F) and the immune associated alloantigen (B-L) in the pattern has been already done with alloantibodies specific to these antigens in crossed immunoelectrophoresis with intermediate gel and line-absorption immunoelectrophoresis (7,14). Further identification was carried out with polypeptide studies of radiolabelled immunoprecipitates, cut out from the agarose gel, in O'Farrell 2-D-gel electrophoresis (to be published). It remains still unclear whether a third peak (marked as X) belongs to the class II alloantigens or represent a more complex protein with binding to selective alloantisera (to be discussed elsewhere). The identification of the fourth peak (marked Y) has not been successfully accomplished as yet.

The binding properties of the 7 lectins to various MHC-antigens are shown in Fig. 1. The results are summarized in Table 1. The crossed immuno-affinoelectrophoresis allows a semiquantitative estimation of the binding, if we assume, that the lectin is in excess in the intermediate gel, which generally should be the case. An exception could be the galactose binding lectins RCA-I and RCA-II, which in the agarose (galactose-polymer) could be partially block-

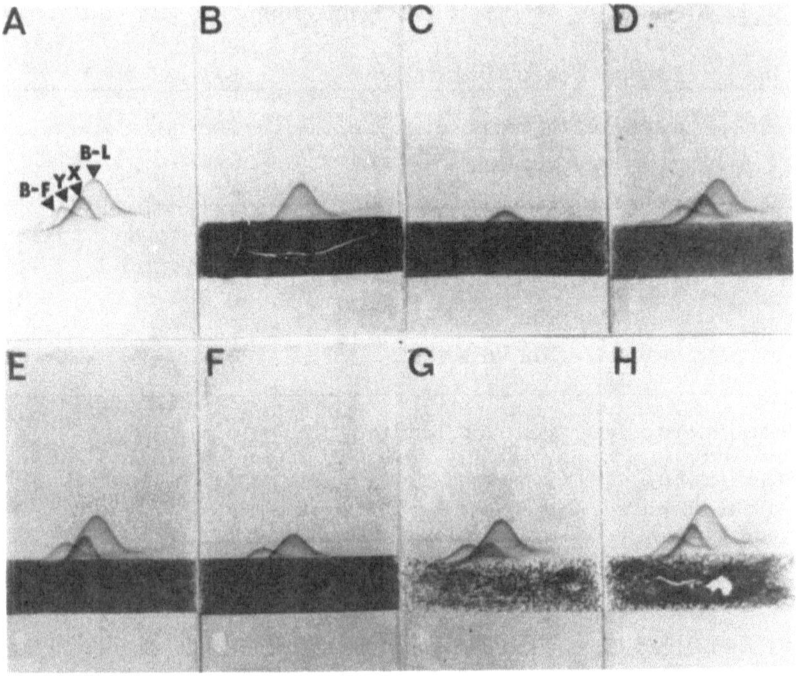

Fig. 1. Crossed immuno-affinoelectrophoresis showing lectin binding properties of chicken lymphocyte MHC-alloantigens. Antigen: NP-40 solubilized chicken (B^6) lymphocyte plasma membrane proteins. Thirty µl (100 µg protein) applied in each well. Antibodies: Rabbit antiserum against LcH-binding glycoproteins of the lymphocyte plasma membrane applied with 6.6 µl/cmm. Plate size 6.25 x 10 cm. The lectin intermediate gels were made with 250 µl packed lectin-Sepharose beads. The height of the intermediate gel was 1.5 cm. Spacing between the lectin intermediate gel and the antibody containing gel consist of 0.5 cm blank agarose gel. Following lectins were tested: A: Sepharose 4B (control), B: Con A-S (200 µg lectin/cmm), C: LcH-S (40 µg/cmm), D: PEA-S (about 50 µg/cmm), E: RCA-I-S (about 50 µg/cmm), F: RCA-II-S (about 50 µg/cmm), G: WGA-S (100 µg/cmm), and H: HPA-S (20 µg/cmm). B-F and B-L indicate the loci of the chicken MHC (B-complex) coding for the alloantigens in the peaks. For indication X and Y see the text.

Table I. Binding of lectins to chicken lymphocyte MHC-antigens.

Lectins	Carbohydrate specificity	B-F	B-L	X
Con A	mannose, glucose	+++[1]	+	+++
LcH	mannose, glucose	++	++	++
PEA	mannose, glucose	-	-	-
RCA-I	galactose	-	-	-
RCA-II	galactose, N-acetyl-galactosamine	+/-[2]	++	+++
WGA	N-acetyl-glucosamine	-	+	++
HPA	N-acetyl-galactosamine	-	-	-

1) Semiquantitative value for binding proportion: complete (+++), partial (++), slight (+), no binding (-).
2) The binding is haplotype dependent. B-F13 showed no or very slight binding, whether B-F6 expressed slight binding to this lectin.

ed. Nevertheless, these lectins retain their binding properties in the immobilized form with Sepharose. LcH bound the B-L antigen nearly complete, but showed only a partial binding to the B-F antigen. On the contrary, Con A revealed only a slight binding proportion of B-L antigen but a complete binding of B-F antigen. These results were used to design a sequential column system of LcH-S followed by Con A-S for fractionation of lymphocyte membrane glycoproteins as described elsewhere (5). The results further indicate, that RCA-II could be useful in isolation of B-L alloantigen free of contaminating B-F antigen. Moreover, WGA-S might be useful to eliminate antigen X from both B-L and B-F alloantigens. An optimal sequential lectin column system for fractionation of various MHC-antigens can consist of WGA-S followed by RCA-II-S and Con A-S. Additionally, LcH-S could be placed in front of Con A-S to absorb unbound B-L antigen to allow pure B-F antigen absorption to Con A. This column system will be used in preparative scale for isolation of MHC-antigens.

Line-absorption immunoelectrophoresis was also used to analyse the lectin binding properties of MHC-antigens. In this method a series of lectins can be screened in one experiment. The binding of lectin to a respective line precipitate antigen is seen as a V-shaped drop on the line. The size of the drop depends on the amount of lectin in the well and the binding proportion of the respective antigen. Since the antigen will be in excess in this technique, this method can only be regarded as qualitative.

Figure 2 illustrates the results of such a study. Identification of various MHC-alloantigen line precipitates was done by applying specific alloantisera in the wells as described elsewhere (15). The results verified the data from Table I. Con A, LcH, and RCA-II coupled to Sepharose and Magnogel reacted with the MHC-antigens. WGA-S gave a slight deflected on the B-L line-precipitate, but a clear drop on the X-antigen line. Among the additional new lectins

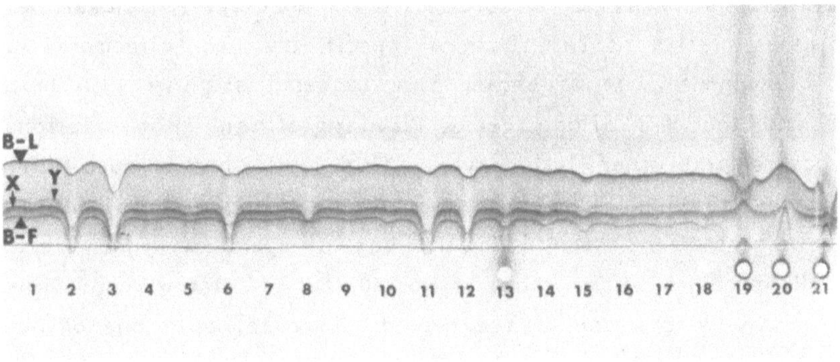

Fig. 2. Line-absorption immunoelectrophoresis for test of lectin binding properties of the chicken lymphocyte MHC-alloantigens. Antigen: Chicken (B^{13}) lymphocyte plasma membrane proteins applied in a 1.5 x 25.5 cm line gel containing 200 µl NP-40 solubilized membrane proteins (about 1 mg protein). Antibodies: Rabbit anti LcH-binding chicken lymphocyte plasma membrane glycoproteins applied as purified Ig-fraction with 4 $\mu l/cm^2$. Plate size 12 x 25.5 cm. In the absorption wells were applied 15 µl samples of packed lectin-beads in following order: Sepharose 4B (no.1), Con A-S, LcH-S, PEA-S, RCA-I-S, RCA-II-S, Lotus A-S, WGA-S, HPA-S (nos. 2-9), then Magnogel coupled samples, goat anti-chicken immunoglobulin light chain (no. 10), Con A-M, LcH-M, Ricin-M, VVA-M, STA--M, PNA-M, and two samples of Limulin-M (nos. 11-18). For identification of the line precipitates following alloantisera were applied: Anti-B13, anti-B-F13, and anti-B-L13 (nos. 19-21). The line precipitates representing the B-F. B-L, X, and Y antigens are indicated.

(Lotus A, VVA, STA, PNA, Limulin) tested in these experiments, STA and Limulin showed reactivity with the chicken MHC-alloantigens. Goat anti-chicken immunoglobulin light chain antibodies coupled to Magnogel were applied as a control for the lectin-Magnogel samples and for detection of eventual immunoglobulins in the pattern. However, this sample showed a slight reaction with the B-F antigen. This could be due to crossreacting antibodies against beta-2-microglobulin, which is the light chain of the histocompatibility antigen (16). The Lotus agglutinin has been described to express binding to B-L analogue products in the mouse serum (17). However, Lotus A-S, which has a specificity for terminal L-fucose residues, did not show any detectable reaction with the chicken lymphocyte membrane MHC-antigens. Vicia Villosa agglutinin has been shown to react specifically with cytotoxic T-lymphocytes (18). In the present studies it seems not to interact significantly with the MHC-antigens. VVA has a specificity for N-acetyl-galactosamine. The potatoes lectin (STA) has a specificity for oligomers of N-acetyl-glucosamine. It is shown here to react slightly with both the B-F and the B-L alloantigens. PNA agglutinate subpopulations of thymocytes and other lymphocytes. This lectin has a specificity for galactose residues (19). No detectable binding to the chicken MHC-alloantigens were seen. Limulin has a specificity for sialic acids. MHC-products have been reported to be sialoglycoproteins (3,20,21), why a reaction was expected. However, only one of our two samples of Limulin showed a very slight reaction and only with the B-F antigen.

In general, studies of lectin binding properties of proteins by electroimmunoprecipitation in agarose has some drawbacks. The electroimmunomigration has to be carried out at a basic pH (pH 8.6), not necessarily optimal for the lectin binding. Moreover all cofactors, e.g. Ca^{2+}, Mg^{2+}, and Mn^{2+}, needed for reactivity of lectins can not be included in the buffer or gel. Mn^{2+} gets precipitated at the basic pH used in electroimmunoprecipitation. However, it has been reported that non-inclusion of cofactors in the electrophoretic experiments (buffer and gel) did not effect the reactivity of the lectins (6,22).

Speculations with respect to the dominant difference in the binding properties of the three mannose binding lectins Con A, LcH, and PEA brought us to consider that the steric configuration around the binding site of the lectin gives each lectin an important individual feature. Furthermore, the fine structures of the carbohydrate moieties of the single glycoproteins express a microheterogeneity, which probably determines the ability of the lectins to interact with the glycoproteins.

A main observation was the proportional or partial binding of individual molecular species to the lectins, probably due to a highly heterogeneity in the terminal glycosylation. The terminal sugars can create new binding sites, but block others existing before addition of e.g. sialic acids and fucose. For preparative fractionation purpose it is important to select a lectin which express as complete a binding as possible. Only in this case the yield after affinity chromatography can be reasonably good. A second problem in this context is the efficiency of the elution with specific sugars, which can be obtained without using denaturating reagents. In studies of human and mouse MHC-alloantigens LcH-S columns have been the most convenient to use, since they were easier to elute than many other lectin columns, especially Con A-S.

The results obtained from the present analytical studies have been successfully used in selecting a suitable lectin column system for preparative isolation of MHC-antigens (5), but also in O'Farrell 2-D-gel electrophoretic technique for preliminary analytical isolation of comparatively pure products to be loaded into the gels (3).

The poor binding of the chicken class I alloantigens to LcH-S, however makes it preferable to use Con A-S for isolation of B-F antigen in the chicken system. We still search for a more optimal lectin for separation of B-F and B-L antigen than the toxic RCA-II (Ricin), why other galactose binding lectins will be studied, such as the Misteltoe lectin described elsewhere in this book (23).

Conclusively, MHC-alloantigens were found to be the major lectin acceptors on the lymphocyte membrane. Similar results have been obtained using SDS-PAGE (24) and 2-D-gel electrophoresis (3). These studies further indicate that Ricin can be used for fractionation of B-F and B-L antigens. An analogues observation has been done for the mouse H-2 antigens (25). The use of lectins for the fractionation of lymphocytes (26) should be seen in the view of these molecular studies.

Acknowledgements

This work has been supported by Danish Medical Research Council, Danish Cancer Society, King Christian Xth Foundation, and The Blood Donors Foundation. We are thankful to Drs. M. Monsigny, S. Olsnes, R. Lotan, and A. Kimura for their gifts of lectins. We thank Misses Hanne Boesen, Lone Malte, and Lone Sørensen for skilful technical assistance, and Ms. Emmy Madsen for valuable secretarial help. S. Bisati was supported by a DANIDA fellowship from Ministry of Foreign Affairs, Government of Denmark.

References

1. Simonsen, M.: In M.B. Zaleski (ed.). Proc. 7th Int. Convocation on Immunology. Karger, Basel. 1981. In press.
2. Bisati, S., Brogren, C.-H., Simonsen, M.: Protides Biol. Fluids. 27, 471-474 (1979).
3. Brogren, C.-H., Bisati, S., Simonsen, M.: Protides Biol. Fluids. 27, 467-470 (1979)
4. Peterson, P.A.: Personal communication (1980)
5. Bisati, S., Mikkelsen, L., Brogren, C.-H.: ibid. Walter de Gruyter, Berlin, 1981.
6. Bøg-Hansen, T.C.: Analyt. Biochem, 56, 480-488 (1973).
7. Bisati, S., Brogren, C.-H., Simonsen, M.: Folia Biologica (Praha) 25, 331-332 (1979).
8. Guesdon, J.-L., Courcon, J., Avrameas, S.: J. Immunol.Methods 21, 59-63 (1978).

9. Blanchet, J.P.: Expt. Cell Research. 84, 159-166 (1974).

10. Brogren, C.-H., Bisati, S.: In preparation.

11. Simonsen, M., Hála, K., Nicolaisen, E.M.: Immunogenetics, 10, 103-112 (1980).

12. Bjerrum, O.J.: Biochem. Biophys. Acta. 472, 135-195 (1977).

13. Harboe, N., Ingild, A.: A Manual of Quantitative Immunoelectrophoresis. Scand. J. Immunol. 2, suppl. 1, 161-164 (1973).

14. Krøll, J., Andersen, M.M.: J. Immunol. Methods. 9, 141-146 (1975).

15. Bisati, S., Simonsen, M., Brogren, C.-H.: In preparation.

16. Gottlieb, A.B., Engelhard, M., Kunkel, H.G., Tanigaki, N., Pressman, D.: J. Immunol. 119, 3998-4001 (1977).

17. Parish, C.R., Jackson, D.C., McKenzie, F.C.: Immunogenetics 3, 455-463 (1976).

18. Kimura, A., Wigzell, H., Holmquist, G., Ersson, B., Carlsson,P.: J. Exp. Med. 149, 473-484 (1979).

19. Novogrodsky, A., Lotan, R., David, A., Sharon, J.: J. Immunol. 155, 1243-1248 (1975).

20. Parham, P., Humphreys, R.E., Turner, M.J., Strominger, J.L.: Proc. Nat. Acad. Sci. U.S.A. 71, 3998-4001 (1974).

21. Pink. J.R.L.: Folia Biologica (Praha) 25, 333-334 (1979).

22. Bøg-Hansen, T.C.: In J.M. Egly (ed.). Affinity Chromatography and Molecular Interactions. INSERM Symposia Series Vol. 86. Paris. 1979, pp. 399-416.

23. Franz, H., Ziska, P.: ibid. Walter de Gruyter, Berlin, 1981.

24. Nilsson, S.F., Waxdal, M.J.: Biochemistry 17, 903-910 (1978).

25. Willoughby, E.M., Turner, M.J., Sanderson, A.R.: Eur. J. Immunol. 8, 628-634 (1978).

26. Waxdal, M.J., Fowlkes, B.J., Sharrow, S.O., Mathieson, B.J.: Protides Biol. Fluids 27, 525-528 (1979).

SEQUENTIAL LECTIN AFFINITY CHROMATOGRAPHY OF SOLUBLE CHICKEN LYMPHOCYTE PLASMA MEMBRANE PROTEINS AND EVALUATION OF SPECIFICITY OF RABBIT ANTISERA AGAINST THE LECTIN BINDING GLYCOPROTEINS.

S. Bisati[+], L. Mikkelsen, C.-H. Brogren.
Institute for Experimental Immunology, University of Copenhagen, DK-2100 Copenhagen Ø, Denmark.

Introduction.

Both B-F (human HLA-A,B,C analogue) and B-L (HLA-D analogue) alloantigens, coded by major histocompatibility complex (MHC), are expressed in chicken lymphocyte plasma membrane. These antigens have been studied mainly by serological and cellular methods, involving use of specific alloantisera (1,2). These approaches result in the identification of alloantigens in intact cells. Later solubilized cell membranes were used to study these antigens in electroimmunoprecipitation methods (3,4). The glycoproteinic nature of these antigens lead to the use of lectins in these studies. Further work demonstrated that major glycoproteins of the chicken lymphocyte membrane are the MHC alloantigens (5,6). In the present study, sequential lectin affinity chromatography were used to prepare solubilized fractions enriched with B-F and B-L alloantigens from chicken lymphocyte plasma membranes. The lectin binding fractions were subsequently used to prepare xenogeneic antisera in rabbits with restricted specificity to the MHC antigens.

[+]Supported by a DANIDA fellowship from Ministry of Foreign Affairs, Government of Denmark.

Materials and Methods.

1) Animals. Two strains of White Leghorn (WL) chickens were used. The WL-B13 and WL-B6 chickens are homozygous for B^{13} and B^{6} MHC haplotypes, according to Briles nomenclature (7). These strains are maintained at our institute in closed flocks and have been derived from outbred White Leghorn birds after selection for respective haplotypes. Three months old New Zealand albino rabbits, weighing 3-4 kg at the beginning af immunization, were supplied by State Serum Institute, Copenhagen.

2) Preparation of plasma membranes. Lymphocytes from about ten (4-6 months old) B^{13} or B^{6} homozygous chickens were used for isolation of plasma membranes. Both spleen and peripheral blood lymphocytes were used. The membranes were prepared according to a modified version of the procedure for erythrocyte membranes as described by Blanchet (8). The Ficoll separated lymphocytes were suspended in hypotonic salt solution and disrupted mechanically by applying a pressure of 40 bar in a French pressure cell. The intact nuclei, resting nuclei and nuclear membrane were removed by differential centrifugation. The enriched plasma membranes were finally collected and washed by centrifugation at 32,000 g for 4 hours.

3) Sequential lectin affinity chromatography. The isolated plasma membrane was solubilized in 10% v/v Nonidet P-40 (NP-40) by mixing with the help of a magnetic stirrer for 3 hours at 4°C. The protein content of the solubilized membrane was estimated according to the procedure of Wang and Smith (9). Two columns of Sepharose bound *Lens culinaris* hemagglutinin (LcH-S, 10 x 1.6 cm) and Concanavalin A (Con A-S, 5 x 1.6 cm) were separately packed and equilibrated with 0.01 M Tris-buffered-saline, pH 7.4, containing 10 mM sodium azide and 2% NP-40 (TBS/NP-40). The immobilized lectins were obtained from Pharmacia Fine Chemical, Uppsala. The two columns were then interconnected to make a sequential column system. The solubili-

zed membrane (B^{13} chickens) was diluted to a final concentration of 2% NP-40. The diluted sample was then passed through the column system in such a way that the membrane material was first absorbed in the LcH-S column and then into the Con A-S column. The material was passed through the column system at a speed of 26 ml per hour and the unbound fraction collected. The columns were then disconnected from each other and washed separately with TBS/NP-40. The two columns were eluted separately with 10% α-methyl glucoside using a reverse flow of the eluant. The various fractionation steps in the whole column system were monitored by fused rocket immunoelectrophoresis by developing the pattern with rabbit antisera against whole lymphocyte plasma membrane. Various lectin binding membrane fractions from the two columns, showing reactivity in the fused rocket immunoelectrophoretic pattern were pooled and concentrated. The pooled samples were first transferred into dialysis tubings and then dehydrated, by keeping the tubings covered with polyethylene glycol (Carbowax 20,000) flakes for 3 hours in cold room.

4) Immunization. The concentrated lectin binding fractions were immunized into rabbit by mixing with equal volumes of Freunds incomplete adjuvant according to the procedure of Harboe and Ingild (10). An amount of 0.6 mg protein of LcH binding fraction, 0.1 mg protein of Con A binding fraction, and 1 mg protein of whole lymphocyte membrane was administered per injection in each animal.

5) Evaluation of the prepared rabbit antisera. The two rabbit antisera were evaluated against solubilized lymphocytes plasma membranes from different haplotypes to detect specificity and titre for MHC antigens and other membrane proteins. Both simple crossed and crossed immunoelectrophoresis with intermediate gel were used in these studies. A reference rabbit antiserum with anti B-F activity was used for confirmation of B-F specificity. The antiserum was raised against the Con A binding

fraction of chicken erythrocyte plasma membrane as described elsewhere (6). The anti B-L specificity was confirmed by the lectin and alloantibody binding properties of the precipitated B-L antigen (Brogren and Bisati, ibid).

6) Electroimmunoprecipitation methods. Crossed immunoelectrophoresis, crossed immunoelectrophoresis with intermediate gels and fused rocket immunoelectrophoresis were performed as described elsewhere (4,11). Agarose of HSA and HSB type (Litex, Denmark) was used in these studies. Nonidet P-40 (0.2% v/v) was present both in the Tris-veronal buffer and in the agarose gel.

Results and Discussion.

Figure 1 shows the fused rocket immunoelectrophoretic analysis of the various fractions obtained by lectin affinity chromatography of soluble lymphocyte membrane proteins in the sequential system. Peak A depicts the membrane proteins which did not bind to column system. The two major peaks B and D represent the glycoside eluted LcH and Con A binding membrane proteins respectively. Peaks C and E additionally eluted LcH and Con A binding proteins. These peaks were obtained by a second elution after an overnight incubation of the two columns with eluate buffer. It can be inferred from the relative area of various peaks that a majority of lymphocyte membrane antigens are glycosylated. The fractions C and E were not used further. Peaks B and D were separately pooled and concentrated. The concentrated samples were evaluated in crossed immunoelectrophoresis for the presence of MHC antigens and other proteins (Fig. 2). The identification of B-F and B-L antigen peaks was done by lectin, specific alloantibody and molecular weight studies (to be published elsewhere). The patterns were developed by rabbit antisera against whole lymphocyte plasma membrane. Comparison of the patterns was made with the pattern obtained with solubilized lymphocyte membrane. A relative enrichment of B-L antigens was seen in the LcH binding fractions of the

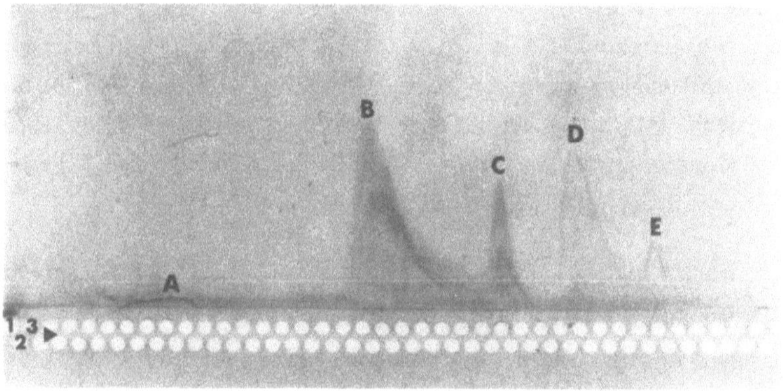

Fig. 1. Fused rocket-immunoelectrophoresis of the various fractions obtained by sequential lectin affinity chromatography of detergent solubilized chicken (B^{13}) lymphocyte plasma membrane proteins. An aliquot of 20 ul of each fraction was applied to the holes in the antibody-free gel of the plate. The upper gel contained rabbit antiserum against chicken (B^{13}) lymphocyte plasma membrane (20 ul/cm^2). Electrophoresis was carried out at 2V/cm for 18 hours. Peak A indicate the unbound proteins whereas peaks B,C and D,E represent glycoside eluted LcH and Con A binding membrane proteins, respectively. Plate size 10 x 20.5 cm.

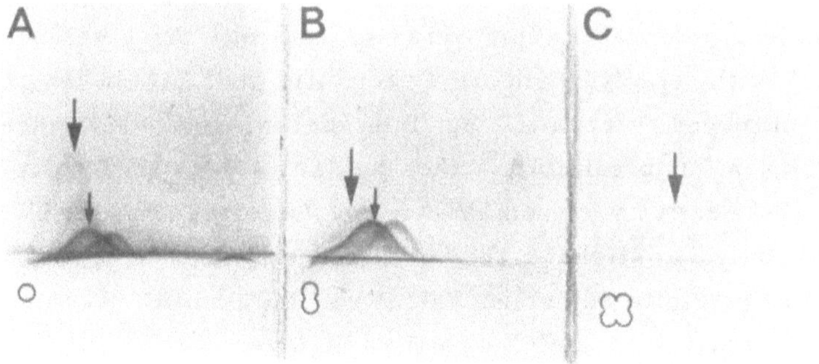

Fig. 2. Comparison of crossed immunoelectrophoresis of detergent solubilized chicken (B^{13}) lymphocyte plasma membrane, LcH and Con A binding membrane proteins. A: Solubilized lymphocyte plasma membrane proteins (10 ul, 25 ug protein). B: LcH binding membrane proteins (25 ul, 35 ug protein). C: Con A binding membrane proteins (50 ul, 5 ug protein). All upper gels contained rabbit antiserum against chicken (B^{13}) lymphocyte plasma membrane (10 ul/cm^2). Small arrows indicate B-L antigens whereas large arrows indicate B-F antigens. Plate size 5 x 7 cm.

membrane. The B-L peak showed a larger area as compared to the B-L peak on the pattern of the lymphocyte membrane. B-F antigen showed a prominent enrichment in Con A binding fraction. However, the two fractions were not completely devoid of either B-F or B-L antigens. Besides other lectin binding proteins in the form of weak precipitates were also seen.

The specificity of the two rabbit antisera was tested against solubilized lymphocyte membranes from B^{13} and B^{6} haplotypes. Rabbit antisera against LcH binding membrane fraction (RaCLM -LcH) precipitated mainly B-L antigen in the crossed immunoelectrophoresis from the both haplotypes. Similarly, rabbit antisera against Con A binding membrane fraction (RaCLM-Con A) precipitated mainly B-F antigen from the two haplotypes. These results indicate that the two antisera are non specific with respect to MHC-alloantigens of different haplotypes.

Figure 3 shows a further confirmation of the specificities of the two rabbit antisera and evaluation of the relative titer by crossed immunoelectrophoresis with intermediate gel. Crossed patterns of the solubilized membrane from B^{6} haplotype were developed against RaCLM-LcH antisera. Normal rabbit sera as control, the reference rabbit antisera with specific anti B-F activity and RaCLM-Con A antisera were employed separately in intermediate gels. Reference sera and RaCLM-Con A precipitated the B-F antigen peaks confirming the anti B-F activity of RaCLM-Con A. Moreover, the RaCLM -Con A showed only a slight lowering effect on the B-L antigen peak which was precipitated prominently by RaCLM-LcH. A third peak was seen to react with RaCLM-Con A in a lower titre.

Electroimmunochemical analysis of the two sera reveal mainly precipitates of MHC-antigens. Both B-F and B-L antigens were precipitated by the two sera, although the specificity for the antigens differ strongly (Fig. 2 and 3). Same is true for other glycoproteins in the pattern. These observations strongly indicate that some glycoproteins, including MHC antigens, were subfractionated on the sequential column system due to microheterogeneity in the carbohydrate moiety. It seems plausible to suggest that microheterogeneity in carbohydrate residues of glycoproteins is a general feature, manifested in lectin-glycoprotein interactions. This phenomenon can lower the yield in purification of a glycoprotein by simple lectin affinity chromatography, but might increase the purity in case of sequential column systems. The analyses showed that the antiserum against the LcH binding fraction precipitated MHC-alloantigens and few other glycoproteins, whereas the antiserum against the Con A binding fraction was found to have a strong specificity for the classical transplantation antigen (B-F).

It can be concluded from these studies that an enrichment of B-F and B-L antigens in the solubilized lymphocyte membrane preparation can be achieved by sequential lectin affinity chromatography. The obtained xenogeneic rabbit antisera are now being used in a systematic analysis of MHC antigens from lymphocytes and serum by electroimmunoprecipitation methods.

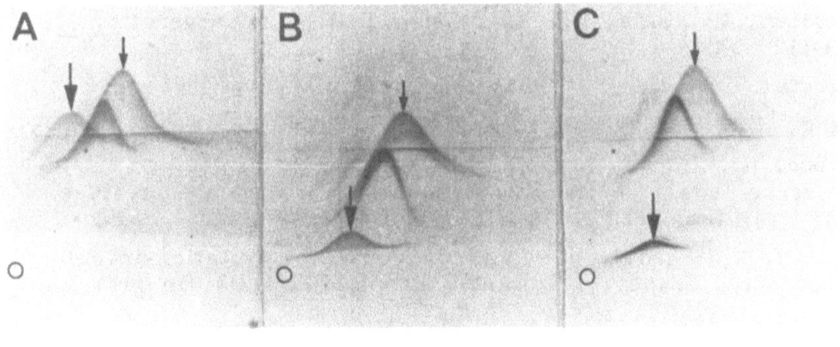

Fig. 3. Crossed immunoelectrophoresis of solubilized chicken (B^6) lymphocyte plasma membrane with rabbit antisera in intermediate gels. A: Intermediate gel with normal rabbit serum (8 µl/cm^2) as control. B: Intermediate gel with rabbit antisera against Con A binding lymphocyte plasma membrane proteins, RaCLM-Con A (8 µl/cm^2). C: Intermediate gel with reference rabbit antiserum with anti-B-F specificity (8 µl/cm^2) (6). An aliqout of 25 µl solubilized membrane (20 µg protein) was used in each application well. All upper gels contained rabbit antisera against LcH binding membrane proteins, RaCLM-LcH (6.6 µl/cm^2). Small arrows indicate B-L, whereas large arrows indicate B-F antigen. Plate size 6.25 x 10 cm.

Acknowledgements

This work has been supported by Danish Medical Research Council, Danish Cancer Society, King Christian Xth Foundation, and The Blood Donors Foundation. We are thankful to Misses Hanne Boesen, Lone Malte, and Lone Sørensen for skilful technical assistance and Mrs. Emmy Madsen for secretarial help.

References

1. Simonsen, M.: Acta pathol. microbiol. Scand. Sect. C, 83, 1-14 (1975).
2. Schou, M.: Tissue Antigens 15, 373-380 (1980).
3. Brogren, C.-H., Bisati, S., Blanchet, J.P.: Folia Biologica (Praha) 25, 328-330 (1979).
4. Bisati, S., Brogren, C.-H., Simonsen, M.: Folia Biologica (Praha) 25, 331-332 (1979).
5. Bisati, S., Brogren, C.-H., Simonsen, M.: Protides Biol. Fluids 27, 471-474 (1979).
6. Brogren, C.-H., Bisati, S., Simonsen, M.: Protides Biol. Fluids 27, 467-470 (1979).
7. Simonsen, M., Hala, K., Nicolaisen, E.M.: Immunogenetics 10, 103-112 (1980).
8. Blanchet, J.P.: Expt. Cell Research 84, 159-166 (1974).
9. Wang, C.S., Smith, R.L.: Analyt. Biochem. 63, 414-417 (1975).
10. Harboe, N., Ingild, A.: In N.H. Axelsen, J. Krøll, and B. Weeke (eds.). A Manual of Quantitative Electrophoresis. Scand. J. Immunol. 2, suppl. 1, 161-164 (1973).
11. Axelsen, N.H. (ed.): Handbook of Immunoprecipitation-in-gel Techniques. Scand. J. Immunol. suppl. 10 (1981), in press.

AFFINITY OF MEASLES VIRUS PROTEINS TO CONCANAVALIN A AS EVALUATED BY CROSSED IMMUNOELECTROPHORESIS

Ib Rode Pedersen

Institute of Medical Microbiology, University of Copenhagen
Juliane Maries Vej 22, DK-2100 Copenhagen Ø, Denmark

Carl H. Mordhorst

Ornithosis Department, State Serum Institute
Amager Boulevard 80, DK-2300 Copenhagen S, Denmark

Introduction

Measles virus is made up of 6 different structural proteins in the molecular weight range of 37,000 - 200,000, and in addition, there is a host specified protein actin, with a molecular weight of 43,000 (1, 2, 3). Two of these proteins are glycoproteins which form projections on the envelope membrane of the virus. One glycoprotein, HA, has hemagglunating activity, while the other, F, can induce fusing of animal cells and cause hemolysis. The viral envelope contains also the membrane protein M. The remaining structural proteins are designated L (large), P (polymerase) and NP (nucleoprotein). The two nucleoproteins have affinity for Concanavalin A (Con A). This property can be used for their identification by crossed immunoelectrophoresis, and it can also be used for their isolation in a batch procedure.

Results

For the production of viral protein, Vero cells in roller bottles were infected with Edmonston B measles virus. Three days later, when they were heavily infected, the cells were

Lectins - Biology, Biochemistry, Clinical Biochemistry, Vol. I

harvested. The viral proteins were extracted from the cells in the presence of the detergent Triton X-100 by use of methods previously described (4). The solubilized viral proteins were then analyzed by crossed immunoelectrophoresis in agarose gels containing anti-measles serum during the second dimension electrophoresis. The antiserum was obtained from rabbits immunized with purified measles virus. The conditions for electrophoresis were similar to those of conventional procedures (4, 5). The result of the crossed immunoelectrophoresis can be seen in Fig. 1 which shows 5 different viral proteins. A similar pattern of precipitation lines is obtained if a reconvalescent serum is used instead of rabbit antiserum in the second dimension gel. The viral glycoproteins, here designated G_1 and G_2, could be identified by their affinity to Con A if this was included in an intermediate gel as shown in Fig. 2. However, it was not possible to distinguish between the two viral glycoproteins. Further identification of the other measles virus proteins was undertaken by crossed immunoelectrophoresis of isolated viral proteins. During this isolation, advantage was taken of the different salt solubility of viral proteins. Thus in succession, the NP nucleoprotein can be isolated due to its insolubility in 1 M KCl and the hydrophobic M membrane protein in distilled water. The remaining non-precipitable glycoproteins and polymerase protein can be separated from each other by binding of the glycoproteins to Con A Sepharose in a batch procedure. In this isolation step, a small volume of G+P proteins was mixed with twice the volume of Con A Sepharose (Pharmacia) in pH 8.6 electrophoresis buffer. The mixture was left at room temperature for 10 min with occasional shaking, whereafter the Sepharose-bound glycoproteins were isolated by centrifugation, leaving the polymerase in the supernatant. The glycoproteins were eluted from the precipitated Sepharose by resuspension in electrophoresis buffer containing 0.3 M α-D-methylmannoside. After a final centrifugation, the glycoproteins were isolated from the supernatant. The eluted glyco-

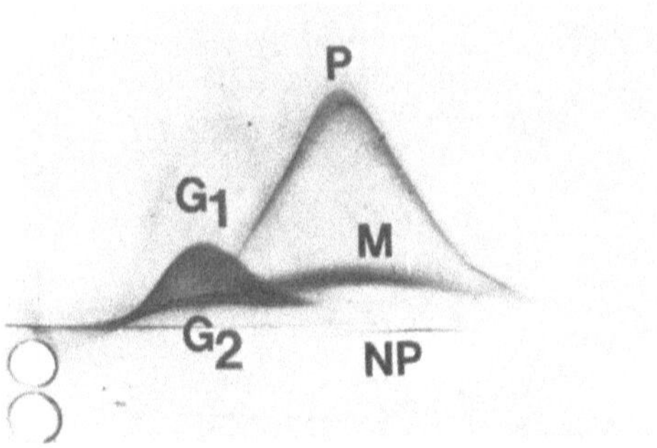

Fig.1. Crossed immunoelectrophoresis of measles virus proteins extracted with Triton X-100 from infected Vero cells.

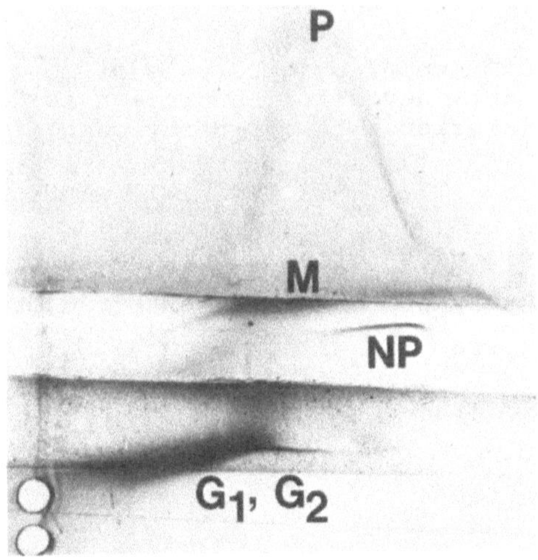

Fig.2. Crossed immunoaffinoelectrophoresis of measles virus proteins extracted with Triton X-100 from infected Vero cells. The intermediate gel Contained Con A in a concentration of 100 $\mu g/cm^2$.

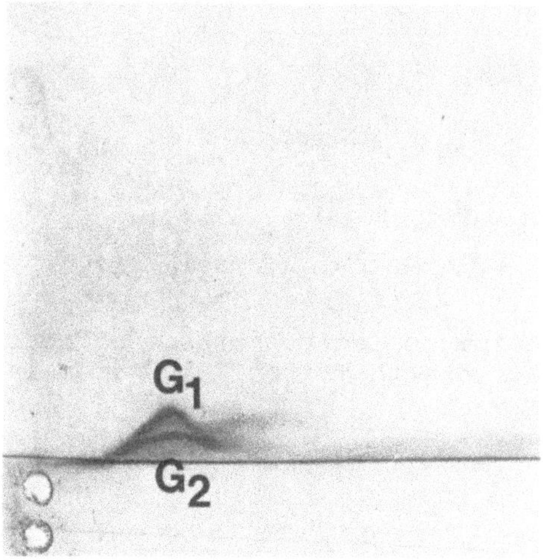

Fig.3. Crossed immunoelectrophoresis of measles virus glycoproteins obtained after absorbtion to Con A Sepharose and elution with α-D-methylmannoside.

proteins were submitted to crossed immunoelectrophoresis which demonstrated that they are obtained free from other viral proteins, as shown in Fig. 3.

Discussion

The experiments performed demonstrate that affinity for Con A can be a useful property for the identification and isolation of measles virus glycoproteins. The successive isolation of viral proteins as described, followed by crossed immunoelectrophoresis, makes it possible to obtain precipitation lines in the agarose gels without cross contamination of viral proteins. These precipitates can then be cut out and used for the immunization of rabbits in order to produce monospecific antisera (6). Thereafter, the monospecific antisera could be used in an enzyme-linked immunosorbent assay (ELISA) as a very sensitive method to follow viral protein synthesis in infected cells. The method could also be used to measure the level of antibodies directed to the different viral components in persons infected or vaccinated with measles virus.

Acknowledgement

The present work was performed with support from the Danish Medical Research Counsil.

References

1. Graves, M.C., Silver, S.M., Choppin, P.W.: Virology 86, 254-263 (1978).
2. Wechsler, S.L., Fields, B.N.: J. Virol. 25, 285-297 (1978).
3. Stallcup, K.C., Wechsler, S.L., Fields, B.N.: Virology 30, 166-176 (1979).
4. Vestergaard, B.F.: Acta path. microbiol. scand. Sect. B 81, 808-810 (1973).
5. Axelsen, N.H., Krøll, J., Weeke, B.: Scand. J. Immunol. 2, Suppl. 1 (1973).
6. Vestergaard, B.F.: Scand. J. Immunol. 4, Suppl. 2, 203-206 (1975).

ISOLATION OF ANTIGENIC SUBUNITS FROM BOVINE VIRAL DIARRHOEA VIRUS BY MEANS OF IMMOBILISED CROTALARIA JUNCEA LECTIN

P. Kårsnäs and T. Kristiansen

Institute of Biochemistry, Biomedical Center, Box 576, S-751 23 Uppsala, Sweden

J. Moreno-Lopez

Department of Virology, College of Veterinary Medicine, Biomedical Center, Box 585, S-751 23 Uppsala, Sweden

Introduction

Bovine viral diarrhoea virus (BVD-virus) is a nonarbo RNA-virus of the family Togaviridae. The nucleic acid is enclosed in a capsid of probably icosahedral symmetry which is surrounded by a lipoprotein envelope carrying surface projections (peplomers), which give the surface a convoluted appearance in the electron microscope. The chemical nature of the peplomers is unkonwn but it is suggested that they consist of glycopeptides (2). Their attachment to the envelope is weak and the virus is mechanically labile due to the easy loss of the peplomers. The lability leads to poor recovery by conventional purification methods such as repeated sucrose gradient centrifugations or ammonium sulphate precipitation (1,2).

By taking advantage of the interaction between the glycoproteins of the envelope and lectins coupled to a matrix we obtained a high recovery of infective virus particles in a single affinity chromatography purification step.

The same technique was applied for isolation of the glycoprotein subunits after disruption of the virion by treatment with a detergent.

Lectins - Biology, Biochemistry, Clinical Biochemistry, Vol. I

Materials and Methods

Preparation of immobilised lectins. The galactose-binding lectin from *Crotalaria juncea* was purified and kindly supplied by Dr. Bo Ersson (3). The lectin was coupled to Sepharose 2B by the CNBr method (4), normally giving 1 to 3 mg coupled lectin per ml sedimented gel.

Preparation of virus. BVD-virus, strain UG 59, was grown on calf testicle cells as reported elsewhere (5) to give a titer of 10^6 PFU/ml. (Plaque assays were carried out in a conventional way.) Cell debris were sedimented by centrifugation at 10,000 g for 30 minutes. The supernatant was then passed through a column, packed with the immobilised lectin, and equilibrated with phosphate buffered saline (PBS, 0.02 M sodium phosphate pH 7.4, 0.9 % sodium chloride). After washing with three column volumes of PBS the adsorbed fraction was eluted with 0.2 % lactose in PBS, with reversed direction of flow.

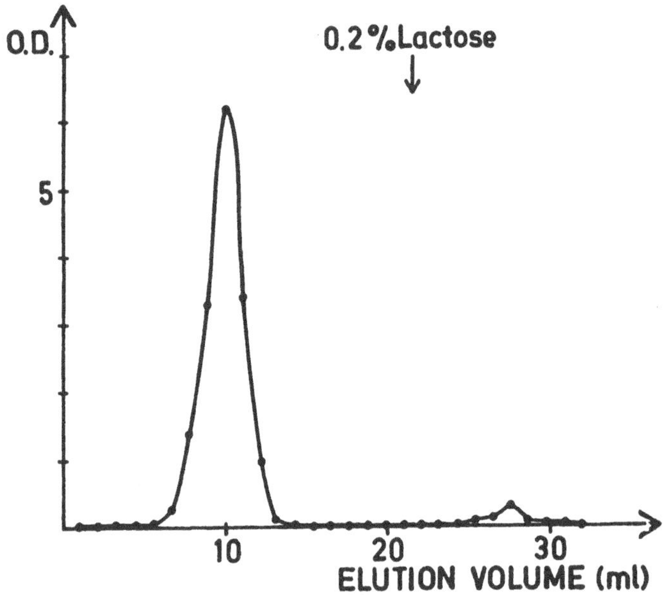

Fig. 1. Affinity chromatography of the extracellular fluid containing BVD-virus. Infectivity is gathered in the second peak.

Subunits. Virus eluted from the lectin column was disrupted by treatment overnight with 1 % Triton X-100. Chromatography of this material on another *Crotalaria juncea* lectin column (3 x 9 cm) was performed, and the adsorbed fraction again eluted with 0.2 % lactose in PBS.

Antisera. Antisera against the virus material treated with Triton X-100 (second peak, Fig. 1) were prepared in rabbits. These sera contain antibodies against viral antigens of the envelope and exposed structures of the capsomer as well as antigens derived from contaminating cell debris.

Immunoelectrophoresis. Crossed and line immunoelectrophoresis, as well as linerocket immunoelectrophoresis were performed in principle as described by Axelsen *et al*. (6). Normally the antibody-containing part of the gel included 200 µl of the whole rabbit antiserum (Figs. 3-5).

Results

Affinity chromatography of virus. The chromatogram of a typical run is shown in Fig. 1. The main part of the material passed the column unretarded. In several runs the desorbed fraction showed between 60 and 70 % of the overall infectivity of the starting material.

Affinity chromatography of subunits. Fig. 2 shows the result of a typical run of the disintegrated viral fraction. A_{230}, A_{280}, and A_{260} were recorded. Based on the ultraviolet adsorption about 40 percent of the entire material is adsorbed to the column. The fact that the size of the elution peaks was constant in several runs suggests that elution was complete in all cases. Glycoconjugates from the envelope but also sugar-containing compounds of the capsomer are to be expected in this peak.

Immunoelectrophoresis. Material from the desorbed peak was tested by crossed immunoelectrophoresis (Fig. 3). The picture reveals clearly three immunoprecipitates corresponding to three different antigenic components. A similar run, performed with antigens prepared as described above from uninfected culture medium, gave a single clearly visible precipitate (Fig.

4), which by tandem crossed immunoelectrophoresis fused with the second peak of Fig. 3. Thus the corresponding antigen originates from the cells. Line immunoelectrophoresis of subunit material normally gave four easily distinguishable precipitates. When antigens from uninfected culture medium were run simultaneously two rockets fused with line number three and four respectively, leaving the remaining lines unaffected (Fig. 5).

Discussion

Affinity chromatography of BVD-virus on immobilised Crotalaria juncea lectin proved to be a suitable technique resulting in a high recovery of infectivity compared to most other reports (1,2,7). It is not possible

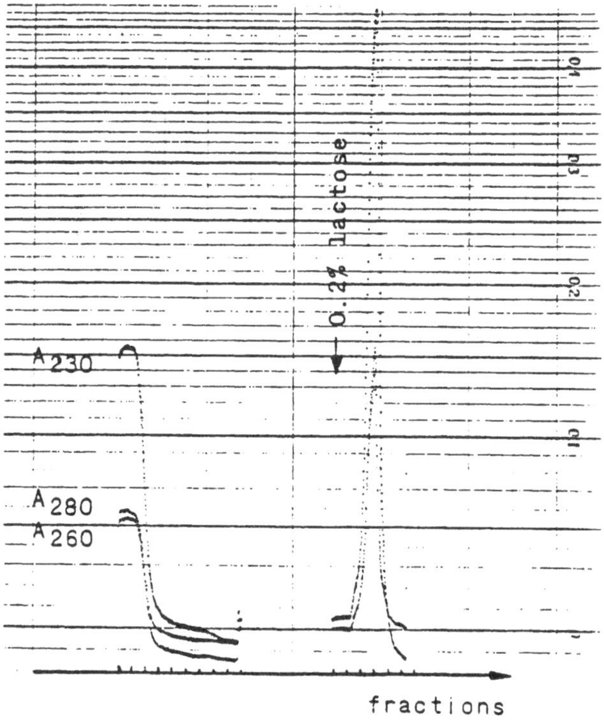

Fig. 2. Affinity chromatography of the virus fraction disintegrated in 1 % Triton X-100. Compounds with terminal galactose are eluted in the second peak.

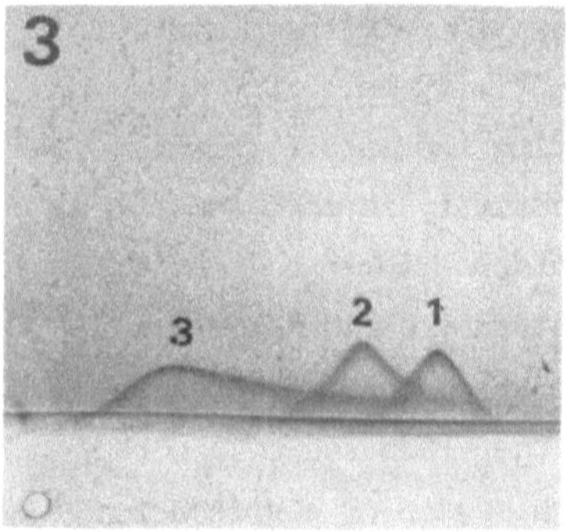

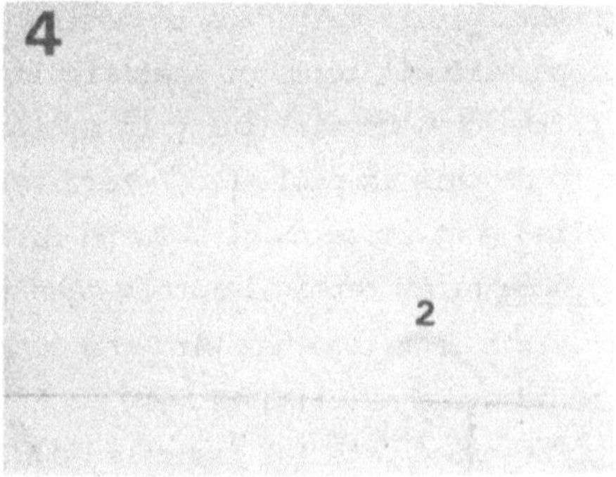

Fig. 3. Crossed immunoelectrophoresis of the proteins desorbed from the Crotalaria juncea lectin column. 5 µl of the antigen was run 1 hr in the first direction with 10 V/cm. The second direction was run overnight with 2 V/cm against 200 µl of the antiserum.

Fig. 4. Crossed immunoelectrophoresis of proteins from noninfected culture medium adsorbed on Crotalaria juncea lectin. The peak was shown to fuse with the second peak of Fig. 3.

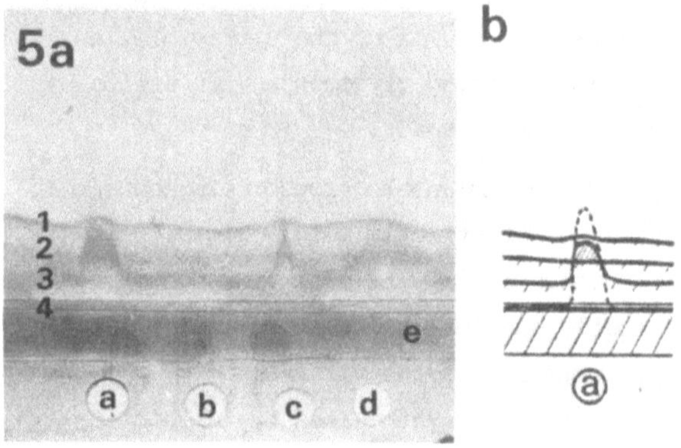

Fig. 5a. Rocket line electrophoresis of proteins from noninfected culture medium (rockets) and purified viral compounds (lines). a. Compounds desorbed from a Crotalaria lectin column after passage of supernatant obtained by centrifugation of noninfected culture medium treated with Triton X-100. Desorption zone was concentrated 25-fold before electrophoresis. b. Washings from concentration membrane used in step a. c. The fraction of supernatant passing through the lectin column and also concentrated 25-fold. d. Washings from concentration membrane used in step c. e. Compounds desorbed after passage of material obtained by treatment of infected culture medium in the same way as described in step a. The rockets of well "a" fuse with line three and four, indicating identity between the corresponding glycoconjugates.

Fig. 5b. Elucidation of the precipitin pattern of well "a" based on the original immunoelectrophoresis plate.

from our present, preliminary data to determine precisely the purity of the virus preparation in terms of specific infectivity (e.g. PFU/mass unit), due to the obvious difficulty in estimating the mass correctly. One attempt to do this is published, where infectivity is compared to the ^{3}H-radiolabelled host proteins (1). There is evidence that the lectin of Crotalaria juncea binds terminal rather than internal chain galactose residues (3). This fact together with the observation that the peak representing nonadsorbed material is very much larger than the peak corresponding to desorbed material indicated a high degree of purification of the active viral particles.

In elution the direction of flow was reversed to provide the shortest migration distance for the strongly adsorbed material bound to the proximal part of the column.

The aim of the affinity chromatography of the disintegrated viral fraction was the isolation of the glycoconjugates of the envelope. The chromatogram (Fig. 2) shows the adsorption of a large fraction of the ultraviolet absorbing material where the galactose-containing glycoconjugates are to be found (UV-absorbance/mass unit was not determined).

Our preliminary investigation by means of immunoelectrophoresis shows that two of the isolated glycoconjugates are of strictly viral origin, which is in agreement with the findings of others (1,2). The dual presence of some antigenic material both in noninfected and infected culture medium is consistent with the fact that the formation of the viral envelope may involve the incorporation of membrane proteins synthesised by the host cell. However, it remains to decide whether or not the two host cell components of fig. 3 are present in the viral envelope as well.

References

1. Pritchett, R.F., Zee, Y.C.: Am. J. Vet. Res. 36;12, 1731-1734 (1975).
2. Matthaeus, W.: Arch. Virol. 59, 299-305 (1979).
3. Ersson. B.: Biochim. Biophys. Acta 494, 51-60 (1977).
4. Axén, R., Porath, J., Ernback, S.: Nature 214, 1302-1304 (1967).
5. Moreno-Lopez, J., Kristiansen, T.: submitted to publication (1980).
6. Axelsen, N.H., Krøll, J., Weeke, D.: Scand. J. Immunol. Suppl. 1 (1973).
7. Maess, J., Reczko, E.: Arch. ges. Virusforsch. 30, 39-46 (1970).

RAPID CHANGES IN THE AFFINITY BETWEEN CON-A-SEPHAROSE AND LYSOSOMAL ENZYMES OF HUMAN SALIVA DURING THE DAY.

Kint, J.A. Kliniek voor Kinderziekten, Rijksuniversiteit Gent, Belgium.

Immobilized lectins can be used as a tool for evaluating the affinity of lectins to the carbohydrate part of a glycoprotein (1). The ratio of bound to not bound glycoprotein after incubation with a constant amount of immobilized lectin can be regarded as a measure of the affinity between both.

Short-time changes in this affinity can be understood as changes in the over all structure of the carbohydrate structure. We wished to examine possible rapid changes in the carbohydrate chains of extra-cellular glycoproteins.

Lysosomal enzymes (e.g. N-acetyl-β-hexosaminidase) were used as models for extra-cellular glycoproteins, because of the sensitivity and the ease of the fluorometric assay of glycosidases. Human saliva was chosen as source of these enzymes. Total saliva was collected at different moments during the day. One to five milliliters were obtained each time. The saliva was centrifuged and the supernatant kept at 4°C until the end of the collection period. At that time 200 µl of each sample was mixed with 200 µl of washed Con-A-Sepharose, diluted 1:4 with phosphate buffered saline. These mixtures were allowed to equilibrate during two hours at room temperature. After centrifugation the activity of N-acetyl-β-hexosaminidase was measured in the supernatant. As can be seen from Fig. 1, four individuals A, B, C and D show a marked change in the Con-A-affinity after breakfast, while person E which did not have a breakfast, did not show such a change. When the experiment was followed for 24 hours a clear circadian rhythm became apparent: Just before each meal the amount of not bound activity was maximal and this value dropped to a minimum after the meal.

Lectins - Biology, Biochemistry, Clinical Biochemistry, Vol. I

More recently Bøg-Hansen and Takeo (2) used affinity electrophoresis for the exact determination of the dissociation constant. It may be interesting to use this method to control the presented data.

1) CHAIKEN, I.M.: Analytical Biochemistry 97, 1-10, 1979.
2) BØG-HANSEN, T.C. and TAKEO, K.: Electrophoresis, 1, 67-70, 1980.

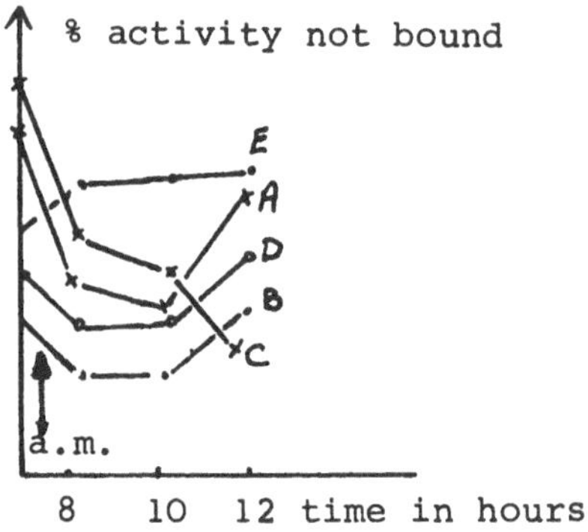

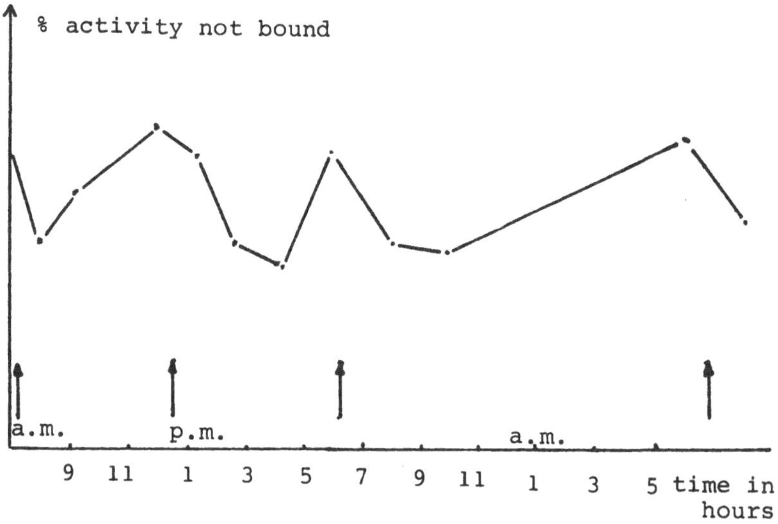

Affinity between N-acetyl-β-hexosaminidase of saliva and conA-Sepharose.
Ordinate: % of enzyme activity not bound to conA-Sepharose.
Abscissa: Time of the day in hours.
A, B, C and D: Individuals who had a breakfast.
E: Individual who did not hav a breakfast.
The arrow indicates the time of the meal.

AUTHOR INDEX

INDEX

FSC
www.fsc.org
MIX
Papier | Fördert
gute Waldnutzung
FSC® C083411

Kolibri 360 GmbH
Ferdinand-Jühlke-Straße 7
99095 Erfurt, Deutschland
produktsicherheit@kolibri360.de